# Lecture Notes in Physics



361

P. Gaigg W. Kummer
M. Schweda (Eds.)



# Physical and Nonstandard Gauges

Proceedings of a Workshop Organized
at the Institute for Theoretical Physics
of the Technical University, Vienna, Austria
September 19–23, 1989

Springer-Verlag
Berlin Heidelberg GmbH

**Editors**

P. Gaigg
W. Kummer
M. Schweda
TU Wien, Institut für Theoretische Physik
Wiedner Hauptstraße 8–10, A-1040 Vienna, Austria

ISBN 978-3-662-13799-4 ISBN 978-3-540-47156-1 (eBook)
DOI 10.1007/978-3-540-47156-1



Originally published by Springer-Verlag Berlin Heidelberg New York in 1990
Softcover reprint of the hardcover 1st edition 1990

2153/3140-543210 – Printed on acid-free paper

# Preface

The development of ordinary and supersymmetric gauge theories and string theories in recent years has brought renewed interest to the wide field of noncovariant gauges, with special emphasis on gauges characterized by one or more gauge vectors fixing a direction in Minkowski space. The use of such gauges has turned out to be crucial for an improved understanding of gauge field phenomena from finite supersymmetric field theories to applications in QCD, like the computation of propagators of the quark - gluon plasma or the investigation of inclusive effects in processes involving hadrons. The basic reason for this usefulness can be traced back to the fact that usually such gauges lead to the reduction of redundant degrees of freedom. However, the price to be paid for these more physical features in formal approaches has been serious complications at the mathematical level, i.e. from actually performing calculations of Feynman graphs within a well-defined, consistent formulation. These problems are almost exclusively tied to the appearance of spurious poles in the propagator of the gauge field. They may even lead to a "nonstandard" renormalization, which means that the standard renormalization approach, based upon local counter terms to individual Feynman graphs, may have to be abandoned. Of course, also in nonstandard renormalization physical observables must be still well-defined in terms of a finite number of local counterterms added to the action.

On the other hand, new original ideas have emerged recently which seem to pave a way around those obstacles. Hence the specialized workshop "Physical and Nonstandard Gauges" was organized at the Institute for Theoretical Physics of the Technical University, Vienna (19 - 23 September 1989) to enable the somewhat scattered group of specialists to present their respective news, to discuss for the first time open problems in a forum of experts, and, if possible, to reach a consensus on open issues.

Scientific contributions to this workshop were related to the broad fields of

1. **Canonical quantization**
2. **Spurious poles and problems with Feynman-integrals in noncovariant gauges**
3. **Renormalization and gauge dependence**
4. **QCD at finite temperature**

Separate short introductions to four sections corresponding to these fields will explain salient features of the respective state of the art, emphasizing solutions and remaining open problems.

The proceedings of the Vienna Workshop on Physical and Nonstandard Gauges 1989 will constitute the first book ever published on the subject of noncovariant gauges. We believe that owing to the increasing importance of the field it may serve as a useful introduction to newcomers. It addresses itself to the whole range of researchers from graduate students to specialists in the broad field of gauge theories, which, to the best of our present knowledge, are the fundamental laws of Nature.

Vienna
March 1990

Peter Gaigg
Wolfgang Kummer
Manfred Schweda

# Acknowledgements

The editors gratefully acknowledge the financial support from the coorganizers (Austrian Academy of Sciences, Federal Ministry of Science and Research) and from the other sponsors (Bundeskammer der Gewerblichen Wirtschaft, CA, Kulturabteilung der Stadt Wien, Österreichische Länderbank AG, and Österreichische Physikalische Gesellschaft). They thank the international advisors G. Leibbrandt (Guelph, Canada) and J.C. Taylor (Cambridge, England). The organization of this workshop would have been impossible without the enthusiastic help of all the members of the Institute for Theoretical Physics of the Technical University, Vienna.

# Contents

# Introduction

Perhaps the most remarkable progress in the last years has been achieved in understanding prescriptions of spurious poles, starting from the basic principles of canonical quantization. The LM-prescription is now well understood in this framework, as testified by the reviews of Cheng and Lazzizzera. Among the special contributions, the work on the Fock-Schwinger gauge (Galvão) should be mentioned, because this gauge violates translation invariance and is thus another example of a noncovariant gauge requiring nonstandard renormalization.

# I

# Canonical Quantization

# Why Canonical Quantization Is the Only Known Way to Quantize Correctly

Hung Cheng

Department of Mathematics, Massachusetts Institute of Technology,
Cambridge, MA. 02139, U.S.A.

The path integral approach to gauge field theories [1] is essentially a Lagrangian approach, and as such, it has many advantages. It is conceptually beautiful and stylistically elegant. It exhibits explicitly the symmetries of the underlying theory, and it treats space and time on equal footing. However, it has at least the following three difficulties:

(a) The transition amplitude is not always given by

$$\int \mathcal{D}\phi e^{i\int L d^4x} . \tag{1}$$

An elementary example can already be found in ordinary quantum mechanics for a particle of time-dependent mass: $L = \frac{1}{2}m(t)\dot{q}^2$. In this case, we all know that a determinant has to be added to the expression in (1). A more modern example can be found in the Skyrme model, the form of the Lagrangian of which can be expressed as

$$L = \frac{1}{2}\dot{\phi}_a M_{ab}\dot{\phi}_b - V(\phi_a, \nabla\phi_a), \tag{2}$$

where the time derviatives of the fields are exhibited explicitly, with $M_{ab}$ dependent on $\phi_a$ and $\nabla\phi_a$. Again, the determinant of $M_{ab}$ must be inserted to (1). Since it depends on $\nabla\phi^a$ but not on $\dot{\phi}^a$, the resulting path integral is not relativistically invariant, even though $L$ is [2].

This is one example in which a theory which is covariant at the classical level cannot be made so at the quantum level. While the necessity of modifying (1) is only too well-known, I bring up this example to prove that we tend to forget to do so. We do not always recognize that (1) is not the starting point of a quantum theory, and that it is derived from the canonical approach, valid only for theories with a Lagrangian of certain specified form. For a Lagrangian not in this form, it is not always clear what modification one must make without going through the canonical derivation. If one chooses to ignore this fact and uses (1) regardless, then the theory obtained may not be unitary, to name one possible difficulty. Indeed, the path integral in (1) may not even be finite.

(b) The expression (1) is not complete. To make it complete, one needs to put in the initial and final wave functions, and integrate (1) over the variables of these wavefunctions. While this presents no hardship in many quantum field theories, it is not so with gauge field theories. This is ironic as the path integral approach first became popular partly due to its unique suitability in handling gauge field theories, or so people thought. The problem is that wave functions in gauge field theories are not normalizable. For example, the path integral (1) with the vacuum wave functions inserted as the initial and the final state wave functions may be meaningless, as these wave functions in a physical theory (defined below) have no inner products. This is not a question of mathematical rigor. The problem is that one may get all kinds of erroneous answers if such a difficulty is ignored.

That the wave functions in (1) are usually left out in most path integrals also means that the Feynman rules derived from these integrals are not yet defined. Specifically, in the path integral approach, the Lagrangian in (1) provides the differential equations satisfied by the propagators. However, such propagators cannot be determined unless we are given, in addition, the boundary conditions they satisfy. These conditions are provided by the initial and final state wave functions.

I have often been asked, "what is wrong with making (1) an ansatz, especially if we go to the Euclidean space and set the boundary values for the fields to be zero at infinity?" The answer is already contained in the question: *it is an ansatz.* Sometimes this ansatz provides the correct answer, but sometimes it does not. I have no objection to using the path integral expression for the cases in which its correctness has been established. But whenever we start with a new theory, it would be unscientific to *assume* that (1) is correct.

(c) The path integral (1) does not incorporate operator ordering. Basically, the path integral is one involving the classical action summed over all paths. The problem is, the classical theory does not always define the quantum theory. An elementary example is the theory for a nonrelativistic particle in two spatial dimensions. In polar coordinates, the classical Hamiltonian is

$$\frac{1}{2}(p_r^2 + \frac{1}{r^2}p_\theta^2).$$

As we all know, the quantum Hamiltonian is not equal to this expression with $p_r$ replaced by $-i\frac{\partial}{\partial r}$. Rather, we must replace $p_r^2$ by $\frac{1}{r}\frac{\partial}{\partial r}r\frac{\partial}{\partial r}$; the order of the operators which we *assume* to be that deduced from the Cartesian coordinates cannot be ignored.

While it is easy to take care of operator ordering in this elementary example, it is not easy to do so in nonabelian gauge field theories. As an example, the formulation of nonabelian field theories in the Coulomb gauge was first correctly done by Schwinger [3,4,5], using the formulation of canonical quantization. The subsequent path integral formulation of the same problem [6] differs in content from Schwinger's formulation: some interaction terms are missing. This difference is due to the failure of the path integral approach to take care of operator ordering, and, as it turns out, the resulting theory does not have relativistic covariance.

With the present prevalence of the path integral formalism, many people are perhaps under the impression that it is difficult to treat nonabelian gauge field theories with the canonical approach. We have found that such is not the case. It is simple to quantize nonabelian field theories canonically by methods that were generally known to Fermi [7] as far back as the 1930s.

Here, let me emphasize an important point. While gauge field theories are mainfestly gauge invariant at the *classical* level, nothing guarantees that they are so at the *quantum* level. Therefore, it is not enough just to derive the Feynman rules in a particular gauge, which is often easy. It is also necessary to prove that the rules in this gauge are *equivalent* to those in other gauges.

For example, it is straightforward to follow the prescription of Faddeev and Popov [1] to the latter, and bingo: we come up with the Feynman rules in the Coulomb gauge. Much more difficult is the task to prove or disprove the validity of these rules. The same comment applies to the Feynman rules in the axial gauge, or any other gauges one cares to propose.

Let me start with the quantum theory in the temporal gauge $A_0^a = 0$, i.e.,

$$H_w = \frac{\boldsymbol{\pi}^{a\,2} + \boldsymbol{B}^{a\,2}}{2}. \tag{3}$$

with the Gauss law imposed as a supplementary condition on the quantum state:

$$G^a |\psi_w >= 0, \tag{4}$$

where

$$G^a \equiv \nabla \cdot \boldsymbol{\pi}^a + g f^{abc} \boldsymbol{A}^b \cdot \boldsymbol{\pi}^c.$$

The Hamiltonian (3) is gauge invariant, as are the quantum states satisfying the Gauss law (4). It also turns out that this theory is unitary and relativistically covariant. We shall call this theory the *physical theory*. We shall define a theory associated with a certain gauge to be correct if and only if it is equivalent to this theory.

It turns out that there are two ways for a gauge theory to be equivalent to the physical theory. The first way is for the theory in question to be the physical theory itself with some superfluous variables eliminated. An elementary example of such a kind of equivalence is that of the three–dimensional Schrödinger equation for a free particle with the corresponding radial Schrödinger equation. The Coulomb gauge theory and the axial gauge theory are related to the physical theory in this way, as we shall show later in this talk.

The Feynman gauge theory is an entire different matter. Indeed, I would like to remind you that the Feynman gauge is somewhat peculiar; it does not even exist in classical electromagnetism. As a matter of fact, a gauge field has only two physical degrees of freedom. In the Feynman gauge, one introduces two extra and independent unphysical degrees of freedom. On top of these unphysical degrees of freedom we add a degree of freedom belonging to the ghost. Thus the space spanned by the quantum states in the Feynman gauge is much larger than that in the Coulomb gauge. Indeed, there is no elementary example to illustrate the relationship between the Feynman gauge theory and the physical theory. Thus I do not consider it to be the end of the story when one is able to prove that the contribution of the extra degrees of freedom to the unitarity condition cancel. In other words, it is not enough to prove that such a theory is unitary with only the physical degrees of freedom included. One must also be able to prove that the theory in the Feynman gauge is *equivalent* to the physical theory. As it turns out, these two field theories are equivalent not in the way we discussed earlier but only in the sense that they give the same on–shell scattering amplitudes, as well as the same matrix elements of gauge–invariant operators. In particular, the quantum states in the Feynman gauge theory do not even form a Hilbert space. This will be discussed in more details in Professor Er–Cheng Tsai's talk.

For the purpose of proving equivalence, we find the following theorem helpful:

**Theorem** *Let the gauge boson propagator be*

$$-i\frac{g_{\mu\upsilon} - a_\mu(k)k_\upsilon - b_\upsilon(k)k_\mu}{k^2 + i\epsilon},$$

*where*

$$a_\mu(k) = -b_\mu(-k),$$

*let the ghost–ghost–gauge boson coupling be*

$$[(a \cdot k) - 1]k^\mu - k^2 a^\mu,$$

*where $k^\mu$ is the outgoing ghost momentum, and let the ghost propagator be*

$$\frac{1}{k^2 + i\epsilon};$$

*then the on–shell scattering amplitude (as well as the matrix elements of gauge invariant operators) is independent of $a_\mu$.*

With this theorem, one is able to play the game of inventing new gauges. Note that this theorem does not distinguish in which of the two ways the equivalence is.

There is one thing to keep in mind when we apply the theorem. The function $a_\mu(k)$ in this theorem may depend on one or more parameters. As one of these parameters takes a special value, the Feynman integrals may become singular. An example is the propagator of the gauge boson in the axial gauge, in which $a_\mu(k) = \dfrac{n_\mu}{k \cdot n} - \dfrac{k_\mu}{2(k \cdot n)^2}$. We may define $(k \cdot n)^{-1}$ as $\lim_{\epsilon \to 0}(k \cdot n + i\epsilon)^{-1}$, or some other way. The point is that the propagator is singular as $\epsilon \to 0$. The correct scattering amplitude is obtained by first choosing a nonzero $\epsilon$, and taking the limit $\epsilon \to 0$ after the integration has been carried out, not before. In other words, the limit $\epsilon \to 0$ should be taken for the Feynman integral, not the integrand. With this point in mind, we were able to prove that the Feynman rules in the Coulomb gauge obtained via path integration is incorrect, and that the Feynman rules in the axial gauge with the principal–value prescription is also incorrect[5]. In particular, there are contributions of ghost–loops in the latter set of Feynman rules.

Since there is not enough time for me to present the diagrammatic proof of the theorem of equivalence here, I shall only mention that there is one point which greatly facilitates the general proof. That point is that it is possible to change the Feynman rules in a gauge to those in another gauge not in a wholesale, but in a piece–meal way. More precisely, one does not have to change the Feynman rules for all lines in a diagram. One may change them one line at a time. This reduces a nightmarish problem in combinatorics to a manageable one.

For the rest of this talk, I would like to give a systematic presentation of canonical quantization, which is essentially a Hamiltonian approach. Let me begin with a nonrelativistic particle in a potential

$$H = \frac{1}{2}p^2 + V(q).$$

The quantum theory is defined by treating $p$ and $q$ as hermitian operators satisfying

$$[p, q] = -i.$$

(Planck's constant is chosen to be unity.) A quantum state is represented by $|\psi>$, and we define the inner product

$$< q|\psi >\equiv \psi(q),$$

where $|q>$ is an eigenstate of the position operator $q$. In this representation

$$p = -i\frac{\partial}{\partial q}.$$

The inner product of two quantum states is given by

$$< \psi_1|\psi_2 >= \int_{-\infty}^{\infty} \psi_1^*(q)\psi_2(q)dq. \qquad (5)$$

I spell out these elementary notations, as we have found it helpful to use the notations $\psi(q)$ generalized to field theories, which we denote as

$$\psi(A_\mu) \equiv < A_\mu|\psi > \qquad (6)$$

a functional of the fields, which we shall call the wave functional. The inner product of two quantum states is therefore given by

$$< \psi_1|\psi_2 >= \int \mathcal{D}A_\mu \psi_1^*(A_\mu)\psi_2(A_\mu). \qquad (7)$$

The integral in (7) is the counterpart of (5). It involves the integration of the field variables at the same time, and is *not* a path integral.

Let us consider the nonabelian field theory described by the Lagrangian density

$$L = -\frac{1}{4} F^a_{\mu\upsilon} F^{\mu\upsilon a}, \tag{8}$$

where

$$F^a_{\mu\upsilon} \equiv \partial_\mu A^a_\upsilon - \partial_\upsilon A^a_\mu - g f^{abc} A^b_\mu A^c_\upsilon .$$

(We may add other Bose-fields or Fermi-fields. But the treatment is almost identical and little insight is gained.) The equations of motion are

$$(D_\mu F^{\mu\upsilon})^a = 0 \ , \tag{9}$$

where $D_\mu$ is the gauge covariant derivative. Because the absence of $\dot{A}^a_0$ in the Lagrangian of (8), the variable conjugate to $A^a_0$ necessary for the procedure of quantization cannot be defined. There are different ways to get around this. For example, we may choose the temporal gauge $A^a_0 = 0$. The Hamiltonian corresponding to (8) is therefore that of the temporal gauge described by (3) and (4).

Alternatively, we may follow Fermi and add a term $-\frac{1}{2}(\partial_\mu A^{\mu a})^2$ to the Lagrangian. Thus

$$L_F \equiv -\frac{1}{4} F^a_{\mu\upsilon} F^{\mu\upsilon a} - \frac{1}{2}(\partial_\mu A^{\mu a})^2, \tag{10}$$

with the equations of motion given by

$$(D_\mu F^{\mu\upsilon})^a = -\partial^\upsilon \ell^a, \tag{11}$$

where

$$\ell^a \equiv \partial^\mu A^a_\mu .$$

The theory defined by (10) is of course different from that defined by (8). However, (9) and (11) differ only by a term $\partial^\upsilon \ell^a$. If we pick the solutions of (11) which obey the additional constraint of

$$\ell^a(x, t) = 0, \quad \text{forall } x \text{ and } t,$$

then these solutions also satisfy (9). Therefore, these solutions are the solutions of (9) in the Lorentz gauge $\partial_\mu A^{\mu,a} = 0$. Futhermore, by applying $D^\upsilon$ to (11), we easily obtain

$$\Box \ell^a = g f^{abc} A^b_\mu \partial^\mu \ell^c ,$$

which is a differential equation for $\ell^a$ and is $2^{nd}$ order in time. Consequently, $\ell^a(x, t)$ vanishes for all times if the initial conditions of

$$\ell^a(x, 0) = 0 \tag{12a}$$

and

$$\dot{\ell}^a(\boldsymbol{x},0)=0 \tag{12b}$$

are satisfied.

To implement these initial condition quantum mechanically we restrict ourselves to quantum states $|\psi>$ which satisfy

$$\ell^a(\boldsymbol{x},a)|\psi>=0 \tag{13a}$$

and

$$\dot{\ell}^a(\boldsymbol{x},0)|\psi>=0. \tag{13b}$$

Setting $v=0$ in (11), we get

$$\dot{\ell}^a=(D_jF^{0j})^a\equiv G^a.$$

We may regard the quantum theory of (10) together with (13) that in the Lorentz gauge.

From (10), we obtain $\pi_0^a$, the variable conjugate to $A_0^a$, as $-\ell^a$. Thus (13a) simply means

$$\pi_0^a|\psi>=0\ . \tag{14}$$

In the notation of thc wavc functional, (14) is

$$\psi=\psi(\boldsymbol{A}^a)$$

independent of $A_0^a$. Also, it is straightforward to obtain the Hamiltonian corresponding to (10). With the condition of (13) utilized, this Hamiltonian is

$$H_F=\frac{\boldsymbol{\pi}^2+\boldsymbol{B}^2}{2},$$

which is the Hamiltonian in the temporal axial gauge. Indeed, (13b) is the Gauss law one imposes on the quantum states in the temporal gauge. Therefore, the quantum theory in the Lorentz gauge is exactly the same as that in the temporal gauge [8].

We make the following observations:

(i) Since the wave functionals in the Lorentz gauge are independent of $A_0^a$, the inner product of

$$\int \mathcal{D}\boldsymbol{A}^a\mathcal{D}A_0^a\psi_1^*(\boldsymbol{A}^a)\psi_2(\boldsymbol{A}^a)$$

is equal to infinity. This is what I meant by the wave functionals being not normalizable.

(ii) If we ignore this difficulty of normalization, we may get any number of wrong results. For example, this ignorance once led us to the erroneous conclusion that there is no ghost in the Feynman gauge.

(iii) While we are not allowed to make inner products of states satisfying (14), we may extract the scattering amplitude from these states by utilizing the fact that their wave functionals are independent of $A_0$ and $A_L$ as $t \to \infty$. In this way, we may obtain the Feynman rule of[9]

$$D_{\mu\upsilon}(k) = -\frac{ig_{\mu\upsilon}}{k^2 + i\epsilon} + \frac{i}{k^2}(ak^{\upsilon}n^{\mu} + bk^{\mu}n^{\upsilon} + ck^{\mu}k^{\upsilon}),$$

where $n = (1, 0, 0, 0)$,

$$\begin{aligned} a(k) &= \frac{|\boldsymbol{k}| - k_0}{2}\left[\frac{1}{k^2 - i\epsilon} - \frac{1}{k^2 + i\epsilon}\right], \\ b(k) &= -a(-k), \\ c(k) &= \frac{1}{2}\left[\frac{1}{k^2 - i\epsilon} - \frac{1}{k^2 + i\epsilon}\right], \end{aligned} \tag{15}$$

and there are no ghosts. The precise form of $D_{\mu\upsilon}$ given above is not important. The interesting point is that by modifying the inner products, we may change the Feynman rules drastically. Comparing with the traditional Feynman rules in the Feynman gauge, we see that the $i\epsilon$ in the denominators of (15) can be of either sign, hence $a, b$ and $c$ do not vanish. In addition there is no ghost. It is perhaps customary to believe that the choice of the initial and the final wave functionals determine only the sign of $i\epsilon$, while the volume of the group element under the gauge–fixing condition determines the ghost. The present example contradicts this belief.

One may think that the inner product being infinite can be easily sidestepped. Since the wave functionals are independent of $A_0$, can't we define the inner product to be

$$\int \mathcal{D}\boldsymbol{A}^a \psi_1^*(\boldsymbol{A}^a)\psi_2(\boldsymbol{A}^a),$$

factoring out the infinity that is the integral of 1 over $A_0$? This is indeed possible, but the resulting inner product is still infinite. The reason is that the wave functionals satisfy the Gauss law, the solution of which is simply

$$\psi(\boldsymbol{A}^a) = \psi(\boldsymbol{A}'^a),$$

where $\boldsymbol{A}^a$ and $\boldsymbol{A}'^a$ and related by a time–independent gauge transformation. Therefore there are infinities arising from the volume of the gauge–equivalent points. This is of course easy to overcome, at least in principle. Let us illustrate the way to overcome this by a simple example. Let us consider a non–relativistic particle in a two–dimensional space with the Hamiltonian $\frac{1}{2}(p_1^2+p_2^2)$. Let the state $|\psi>$ of interests satisfy the supplementary condition

$$(q_1p_2 - q_2p_1)|\psi >= 0.$$

Then

$$\psi(q_1, q_2) = \psi(\sqrt{q_1^2 + q_2^2}),$$

i.e., it is rotationally symmetric. Therefore, to know $\psi$ for all $q_1$ and $q_2$, we need only to specify $\psi(q_1, 0)$ for $q_1 \geq 0$. This is to say that we need only to choose the "gauge" $q_2 = 0$. The Hamiltonian in this "gauge " is simply that of the $S$-wave in the polar coordinate

$$H = -\frac{1}{2}\frac{1}{r}\frac{\partial}{\partial r} r \frac{\partial}{\partial r} \tag{16}$$

(Note that $|q_1|$ at $q_2 = 0$ is equal to $r$). The inner product is defined with the angular volume factored out, i.e., $\int_0^\infty \psi_1^*(r)\psi_2(r) r dr$. We note that the factor $r$ is also the Jacobian $J$ for the transformation from the cartesian coordiantes to the polar coordinates. Thus (16) is simply in the form of

$$-\frac{1}{2}\frac{1}{J}\frac{\partial}{\partial q_1} J \frac{\partial}{\partial q_1}.$$

We may go through a similar procedure for the gauge field theory. For example, we may choose the Coulomb gauge $A_L^a = 0$. The Hamiltonian in the Coulomb gauge can then be obtained as

$$H_c = \frac{1}{2}\int d^3x \left[\frac{1}{J}\boldsymbol{\pi}_T^a \cdot J\boldsymbol{\pi}_T^a + \boldsymbol{B}^a \cdot \boldsymbol{B}^a - \frac{1}{J}(\boldsymbol{D}\cdot\boldsymbol{\pi}_T)^T J (\frac{1}{\boldsymbol{D}\cdot\nabla})^T \nabla^2 \frac{1}{\boldsymbol{D}\cdot\nabla}(\boldsymbol{D}\cdot\boldsymbol{\pi}_T)\right] \tag{17}$$

where $A_L$ is set to zero, and where $J$ is the Jacobian of the transformation to the Coulomb gauge. If we ignore, incorrectly, the operator ordering, we may cancel the factors of $J$ in (17) and $H_c$ is reduced to the classical Hamiltonian. The expression in (17) was first obtained by Schwinger [3].

In the classical theory, the Coulomb gauge formulation is obtained by setting $A_L = 0$, while the temporal gauge formulation is obtained by setting $A_0 = 0$. It is therefore a little surprising that in the quantum theory, the Coulomb gauge formulation is obtained by first setting $A_0 = 0$ (and then setting $A_L = 0$). Actually, people already used this procedure in the Coulomb gauge formulation of QED, in which the Hamiltonian has the instantaneous Coulomb interaction term $-\frac{1}{2}\rho\frac{1}{\nabla^2}\rho$, where $\rho$ is the charge. People then identified the factor $-\frac{i}{\nabla^2}$ as the propagator of $A_0$, although $A_0$ does not appear in the Hamiltonian. The same procedure is followed in the Coulomb gauge formulation of nonabelian gauge field theories. The fact is that what we call a gauge in association with a set of Feynman rules is *different* from that in the classical theory. In the latter, it is a condition on the gauge fields ($A_0 = 0$ or $A_L = 0$, or others). In the former it merely denotes the form we express the gauge boson propagator so that the Feynman rules are equivalent. As we have seen, this form depends on more than one thing. For example, the fact that $|\psi>$ satisfies $(\partial_\mu A^\mu)|\psi>=0$ does not imply that the propagator $D_{\mu\upsilon}$ satisfies $\partial^\mu D_{\mu\upsilon} = 0$. One of the reasons is that $D_{\mu\upsilon}$ is not necessarily equal to $<0|TA_\mu A_\upsilon|0>$, where $|0>$ is the vacuum state. Indeed, in the case of the vacuum state being not normalizable, vacuum expectation values are not defined.

We have thus seen that the Coulomb gauge theory is also the same as the temporal gauge theory. The inner product in the Coulomb gauge is defined to be that integrated over $A_T$ only, and is finite. The Hamiltonian in the Coulomb gauge is hermitian, thus the scattering amplitudes in the Coulomb gauge are unitary. Since the temporal gauge Feynman rules and the Lorentz gauge Feynman rules give the same on–shell scattering amplitudes, they satisfy the unitarity requirement as well. No further proof of unitarity is needed. In particular, one does not have to prove that cancellations between ghosts and the unphysical polarization state – a major problem in other formulations.

The spatial axial gauge formulation can be treated similarly. We start with the temporal gauge formulation defined by (3) and (4), and we use (4) to eliminate $A_3$. In this way we obtain the Hamiltonian in the axial gauge to be

$$\begin{aligned} H_a =& \frac{1}{2}\int d^3x\,\{\,[\pi_1^a\pi_1^a + \pi_2^a\pi_2^a + \boldsymbol{B}^a\cdot\boldsymbol{B}^a]_{A_3^a=0} \\ & - (D_1\pi_1 + D_2\pi_2)^\dagger \frac{1}{\frac{\partial^2}{\partial z^2} - \lambda^2}(D_1\pi_1 + D_2\pi_2)\,\}\,, \end{aligned} \tag{18}$$

where $\lambda$ is infinitesimal. Feynman rules are easily deduced from (18). In particular, we identify $-\frac{1}{\nabla^2}(\nabla_\perp\cdot\dot{A}_\perp^a)$ in (18) as $A_0^a$, and the gauge boson propagator is given by [10]

$$D_{\mu\upsilon} = -i\left[g_{\mu\upsilon} - \frac{k_3}{k_3^2+\lambda^2}(N_\mu k_\upsilon + N_\upsilon k_\mu) - \frac{k_\mu k_\upsilon}{k_3^2+\lambda^2} + \frac{\lambda^2}{k_3^2+\lambda^2}N_\mu N_\upsilon\right]/(k^2+i\epsilon), \tag{19}$$

where $N_\mu = (0, \boldsymbol{e}_3)$, and $\epsilon$ and $\lambda$ are both infinitesimal.

We may also obtain the quantum theory in the light–cone gauge by simply setting $A^0 = A^3$ in (8). The Hamiltonian is

$$H_L = H_\omega + A^{3,a}G^a, \tag{20}$$

with the Gauss law (4) imposed on the quantum states. The Feynman rules can be deduced [10] from (20).

Finally, we discuss the Feynman rules in the temporal axial gauge. As we have mentioned, the inner product between two quantum states satisfying (4) does not exist. Since (4) is reduced to $\nabla\cdot\boldsymbol{\pi}^a|\psi>=0$ as $g$ is turned off, the wave functionals satisfying (4) depend on $\boldsymbol{A}_T$ but not on $A_L$ as $t\to\infty$. We may therefore extract the scattering amplitudes from the wave functional at $A_L = 0$. The unperturbed Hamiltonian of the longitudinal mode is simply $\frac{1}{2}\pi_L^2$. This is just like the Hamiltonian of a free non–relativisitc particle $H_0 = \frac{1}{2}p^2$, for which $q(t) = q + pt$. The Green's function in the latter theory will be defined to be

$$< q|Tq(t_1)q(t_2)|0 > |_{q=0} , \tag{21}$$

where $|0>$ is the ground state of $\frac{1}{2}p^2$, i.e.,

$$\psi_0(q) \equiv < q|0 >= 1,$$

which is not normalizable. As we have just explained, we may define the propagator as (21), which can be easily calculated to be

$$< q|Tq(t_1)q(t_2)|0 > |_{q=0} = it_>.$$

where $t_>$ is the larger of $t_1$ and $t_2$. Similarly, the propagator of the longitudinal gauge boson is [11]

$$D_{LL}(\boldsymbol{x}_1, \boldsymbol{x}_2) = -i\delta^3(\boldsymbol{x}_1 - \boldsymbol{x}_2)t_>,$$

while the propagator of the transverse gauge boson is

$$D^T_{ij}(x_1, x_2) = i\int \frac{d^4k}{(2\pi)^4}\frac{e^{-ik\cdot(x_1-x_2)}}{k^2+i\epsilon}(\delta_{ij} - \frac{k_ik_j}{\boldsymbol{k}^2}).$$

The propagator $D_{LL}$ is not a function of $t_1$ $t_2$, and hence cannot be transformed into the momentum space. To obtain the Feynman rules in the temporal gauge in the momentum space, we may add a term $\frac{1}{2}\lambda^2\boldsymbol{A}^2$ to (3) and take the limit $\lambda \to 0$. The propagator obtained is [10]

$$D_{\mu\upsilon} = -\frac{i}{k^2+i\epsilon}\left[g_{\mu\upsilon} - \frac{k_0(n_\mu k_\upsilon + n_\upsilon k_\mu)}{k_0^2 - \lambda^2 + i\epsilon} + \frac{k_\mu k_\upsilon}{k_0^2 - \lambda^2 + i\epsilon} + \frac{\lambda^2 n_\mu n_\upsilon}{k_0^2 - \lambda^2 + i\epsilon}\right], \tag{22}$$

where $n = (1, 0, 0, 0)$. The propagator (22) cannot be transformed into the configuration space, as this transform blows up as $\lambda \to 0$.

## References

[1] L.S. Faddeev and V.N. Popov, Phys. Lett. **25B** (1967) 29.

[2] H. Cheng, K.L. Chang, E.C. Tsai, and X. Zhu, Difficulty with the Quantum Skyrme Model (unpublished).

[3] J. Schwinger, Phys. Rev. **127** (1962) 324 and **130** (1963) 402.

[4] N.H. Christ and T.D. Lee, Phys. Rev. **D22** (1980) 939.

[5] H. Cheng and E.C. Tsai, Phys. Rev. Lett. **57** (1986) 511.

[6] E.S. Abers and B.W. Lee, Phys. Report **9C** (1973) 5.

[7] E. Fermi, Review of Modern Physics **4** (1932) 105.

[8] H. Cheng and E.C. Tsai, Chinese J. Phys. **25** (1987) 95.

[9] H. Cheng and E.C. Tsai, Phys. Rev. **D34** (1986) 3858. See H.S. Chan and M. Halpern, Phys. Rev. **D33** (1986) 540, for other ghostless Feynman rules.

[10] H. Cheng and K. Olaussen (in preparation).

[11] S. Caracciolo, G. Curci, and P. Menotti, Phys. Lett. **113B** (1982) 311.

# Canonical Quantization of Yang-Mills Theories in Axial and Planar Gauges

I Lazzizzera

Dipartimento di Fisica dell'Universitá di Trento
Istituto Nazionale di Fisica Nucleare, Gruppo Collegato di Trento
38050 – Povo (Trento) – Italy

## 1 Introduction

After Leibbrandt [1] and Mandelstam [2] proposed their well known (LM) prescription to handle the spurious poles of the Feynman propagator of the gauge bosons for Yang-Mills theories in the light-cone gauge, Bassetto et al. obtained it from first principles by canonical quantization [3]. Several authors then explored its possible generalizations off the light-cone region [4-9], supported later by the derivation through canonical quantization procedures: in [10-12] the method of [3] was extended to the homogeneous axial and the planar gauges; in [13,14] a new limiting conception of the axial gauges was employed, and in [15] a second constant (so called dual) gauge vector was included in the starting Lagrangean. Strong support also came from the proof of time exponentiation of the Wilson loop in perturbative calculations [16].
Here I will briefly review the canonical quantization of the non-pure space-like axial and planar gauges, recalling that otherwise a completely different theory applies, due to the fact that Gauss' law does follow from the equations of motion, provided that suitable asymptotic conditions in space-like directions are guaranteed for the four potential(s). (see [17]). I will present a unified scheme of quantization according to the BRST formalism [18-21], the use of which appears to be mandatory in inhomogeneous gauges [12] as far as the structure of the physical state-vector space is concerned. It will be shown that such a formalism definitively induces on some crucially important, but unphysical fields the special decomposition in creation and annihilation operators, which eventually produces the LM prescription from first principles. It will also be proved algebraically that even in the inhomogeneous axial gauges the Faddeev-Popov ghosts decouple.

## 2 The BRST Quantization

We start from the Lagrangian

$$\mathcal{L} = -\frac{1}{4} G_{\mu\nu} \cdot G^{\mu\nu} + \bar{\psi} \cdot (i \not\partial - m)\psi + g\, \bar{\psi}\gamma^\mu \tau \psi \cdot A_\mu + \frac{1}{2n^2}(\partial_\mu \omega) \cdot (\partial^\mu \omega) + \\ + \lambda' \cdot (\xi\omega - nA) - i\bar{C} \cdot nDC \,, \tag{2.1}$$

where $n_0 \neq 0$. Of course we also have $n^2 \neq 0$: however, the light-cone gauge is obtained from (2.1) after transforming $\omega$ into $n^2\omega$ and performing the limt $n^2 \to 0$ afterwards. All the non light-cone homogeneous axial gauges are obtained for the real constant parameter $\xi$ equal to zero. $\xi = \pm 1$ gives the planar gauges. It is understood that

$$G_{\mu\nu} = \partial_\mu A_\nu - \partial_\nu A_\mu + g\, A_\mu \times A_\nu \,. \tag{2.2}$$

The $\tau^a$'s form a basis in the fundamental representation of the internal symmetry (semi-simple compact) Lie group; $C$ and $\bar{C}$ are real Grassman fields in the adjoint representation of that group and have canonical dimensions 1 and 2 respectively. $\omega$ belongs to the adjoint representation, too.

$$D_\mu{}^{ab} = \partial_\mu \delta^{ab} + g f^{acb} A^c_\mu \,, \tag{2.3a}$$

$$\nabla_\mu = \partial_\mu - ig\, \tau \cdot A_\mu \,. \tag{2.3b}$$

By construction our Lagrangian is (quasi-) invariant under the following transformations, including, of course, the BRST and the ghost number conservation:

$$\left.\begin{aligned} s(A_\mu) &= D_\mu C \,, & s(\psi) &= ig\, C \cdot \tau\psi \,, \\ s(C) &= -\frac{1}{2} g\, C \times C \,, & s(\bar{C}) &= i\lambda' \,, \\ s(\lambda') &= 0 \,, & s(\omega) &= 0 \,; \end{aligned}\right\} \tag{2.4}$$

$$\left.\begin{aligned} C' = C \exp\{\epsilon\} \,, \quad \bar{C}' = \bar{C} \exp\{-\epsilon\} \\ A'_\mu = A_\mu \,, \quad \psi' = \psi \,, \quad \omega' = \omega \,; \end{aligned}\right\} \tag{2.5}$$

$$\left.\begin{aligned} &\hat{s}(A_\mu) = \hat{s}(\psi) = \hat{s}(\lambda') = \hat{s}(\omega) = 0 \,, \\ &\hat{s}(\bar{C}) = \frac{1}{2}\bar{C} \times \bar{C} \,, \qquad \hat{s}(C) = \bar{C} \times C \,; \end{aligned}\right\} \tag{2.6}$$

$$\left.\begin{aligned} \delta_Z(A_\mu) &= D_\mu(\bar{C} \times C) \,, & \delta_Z(\psi) &= g\bar{C} \times C \cdot \tau\,\psi \,, \\ \delta_Z(\lambda') &= 0 \,, & \delta_Z(\omega) &= 0 \,, \\ \delta_Z(C) &= i\,\lambda' \times C \,, & \delta_Z(\bar{C}) &= i\,\lambda' \times \bar{C} \,; \end{aligned}\right\} \tag{2.7}$$

$$\left.\begin{aligned} &A_\mu \to A'_\mu = \Theta^{-1} A_\mu \Theta + \frac{1}{g}\, \Theta^{-1} \partial_\mu \Theta \,, \\ &\text{with} \quad \begin{cases} \Theta = exp\left\{\zeta g L^3\, \lambda\right\} \,, \\ \lambda = \lambda' - ig\, \bar{C} \times C \,. \end{cases} \end{aligned}\right\} \tag{2.8}$$

L is an arbitrary constant parameter of dimension -1. The transformations (2.4) and (2.6) are nilpotent. The conserved Noether charges are, respectively:

$$Q_B = -n_0 \int d^3x\, C \cdot \left[D_k G_{0k} + g\,\bar{\psi}\tau\psi + \frac{i}{2} g\,\bar{C} \times C\right] \equiv$$
$$\equiv -n_0 \int d^3x\, C \cdot \left[\lambda + \frac{i}{2} g\,\bar{C} \times C\right] , \tag{2.9}$$

$$Q_g = i\, n_0 \int d^3x\, \bar{C} \cdot C \,, \tag{2.10}$$

$$Q_{\hat{B}} = n_0 \frac{i}{2} \int d^3x\, \bar{C} \cdot \bar{C} \times C \,, \tag{2.11}$$

$$i\,Q_Z = -n_0 \int d^3x\, \bar{C} \times C \cdot \left[\lambda + \frac{i}{2} g\,\bar{C} \times C\right] \equiv \hat{s}(Q_B) \,, \tag{2.12}$$

$$Q_{\mathcal{L}} = \frac{1}{2} \int d^3x\, \lambda \cdot \lambda \,. \tag{2.13}$$

We have the following Euler-Lagrange equations:

$$\xi\omega - nA = 0 \,, \tag{2.14a}$$
$$D_\mu G^{\mu\nu} + g\,\bar{\psi}\gamma^\nu\tau\psi = n^\nu \lambda' - ig\, n^\nu \bar{C} \times C = -in^\nu \left(s(\bar{C}) - ig\, \hat{s}(C)\right) \,, \tag{2.14b}$$
$$(i\gamma_\mu \nabla^\mu - m)\psi = 0 \,, \tag{2.14c}$$
$$nD\bar{C} = 0 \,, \qquad nDC = 0 \tag{2.14d}$$
$$\frac{1}{n^2}\partial^2\omega = \xi\lambda' = -i\xi\, s(\bar{C}) \,. \tag{2.14e}$$

It is straightforward to settle the canonical structure of the theory, using the Dirac procedure: we obtain the independent couples of canonical conjugate coordinates as follows:

$$(A^a_k,\, G^a_{0k}) \,, \tag{2.15a}$$

$$(\psi,\, i\bar{\psi}\gamma_0) \,, \tag{2.15b}$$

$$(C^a,\, -i\bar{C}^a) \,, \tag{2.15c}$$

$$\left(\omega^a,\, \pi^a \equiv \frac{1}{n^2}\partial_0\omega^a\right) \,. \tag{2.15d}$$

The total classical Hamiltonian, e.g. with $\xi = 1$, reads:

$$\mathcal{H} = \frac{1}{2} G_{0k} \cdot G_{0k} + \frac{1}{4} G_{ij} \cdot G_{ij} + \bar{\psi}(i\gamma_k \partial_k + m)\psi + g\, A_k \cdot \bar{\psi}\gamma_k\tau\psi -$$
$$-(\xi\omega + n_k A_k) \cdot (D_j G_{0j} + g\bar{\psi}\gamma_0\tau\psi + ig\bar{C} \times C) +$$
$$+ \frac{n^2}{2}\pi \cdot \pi - \frac{1}{2n^2}\omega \cdot \Delta\omega - i\bar{C} \cdot n_k\partial_k C + ign_k A_k \cdot \bar{C} \times C \,. \tag{2.16}$$

The energy-momentum tensor $\mathcal{T}^{\mu\nu}$ and the "angular-momentum" tensor $\mathcal{M}^{\mu\nu\rho}$, disregarding the fermions sector for simplicity, have the following form:

$$\mathcal{T}^{\nu\mu} = -G^{\mu\lambda}\cdot\partial^\nu A_\lambda + \frac{1}{4}g^{\nu\mu}G^{\sigma\rho}\cdot G_{\sigma\rho} + \frac{1}{n^2}\partial^\nu\omega\cdot\partial^\mu\omega - \frac{1}{2n^2}g^{\nu\mu}\partial_\rho\omega\cdot\partial^\rho\omega - in^\mu\bar{C}\cdot\partial^\nu C\,,$$
$$\mathcal{M}^{\mu\nu\rho} = x^\rho\mathcal{T}^{\nu\mu} - x^\nu\mathcal{T}^{\rho\mu} + G^{\mu\nu}\cdot A^\rho - G^{\mu\rho}\cdot A^\nu\,;$$

they obey the equations

$$\partial_\mu\mathcal{T}^{\nu\mu} = 0\,, \tag{2.17}$$
$$\partial_\mu\mathcal{M}^{\mu\nu\rho} = -i\,s\left(\bar{C}\cdot(n^\nu A^\rho - n^\rho A^\nu)\right)\,. \tag{2.18}$$

The generators of the Poincaré transformations, in turn, realize the following algebra of Dirac brackets:

$$\{P^\mu,\,P^\nu\} = 0\,, \tag{2.19a}$$
$$\{P_0,\,J_l\} = 0\,, \tag{2.19b}$$
$$\{P_l,\,J_j\} = \varepsilon_{ljk}P_k\,, \tag{2.19c}$$
$$\{P_0,\,K_l\} = P_l + i\int d^3x\,s\left(\bar{C}\cdot(n_lA_0 - n_0A_l)\right)\,, \tag{2.19d}$$
$$\{P_j\,K_l\} = -g_{jl}P_0\,, \tag{2.19e}$$
$$\{J_l,\,J_j\} = \varepsilon_{ljk}\,J_k\,, \tag{2.19f}$$
$$\{J_l,\,K_j\} = \varepsilon_{ljk}\,K_k - \varepsilon_{lrk}n_r\int d^3x\,x_js(\bar{C}\cdot A_k)\,, \tag{2.19g}$$
$$\{K_l,\,K_j\} = -\varepsilon_{ljk}\,J_k + i\int d^3x\;s\left(\bar{C}\cdot(x_lA_j - x_jA_l - (x_ln_j - x_jn_l)A_0)\right)\,. \tag{2.19h}$$

Evidently the classical physical states, for which both the Gauss' law and the Poincaré covariance hold, are described by those solutions of the equations of motion characterized by the vanishing of the fields $\omega$, $C$ and $\bar{C}$.
Noticed that a residual gauge invariance allows free choice of an initial condition for the first-order-in-time partial differential equation for the longitudinal field $\partial_kA_k$, in order to determine it without influencing the definition of any specific physical configuration. Indeed, from the relation

$$n^\nu G_{\mu\nu} = \xi\,D_\mu\omega - n\partial A_\mu\,, \tag{2.20}$$

the integral equation

$$\partial_kA_k = \mathcal{P}(x - \breve{n}(x_0 - t_0)) + \int_{t_0}^{x_0} dt\;(n^\nu\partial_kG_{\nu k} + \xi\partial_kD_k\omega)\,(x - \breve{n}(x_0 - t)) \tag{2.21a}$$

is obtained, where

$$n\partial\mathcal{P} = 0\,, \tag{2.21b}$$

and

$$\breve{n}^\mu = \frac{n^\mu}{n_0}\,. \tag{2.21c}$$

But, while completely fixing $\mathcal{P}$, any initial conditions on $\partial_kA_k$ can be changed to any suitable value by a time-independent gauge transformation, which has to be understood, in turn, as an initial condition for picking up the consequent residual gauge transformation from the defining equation $nD\Theta = 0$. Moreover,it is noticeable that the four-vector $\breve{n}$ is present in (2.21). This means that an additional violation of the Lorentz covariance occurs. Certainly these two facts can be considered as the respective explanations of the structure of the space of the quantum physical states and of the further violation of the Lorentz covariance in gauge-dependent quantities, like the gauge boson propagator.

The canonical quantization is straightforward. Assuming that for any one of the independent fields

$$\chi = A_k,\, \psi,\, \omega,\, C,\, \bar{C} \tag{2.22a}$$

we have

$$[iQ_B,\, \chi]_\pm = s(\chi)\,, \qquad [iQ_{\hat{B}},\, \chi]_\pm = \hat{s}(\chi)\,, \tag{2.22b}$$

then it can be verified that $Q_B$ and $Q_{\hat{B}}$ are nilpotent. Moreover, introducing the new notations

$$\mathbf{u} \equiv L^3 Q_Z\,, \tag{2.23a}$$

$$\mathbf{n} \equiv i\, Q_g\,, \tag{2.23b}$$

$$\mathbf{q}_+ \equiv L Q_B\,, \qquad \mathbf{q}_- \equiv L^2 Q_{\hat{B}}\,, \tag{2.23c}$$

it can be seen that the following (anti-)commutators algebra holds:

$$[\mathbf{q}_+,\, \mathbf{q}_-]_+ = \mathbf{u}\,, \tag{2.24a}$$

$$[\mathbf{n},\, \mathbf{q}_+] = \mathbf{q}_+\,, \tag{2.24b}$$

$$[\mathbf{n},\, \mathbf{q}_-] = -\mathbf{q}_-\,, \tag{2.24c}$$

with the the remaining (anti)commutators vanishing, while

$$[Q_{\mathcal{L}},\, Q_r] = 0 \qquad (r = B,\, \hat{B},\, Z,\, g)\,. \tag{2.25}$$

Thus, the operators defined in (2.23) in our space of state-vectors with indefinite inner product realize the well known **Z** graded algebra called End(V) in [22], with $V = V_0 \oplus V_1$, both $V_0$ and $V_1$ being one-dimensional vector spaces over the complex numbers. This algebra is isomorphic to the Jordan-Wigner algebra of the Fermi-Dirac one-particle creation and annihilation operators.

We see from eq (2.11) that in the subspace $\mathcal{V}'_p$ of the total space $\mathcal{V}$ with indefinite inner product, defined by

$$\mathcal{V}'_p = \left\{ |F\rangle \in \mathcal{V} :\ Q_r|F\rangle = 0 \qquad (r = B,\, \hat{B},\, Z,\, g,\, \mathcal{L}) \right\}. \tag{2.26}$$

the strict Gauss' law holds, namely:

$$\langle F'|D_k G_{0k} + g\,\bar{\psi}\gamma_0 \tau \psi|F\rangle = 0 \qquad \forall \quad |F\rangle,\, |F'\rangle \in \mathcal{V}'_p\,. \tag{2.27}$$

In addition from the quantum variant of eq (2.17-19) and the commutation relations

$$\left.\begin{array}{c} [Q_r,\, P_\mu] = [Q_r,\, \boldsymbol{J}] = [Q_r,\, \boldsymbol{K}] = 0 \\ (r = B,\, \hat{B},\, Z, g,\, \mathcal{L})\,, \end{array}\right\} \tag{2.28}$$

we deduce that Poincaré covariance is consistently recovered in such a physical sector. It is worthwhile to stress this fact for the inhomogeneous gauges, since for the homogeneous ones it is obvious that the well known decoupling theorem concerning the Faddeev-Popov ghosts is established here on an algebraic basis through (2.27). The same theorem can be proved as well within a path-integral formalism without actual perturbative calculations, thus independent of the regularization scheme. Indeed we can write the Faddeev-Popov term of the Lagrangean as

$$-i\bar{C}\cdot D_0 C = -i\bar{C}\cdot\partial_0 C + ig\,A_0\cdot\hat{s}(C)\,. \tag{2.29}$$

Then, after the change of integration variables:

$$C \longrightarrow \theta\,L^2\,\hat{s}(C)\,, \tag{2.30}$$

$\theta$ being an arbitrary Grassman constant parameter, we get

$$-i\bar{C}\cdot D_0 C \longrightarrow -i\theta L^2\bar{C}\cdot\partial_0(\bar{C}\times C) - ig\theta L^2\,A_0\cdot\hat{s}^2(C) = -i\theta L^2\bar{C}\cdot\partial_0(\bar{C}\times C)\,, \tag{2.31}$$

due to the nilpotency of $\hat{s}$. Since the above transformation neither involves gluons nor fermions, we can eventually cast the functional integral over the Faddeev-Popov ghost fields in a trivial factor.

The above definition of the physical subspace can be even more restricted taking advantage of the fact that in $\mathcal{V}'_p$ the field $\omega$ obeys a free evolution equation. As a matter of fact, calling $\omega_\circ$ the restriction of $\omega$ to $\mathcal{V}'_p$, since

$$\langle F'|\,\partial^2\omega_\circ\,|F\rangle = \langle F'|\,\left[Q_B,\bar{C}\right]\,|F\rangle = 0, \qquad \forall\;\;|F\rangle,\,|F'\rangle\in\mathcal{V}'_p\,, \tag{2.32}$$

the annihilation operator

$$\omega_{\circ(-)}[\phi] = \int\frac{d^3k}{(2\pi)^{3/2}}\tilde{\phi}(\boldsymbol{k})\int d^3x\,\exp\left\{i(|\boldsymbol{k}|x_0 - \boldsymbol{k}\boldsymbol{x})\right\}\overleftrightarrow{\partial_0}\omega_\circ(x)\,, \tag{2.33}$$

can be introduced, $\tilde{\phi}$ being a suitable arbitrary test function, so that it can be consistently defined:

$$\mathcal{V}_p = \left\{|\mathcal{F}\rangle\in\mathcal{V}'_p:\;\omega_{\circ(-)}|\mathcal{F}\rangle = 0\right\}\,. \tag{2.34}$$

Our theory is established on the same footing as those in covariant gauges: Conjecturing that $\mathcal{V}_p$ has positive-semidefinite metric, calling $\mathcal{V}_\circ$ the subspace of the neutral state-vectors of $\mathcal{V}_p$, the physical Hilbert space will be defined as the complete quotient space $\overline{\mathcal{V}_p/\mathcal{V}_\circ}$, where the (formally) unitary physical S-Matrix can also be well defined [18-21].
Of course, for homogeneous gauges the fields $C$, $\bar{C}$, $\omega$ and $\lambda$ are truly free, obeying the equations:

$$n\partial\,\lambda = 0\,,\qquad n\partial\,C = 0\,,\qquad n\partial\,\bar{C} = 0\,,\qquad \partial^2\omega = 0\,, \tag{2.35}$$

Then we can introduce the decompositions [3,10-12]:

$$\lambda(x) = \frac{1}{(2\pi)^{3/2}}\int\frac{d^3k}{\sqrt{|\boldsymbol{k}|}}\,\theta(-\check{n}^*\boldsymbol{k})\cdot$$
$$\cdot\left[\exp\{-i(\check{n}\boldsymbol{k}\,x_0 - \boldsymbol{k}\boldsymbol{x})\}\,\tilde{\lambda}_{(-)}(\boldsymbol{k}) + \exp\{i(\check{n}\boldsymbol{k}\,x_0 - \boldsymbol{k}\boldsymbol{x})\}\,\tilde{\lambda}_{(+)}(\boldsymbol{k})\right]\,, \tag{2.36a}$$

$$C(x) = \frac{1}{(2\pi)^{3/2}}\int\frac{d^3k}{\sqrt{|\boldsymbol{k}|}}\,\theta(-\check{n}^*\boldsymbol{k})\cdot$$
$$\cdot\left[\exp\{-i(\check{n}\boldsymbol{k}\,x_0 - \boldsymbol{k}\boldsymbol{x})\}\,\tilde{C}_{(-)}(\boldsymbol{k}) + \exp\{i(\check{n}\boldsymbol{k}\,x_0 - \boldsymbol{k}\boldsymbol{x})\}\,\tilde{C}_{(+)}(\boldsymbol{k})\right]\,, \tag{2.36b}$$

$$\bar{C}(x) = \frac{-1}{(2\pi)^{3/2}} \int d^3k \sqrt{|k|}\, \theta(-\check{n}^* k) \cdot$$
$$\cdot \left[\exp\{-i(\check{n}k\, x_0 - kx)\}\, \tilde{\bar{C}}_{(-)}(k) + \exp\{i(\check{n}k\, x_0 - kx)\}\, \tilde{\bar{C}}_{(+)}(k)\right] , \qquad (2.36c)$$

where

$$\left.\begin{array}{lll} n^* = -n & \text{for} & n \neq 0\,, \\ n^* \neq 0 \text{ arbitrary} & \text{for} & n = 0\,. \end{array}\right\} \qquad (2.37)$$

The Fourier-transforms $\tilde{\lambda}_{(\pm)}(k)$, $\tilde{C}_{(\pm)}(k)$, $\tilde{\bar{C}}_{(\pm)}(k)$ are supposed to have support in the half-space $\check{n}^* k < 0$ and the conjugation properties

$$\tilde{\lambda}_{(\pm)} = \left(\tilde{\lambda}_{(\mp)}\right)^\dagger , \qquad \tilde{C}_{(\pm)} = \left(\tilde{C}_{(\mp)}\right)^\dagger , \qquad \tilde{\bar{C}}_{(\pm)} = \left(\tilde{\bar{C}}_{(\mp)}\right)^\dagger , \qquad (2.38)$$

and all (anti-) commute, with the exception of

$$\left[\tilde{C}^a_{(\mp)}(p),\, \tilde{\bar{C}}^b{}_{(\pm)}(q)\right]_+ = \delta_{ab}\, \theta(-\check{n}^* p)\, \delta(p - q)\,. \qquad (2.39)$$

They can be interpreted as creation (- sign) or annihilation (+ sign) operators. Thus, since now $\omega \equiv \omega_o$, in the vector space

$$\left\{|v\rangle \in \mathcal{V} : \omega_o|v\rangle = 0,\ \tilde{C}_{(-)}|v\rangle = 0,\ \tilde{\bar{C}}_{(-)}|v\rangle = 0\right\} \equiv \mathcal{V}_A \otimes |0\rangle_{FP}\,, \qquad (2.40)$$

which is isomorphic to the complete space of state-vectors in a formalism without the Faddeev-Popov ghosts, the theory developed in [3,10,11] is recovered. As a matter of fact, calling $\{\Psi_k\}$ a suitable wave packet system, we have

$$\begin{aligned} Q_B &= -n_0 \int d^3x : [C \cdot \lambda - g\, C \cdot Q_{\hat{B}} C] : \equiv \\ &\equiv -n_0 \sum_k \big(C_{(+)}[\Psi_k] \cdot \lambda_{(-)}[\Psi_k] + \lambda_{(+)}[\Psi_k] \cdot C_{(-)}[\Psi_k] - \\ &\quad - g\, C_{(+)}[\Psi_k] \cdot Q_{\hat{B}} C_{(-)}[\Psi_k]\big)\,, \end{aligned} \qquad (2.41)$$

so that $\forall\, |f\rangle \in \mathcal{V}_A$ the condition:

$$0 = Q_B\, (|f\rangle \otimes |0\rangle_{FP}) = -n_0 \sum_k (\lambda_{(-)}[\Psi_k]|f\rangle \otimes C_{(+)}[\Psi_k]|0\rangle_{FP}) \qquad (2.42a)$$

implies:

$$\lambda_{(-)}[\Psi_k]|f\rangle = 0 \quad \forall k\,, \qquad (2.42b)$$

due to the linear independence of the system $\{C_{(-)}[\Psi_k]|0\rangle_{FP}\}$.

It is worthwhile to stress that a decomposition like (2.36a) is the only way in order to separate a single hermitian field into two mutually hermitian conjugate operators. Actually, for abelian theories a different way for introducing creation and annihilation operators from the field $\lambda$ is open, due to the fact that [11,12] a field $\mathcal{P}$, analogous to that of eq.(2.21) which is the conjugate of $\lambda$, is explicitly known, such that

$$[\lambda(t, x),\, \mathcal{P}(t, y)] = -i\, \Delta\delta(x - y)\,. \qquad (2.43)$$

With the definitions:

$$\left.\begin{aligned} a(x) &= \frac{1}{\sqrt{2}}\left(\lambda(x) + i\sqrt{-\Delta}\,\mathcal{P}\right), \\ a(x)^{\dagger} &= \frac{1}{\sqrt{2}}\left(\lambda(x) - i\sqrt{-\Delta}\,\mathcal{P}\right), \end{aligned}\right\} \tag{2.44}$$

$a$ and $a^{\dagger}$ are respectively creation and annihilation operators and

$$\lambda(x) = \frac{1}{\sqrt{2}}\left(a(x) + a^{\dagger}(x)\right). \tag{2.45}$$

But then the BRST charge:

$$Q_B = -i\int d^3x\ :\mathcal{C}\cdot\lambda: \tag{2.46}$$

turns out to unavoidably contain a term with a product of two creation operators, which apparently renders impossible the requirement that it strongly annihilates the physical state-vectors and the vacuum in particular. Of course one could elude the difficulty, avoiding the use of the BRST formalism and defining the physical state-vectors as those annihilated by the operator $a_{(-)}$, singling out a subspace where the Gauss' law is (weakly) restored. But it has been shown [12] that then problems arise trying to recover Poincaré covariance, seemingly due to the necessary (weak) vanishing of the field $\mathcal{P}$ which causes the weak breaking of the residual gauge invariance.

## 3 The Feynman Propagator of the Gauge Boson

We turn to the interaction representation $\mathring{\mathcal{O}}$ of each Heisenberg operator $\mathcal{O}$, introducing moreover the notations

$$U_\mu = \mathring{A}_\mu\,, \tag{3.1}$$

$$F_{\mu\nu} = \partial_\mu U_\nu - \partial_\nu U_\mu\,, \tag{3.2}$$

$$\mathring{\lambda}_\circ = \frac{1}{n_0}\partial_k F_{0k}\,. \tag{3.3}$$

$$Q_B = -n_0\int d^3x\,\mathring{\mathcal{C}}\cdot\mathring{\lambda}_\circ\,, \tag{3.4}$$

Disregarding the fermionic sector for simplicity, we have the equations

$$\partial_\mu F^{\mu\nu} = n^\nu\mathring{\lambda}_\circ\,, \tag{3.5a}$$

$$\partial^2\mathring{\omega} = n^2\xi\mathring{\lambda}_\circ\,, \tag{3.5b}$$

$$\xi\mathring{\omega} = nU\,; \tag{3.5c}$$

$$n\partial\mathring{\lambda}_\circ = 0\,,\qquad n\partial\mathring{\mathcal{C}} = 0\,,\qquad n\partial\mathring{\bar{\mathcal{C}}} = 0\,. \tag{3.6}$$

We also define

$$n_0\mathring{\lambda}_\circ = \Delta_{\mathrm{T}}\Phi\,, \tag{3.7a}$$

where

$$\Delta_{\mathrm{T}} \equiv \Delta - (\check{n}\partial)^2 \,, \tag{3.7b}$$

and introduce the following general solution of (3.5):

$$\partial U(x) = \Pi(x) + (1-\xi^2)\,\check{n}^2 \check{n}_i \partial_i \Phi(x) - (1-\xi^2)\,\check{n}^2\,(x_0 - t_0)\,\Delta_{\mathrm{T}}\Phi(x) \,, \tag{3.8a}$$

where

$$n\partial\Pi = 0 \,; \tag{3.8b}$$

$$U_i^{\perp} = \mathcal{U}_i - \check{n}_i \Phi + \check{n}_j \partial_j \,\partial_i \Delta^{-1} \Phi \,, \tag{3.8c}$$

with

$$\partial^2 \mathcal{U}_i = 0 \,, \qquad \partial_i \mathcal{U}_i = 0 \,; \tag{3.8d}$$

$$\check{\omega} = -n_0\,\xi\,\check{n}^2 \Phi + \varphi \,, \tag{3.8e}$$

$$\partial^2 \varphi = 0 \,. \tag{3.8f}$$

Consequently we can write

$$U_j = \Lambda_j + \Gamma_j + \Sigma_j \,, \tag{3.9a}$$

where

$$\Lambda_i(x) = \mathcal{U}_i + (\partial_0 + \check{n}_k \partial_k)\Delta_{\mathrm{T}}^{-1} \check{n}_j \partial_i \mathcal{U}_j(x) \,, \tag{3.9b}$$

$$\Gamma_i(x) = -\check{n}_i \Phi(x) + (1-\xi^2)\,\check{n}^2 (x_0 - t_0) \partial_i \Phi + \frac{\partial_i}{\Delta_{\mathrm{T}}}\left[-\Pi(x) + (1-\xi^2)\,\check{n}^2 n_j \partial_j \Phi\right] \,, \tag{3.9c}$$

$$\Sigma_i(x) = \frac{\xi}{n_0}(\partial_0 + \check{n}_k \partial_k)\Delta_{\mathrm{T}}^{-1} \partial_i \varphi \,. \tag{3.9d}$$

We notice that the operator $\Delta_{\mathrm{T}}$ is elliptic for $n^2 > 0$ and hyperbolic for $n^2 < 0$. Correspondingly the symbol $\Delta_{\mathrm{T}}^{-1}$ is supposed to be defined by suitable choice of one of the fundamental solutions of $\Delta_{\mathrm{T}}$ itself, picking up in so doing some particular, but physically irrelevant, asymptotic behavior of the four-potentials along the space-like directions.
It can be proved that

$$(\mathcal{U}_j,\, \partial_0 \mathcal{U}_j) \,, \tag{3.10a}$$

$$(\Phi,\, \Pi) \,, \tag{3.10b}$$

$$(\varphi,\, (1/n^2)\partial_0 \varphi) \tag{3.10c}$$

are independent couples of conjugate variables, such that we have

$$[\Phi(t,\, \boldsymbol{x}),\, \Pi(t,\, \boldsymbol{y})] = -i\,\delta(\boldsymbol{x} - \boldsymbol{y}) \,, \tag{3.11}$$

$$[\varphi(t,\, \boldsymbol{x}),\, \partial_0 \varphi(t,\, \boldsymbol{y})] = i\,n^2\,\delta(\boldsymbol{x} - \boldsymbol{y}) \,, \tag{3.12}$$

$$[\mathcal{U}_j(t,\, \boldsymbol{x}),\, \partial_0 \mathcal{U}_l(t,\, \boldsymbol{y})] = i\left(\delta_{jl} - \frac{\partial_j \partial_l}{\Delta}\right)\delta(\boldsymbol{x} - \boldsymbol{y}) \,. \tag{3.13}$$

We notice, on the basis of arguments produced in [23], assuming linearization of the asymptotic dynamics in perturbation theory, that the asymptotic physical state-vectors are subjected to the condition

$$Q_B\,|F_{as}\rangle = 0\,. \tag{3.14}$$

Consequently, recalling the arguments at the end of the preceding section, used now for the consistent realization of (3.14), we conclude that the quantization-scheme with two gauge vectors(2.36) ***should*** be used in our theory, playing there a privileged, if not an exclusive role:

$$\Phi(x) = \frac{1}{(2\pi)^{3/2}}\int\frac{d^3k}{\sqrt{|\boldsymbol{k}|}}\,\theta(-\check{n}^*\boldsymbol{k})\cdot$$
$$\Big[\exp\{-i(\check{n}k\,x_0 - \boldsymbol{kx})\}\,\tilde{\Phi}_{(-)}(\boldsymbol{k}) + \exp\{i(\check{n}k\,x_0 - \boldsymbol{kx})\}\,\tilde{\Phi}_{(+)}(\boldsymbol{k})\Big]\,, \tag{3.15a}$$

$$\Pi(x) = \frac{i}{(2\pi)^{3/2}}\int d^3k\,\sqrt{|\boldsymbol{k}|}\,\theta(-\check{n}^*\boldsymbol{k})\cdot$$
$$\cdot\Big[\exp\{-i(\check{n}k\,x_0 - \boldsymbol{kx})\}\,\tilde{\Pi}_{(-)}(\boldsymbol{k}) - \exp\{i(\check{n}k\,x_0 - \boldsymbol{kx})\}\,\tilde{\Pi}_{(+)}(\boldsymbol{k})\Big]\,. \tag{3.16b}$$

The Fourier transform $\tilde{\Phi}_{(\pm)}(\boldsymbol{k})$ and $\tilde{\Pi}_{(\pm)}(\boldsymbol{k})$ have support in the half-space $\check{n}^*\boldsymbol{k} < 0$, the conjugation properties

$$\tilde{\Phi}_{(\pm)} = \left(\tilde{\Phi}_{(\mp)}\right)^\dagger\,, \qquad \tilde{\Pi}_{(\pm)} = \left(\tilde{\Pi}_{(\mp)}\right)^\dagger\,, \tag{3.16}$$

and the commutation properties:

$$\left[\tilde{\Phi}_{(-)}(\boldsymbol{k}),\,\tilde{\Pi}_{(+)}(\boldsymbol{k}')\right] = \left[\tilde{\Pi}_{(-)}(\boldsymbol{k}),\,\tilde{\Phi}_{(+)}(\boldsymbol{k}')\right] = \delta(\boldsymbol{k}-\boldsymbol{k}')\,, \tag{3.17}$$

the remaining commutators vanishing.

The generalized Leibbrandt-Mandelstam prescription for the spurious poles exhibited by the Feynman propagator of the gauge bosons emerges computing the vacuum expectation value of Dyson's T-product:

$$\begin{aligned}
\langle 0|\mathbf{T}\Phi(x)\Pi(x')|0\rangle &= -\langle 0|\mathbf{T}\Pi(x)\Phi(x')|0\rangle = \\
= -\,i\,\theta(x_0 - x_0')\int\frac{d^3k}{(2\pi)^3}\,&\theta(\check{n}k)\,\exp\{-i\,\check{n}k\,(x_0 - x_0') + i\,\boldsymbol{k}(\boldsymbol{x}-\boldsymbol{x}')\} + \\
+\,i\,\theta(x_0' - x_0)\int\frac{d^3k}{(2\pi)^3}\,&\theta(\check{n}k)\,\exp\{-i\,\check{n}k\,(x_0 - x_0') + i\,\boldsymbol{k}(\boldsymbol{x}-\boldsymbol{x}')\} = \\
= \int\frac{d^4k}{(2\pi)^4}&\left[\frac{\theta(\check{n}k)}{\check{n}k + i0^+} + \frac{\theta(-\check{n}k)}{\check{n}k - i\,0^+}\right]e^{-i\,k(x-x')} = \\
= \int\frac{d^4k}{(2\pi)^4}&\frac{e^{-i\,k(x-x')}}{\check{n}k + i\,0^+ sign(\check{n}k)} \equiv \mathcal{D}_{\check{n}}(x-x')\,,
\end{aligned} \tag{3.18}$$

such that

$$[\check{n}\partial]\,\mathcal{D}_{\check{n}}(x) = -i\,\delta(x)\,. \tag{3.19}$$

Doing calculations along the lines of [3,10-12], we eventually obtain

$$\langle 0|\mathbf{T}U_\mu(x)U_\nu(x')|0\rangle \equiv D_{\mu\nu}(x-x') = i\int\frac{d^4k}{(2\pi)^4}\,\frac{e^{-i\,k(x-x')}}{k^2 + i0^+}\,d_{\mu\nu}(k)\,, \tag{3.19a}$$

$$d_{\mu\nu}(k) = -g_{\mu\nu} + \frac{n_\mu k_\nu + n_\nu k_\mu}{[nk]} + (1-\xi^2)n^2 \frac{k_\mu k_\nu}{n_0} \frac{\partial}{\partial k_0} \frac{1}{[nk]} . \qquad (3.19b)$$

**Acknowledgements** It is a pleasure to express to the organizers and the financial supporters my gratitude for the opportunity given to me of participating at this workshop and for the very kind hospitality found.

## References

[1] S. Mandelstam, Nucl. Phys. **B213** (1983) 149.
[2] G. Leibbrandt, Phys. Rev. **D29** (1984) 1699.
[3] A. Bassetto, M. Dalbosco, I. Lazzizzera and R. Soldati, Phys. Rev. **D31** (1985) 2012.
[4] G. Leibbrandt, Nucl. Phys. **B310** (1988) 405.
[5] P. Gaigg, M. Kreuzer, M. Schweda and O. Piguet, J. Math. Phys. **28** (1987) 2781.
[6] P. Gaigg and M. Kreuzer, Phys. Lett. **205B** (1988) 530;
P. Gaigg, M. Kreuzer and G. Pollak, Phys. Rev. **D30** (1988) 2559.
[7] G. Nardelli and R. Soldati, Phys. Lett. **206** (1988) 495;
G. Nardelli, Phys. Rev. **D40** (1989) 2662
[8] G. Leibbrandt and S. L. Nyeo, Mod. Phys. Lett. **A3** (1988) 1085.
[9] G. Pollak, Phys. Rev. **D40** (1989) 2027.
[10] I. Lazzizzera, Phys. Lett. **210B** (1988) 188.
[11] I. Lazzizzera, Trento preprint UTF-146 (1988), Il Nuovo Cim. **A** in press.
[12] I. Lazzizzera, Trento preprint UTF-100 (1080).
[13] P. V. Landshoff, Phys. Lett. **227** (1989) 427.
[14] P. V. Landshoff and J. C. Taylor, preprint DAMTP 89.
[15] A. Burnel, Phys. Rev. **D40** (1989) 1221.
[16] H. Hüffel, P. V.Landshoff and J. C.Taylor, Phys. Lett. **217B** (1989) 147.
[17] A. Bassetto, I. Lazzizzera and R. Soldati, Nucl. Phys. **B236** (1984) 319.
[18] C. Becchi, A. Rouet and R. Stora, Ann. Phys. **98** (1976) 287.
[19] G. Curci and R. Ferrari, Il Nuovo Cim. **35A** (1976) 273.
[20] T. Kugo and I. Ojima, Suppl. Prog. Theor. Phys. **66** (1979) 1.
[21] L. Bonora, P. Pasti and M. Tonin, Nuovo Cim. **64A** (1981) 307.
[22] P. Jordan and E. Wigner, Z. Phys. **47** (1928) 631;
L. Corwin, Y. Ne'eman and S. Sternberg, Rev. Mod. Phys. **47** (1975) 573.
[23] A. A. Slavnov, Phys. Lett. **217B** (1989) 91.

# BRST-Approach to Noncovariant Gauges

A. Burnel

Université de Liège, Institut de Physique au Sart Tilman (B.5),
B-4000 Liège 1, Belgium

## 1 INTRODUCTION

The aim of this contribution is to discuss the motivations for a consistent universal description of most of the known linear gauges used in Yang-Mills theory. Due to the lack of space, I restrict myself to the leading ideas leaving the (elementary) technical details for the published [1]-[4] or forthcoming [5] papers.

## 2 STANDARD SCHEME FOR QUANTIZING A GAUGE THEORY

In order to set my conventions of language, let me first recall the three different ways of fixing a gauge [1] in the framework of the Hamiltonian analysis of constrained systems. To be precise, I illustrate them directly in the case of Yang-Mills theory. I drop the internal symmetry indices because they do not play any role in the discussion. For the same reason, matter fields are not introduced. The starting Lagrangian is

$$\mathcal{L} = -\frac{1}{4} F_{\mu\nu}F^{\mu\nu} \ . \tag{1}$$

It gives rise to two first-class constraints

$$\pi^0 = 0 \ \dot{\Rightarrow} \ D^k\pi^k = 0 \ , \tag{2}$$

where the $\pi$'s are the variable canonically conjugate to the A's and where $\dot{\Rightarrow}$ means that the second equation can be obtained from the first one by time differentiation in the Hamiltonian formalism.

The three different gauge choices are

$$\text{class} \quad \text{I :} \quad f(A_k,\pi^k) = 0 \ \dot{\Rightarrow} \ g(A_0,A_k,\pi^k) = 0 \ \dot{\Rightarrow} \ \partial_0 A_0 = h(A_0,A_k,\pi^k) \ , \tag{3}$$

class II : $$g(A_0,A_k,\pi^k) = 0 \doteq \partial_0 A_0 = h(A_0,A_k,\pi^k) \ , \tag{4}$$

class III : $$\partial_0 A_0 = h(A_\mu,\pi^\mu) \ . \tag{5}$$

In class I gauges, all the constraints can be realized strongly and all the degrees of freedom are physical. In class II gauges, the Gauss law cannot be satisfied in the general operator form but is imposed as a condition restricting the full states space to the subspace of physical states (Dirac quantization). In class III gauges, one is left with one additional unphysical degree of freedom accompanied by Faddeev-Popov ghosts. The quantization scheme is the BRST one with physical states defined as cohomology classes of the BRST operator.

## 3 NON-COVARIANT CLASS I GAUGES

Two class I gauges are used in perturbative calculations of Yang-Mills theory, the pure axial gauge and the Coulomb gauge characterized respectively by

$$A_3 = 0 \doteq \pi^3 - \partial^3 A_0 = 0 \ , \tag{6}$$

$$\partial_k A_k = 0 \doteq \partial^k \pi^k - \partial^k D^k A_0 = 0 \ . \tag{7}$$

Both suffer from serious troubles. First of all, the solution of the second of eqs.(6) and (7) with respect to $A_0$ is ambiguous. Moreover, even if this difficulty is resolved, the theory remains inconsistent. I refer to Soldati's [6] and Bassetto's [7] contributions for the positivity problem in the axial gauge while noting that the operator ordering in the Coulomb gauge requires additional arguments.

In working with these gauges, two different attitudes can be adopted. The first one is to assume that the gauge condition is frame independent but that $A_\mu$ transforms as a four-vector plus an additional gauge term. I am not sure that this is consistent because such a transformation is a gauge transformation on physical degrees of freedom which, in principle, are gauge invariant. In any case, this way of using a class I gauge is unwarranted in view of the Singer theorem stating (essentially) that, in fact, there is no class I gauge in Yang-Mills theory.

Another attitude is to regard $A_\mu$ as a honest four-vector and the gauge condition as being frame-dependent. This gives its full meaning to the word "noncovariant" which means that not all the coordinate frames are equivalent. Actually, one encounters <u>singular</u> [4] and <u>preferred</u> frames. Let me begin by illustrating what is meant by a singular frame, the definition of preferred frames being differed to the next section.

In general, the pure axial gauge can be written $n \cdot A = 0$. When $n_0 \neq 0$, it involves $A_0$ and is therefore a class II gauge condition. It is only in the frames where $n_0 = 0$ that it is class I. In the same way, the Coulomb gauge can be written as

$n\cdot\partial\; n\cdot A - \partial\cdot A = 0$. When $n_0 \neq 1$, it is a class III condition which becomes class I only when $n = (1,0,0,0)$. In both cases, the class I character is a property of the coordinate frame rather than of the gauge itself. These coordinate frames involve less degrees of freedom than the general frames. This can obviously be misleading. Therefore, I call them singular frames. Now, the pure axial gauge and the Coulomb gauge become compatible with the Singer theorem provided one does not make the analysis in a singular frame.

## 4 THE COULOMB GAUGE AND THE EVOLUTION OF UNPHYSICAL DEGREES OF FREEDOM

Let me introduce the Coulomb gauge condition inside the Lagrangian with the help of a Lagrange multiplier $S$ corresponding to the Nakanishi-Lautrup field.

$$\mathcal{L}_C = -\frac{1}{4} F_{\mu\nu}F^{\mu\nu} - S(n\cdot\partial\; n\cdot A - \partial\cdot A) \; . \tag{8}$$

One easily gets, from the free field equations,

$$[(n\cdot\partial)^2 - \Box]S = 0 \; . \tag{9}$$

The evolution operator for the unphysical degrees of freedom is a parabolic operator for which it is well-known that both the Cauchy problem and the boundary value problem on a closed surface are ill-defined. The fundamental problem in a field theory, at least for scattering, consists in finding the propagator even if we are, in general, unable to solve it rigorously. The propagator results from the commutation relations for any time which are obtained from a Cauchy problem with equal-time commutators as initial data. Therefore, in the Coulomb gauge, the field theoretical problem is ill-defined. This may explain some of the difficulties met in the gauge [8].

In a covariant theory, the evolution operator $\Box$ is hyperbolic in Minkowski space and elliptic in Euclidean spaces, both cases being related by a Wick rotation. In order to have a chance to describe consistently the noncovariant gauges, a similar situation should be required but, owing to noncovariance, Wick rotation is possible only in some particular frames, which I call preferred frames.

A way to treat the Coulomb gauge consists to regularize the operator $(n\cdot\partial)^2 - \Box$ by $(n\cdot\partial)^2 - \alpha\Box$. This new operator is elliptic for $0 < \alpha < 1$, hyperbolic for $\alpha > 1$ and parabolic (and therefore pathological) for $\alpha = 0$ and $\alpha = 1$. The frame in which $n = (1,0,0,0)$ is a preferred frame. Originally due to Frenkel and Taylor [9], this gauge was rediscovered independently by myself [2] and used by Nyeo [10]. It offers a regularization for the static temporal gauge too and is used intensively in finite temperature calculations (see G. Kunstatter's talk [11]).

Its main properties are :

- for $\alpha > 1$, there is an unphysical pole in the propagator which is regularized by the causal $i\varepsilon$ prescription like in relativistic gauges ;
- for $0 < \alpha < 1$, there are only physical real poles in the propagator. This is what

can be called a physical gauge. The unphysical degrees of freedom including the never decoupling Faddeev-Popov ones are associated with complex poles at $(n\cdot k)^2 - \alpha k^2 = 0$.

## 5 THE LEIBBRANDT-MANDELSTAM PRESCRIPTION FOR AXIAL GAUGES

In the usual formulation of axial gauges, the Lagrangian is

$$\mathcal{L}_A = -\frac{1}{4} F_{\mu\nu}F^{\mu\nu} - S\, n\cdot A \qquad (10)$$

where, in a general frame, $n_0 \neq 0$. It leads to $n\cdot\partial\, S = 0$, which shows that the evolution of the unphysical degree of freedom is governed by a first-order equation. For this part of the discussion, it is sufficient to consider only the temporal gauge. The solution of the Cauchy problem leads to the principal value prescription for the unphysical pole at $n\cdot k = 0$. This is known to be unsatisfactory. This is however not Yang-Mills because the Gauss law is not implemented. With the implementation of the Gauss law, the propagator is CCM [16] with its unelegant breaking of time-reversal invariance. The situation is even obscured by two contributions to this Workshop. Indeed, Lazzizzera [12] is able to get the Leibbrandt-Mandelstam prescription in a Hamiltonian formalism based on eq.(10). On the other hand, Kalloniatis [13] gets a prescription similar but not identical to the CCM one by going from the Coulomb to the temporal gauge. On the basis of these results, it seems that, in the temporal gauge, there is no definite prescription conciliating the Lagrangian and Hamiltonian formalisms.

One knows however that the Leibbrandt-Mandelstam prescription seems to resist all the tests which were applied to it. For any $n^2$, it consists of

$$\frac{1}{n\cdot k} \rightarrow \frac{n^*\cdot k}{n\cdot k\; n^*\cdot k + i\varepsilon} \quad . \qquad (11)$$

If we look at the denominator, we again recognize the causal $i\varepsilon$ prescription associated with a hyperbolic second-order operator. The Lagrangian [3] giving rise to it is

$$\mathcal{L}_{LM} = -\frac{1}{4} F_{\mu\nu}F^{\mu\nu} - n^*\cdot\partial\, S\, n\cdot A \quad . \qquad (12)$$

Let me summarize the main properties of this theory.

a) One also meets here singular and preferred frames. For $n^2 = 1$, for instance, the choice $n^* = (0,0,0,1)$ and $n = (1,0,0,0)$ leads to a singular frame while a preferred frame is, for any $n^2$, $n = (n_0,\vec{n})$, $n^* = (n_0,-\vec{n})$. By comparing the values of $(n^*)^2$ in a preferred and in a singular frame, it is easy to see that $n^*$ is not a four-vector. Let me remark that the use of a singular frame for performing integration after the quantization has been achieved in a preferred frame is not inconsistent. This does give the correct answer for the Wilson loop [14].

b) The variation of the Lagrangian with respect to $S$ leads to $n^*\cdot\partial\, n\cdot A = 0$. A priori, this is not the axial gauge $n\cdot A = 0$ but it reduces to it if we take the solution which vanishes asymptotically, as is allowed with a hyperbolic operator.

c) To discuss the propagator, it is useful though not necessary to add an inhomogeneous term $-\frac{1}{2}aS^2$ to (12). The propagator is then

$$D_{\mu\nu}(k) = \frac{-i}{k^2+i\varepsilon}\left\{g_{\mu\nu} - \frac{n_\mu k_\nu + n_\nu k_\mu}{n\cdot k\; n^*\cdot k+i\varepsilon}\, n\cdot k - \frac{k_\mu k_\nu[ak^2 - n^2(n^*\cdot k)^2]}{(n\cdot k\; n^*\cdot k+i\varepsilon)^2}\right\}, \qquad (13)$$

where the $i\varepsilon$ is the causal prescription associated with the hyperbolic $\Box$ and $n^*\cdot\partial\, n\cdot\partial$ operators. One should remark that, for $a \neq 0$, one has, up to four-divergences

$$\mathcal{L} = -\frac{1}{4}A_\mu(D^{-1})^{\mu\nu}A_\nu, \qquad (14)$$

when the $S$ field is removed with the help of field equations. This familiar property does not hold in the Lazzizzera's approach to the Leibbrandt-Mandelstam prescription. However, the term $ak^2$ can generate wrong divergences, as pointed out by G. Pollak. The addition of an inhomogeneous term should therefore be regarded only as an easy means of getting the propagator through eq.(14) but not as a gauge parameter. If one wants a gauge parameter with the Leibbrandt-Mandelstam prescription, one may interpolate between axial and planar gauges (see below with the parameter a').

d) For nonabelian theories, the dimensionally regularized ghost loops vanish. The proof goes like that for the formal poles.

## 6 GENERAL LINEAR GAUGE

The previous two examples show that the usual noncovariant gauges can be considered very usefully as class III gauges. The formulation can be extended to any linear gauge by considering the Lagrangian

$$\mathcal{L} = -\frac{1}{4}F_{\mu\nu}F^{\mu\nu} + Q(C^{\mu\nu}\partial_\mu \xi A_\nu + \frac{a}{2}S\xi - \frac{a'}{2}S'C^\mu\partial_\mu\xi + \partial_\mu u C'^{\mu\nu}\partial_\nu\xi' + uC^\mu\partial_\mu\xi), \qquad (15)$$

where the gauge breaking term is a BRST transform leading therefore to BRST invariance. This BRST invariance may even be extended to the various gauge parameters $C^{\mu\nu}$, a, a', $C'_{\mu\nu}$, $C_\mu$. Of course, one restricts $C^{\mu\nu}$ and $C'_{\mu\nu}$ by the condition that $\Box_C = C^{\mu\nu}\partial_\mu\partial_\nu$ and $\Box_{C'} = C'_{\mu\nu}\partial^\mu\partial^\nu$ are elliptic or hyperbolic operators in order to be able to introduce a preferred frame. The last three terms may look strange but it is the only consistent way I know of for incorporating the planar gauge. Let me quote only some particular examples :

a) relativistic gauges : $C_{\mu\nu} = g_{\mu\nu}$ , $C_\mu = C'_{\mu\nu} = 0$ ;

b) regulated static temporal and Coulomb gauges

$$C_{\mu\nu} = n_\mu n_\nu - \alpha g_{\mu\nu} \quad , \quad n^2 = 1 \quad , \quad \alpha > 0 \ (\neq 1) \quad , \quad C_\mu = C'_{\mu\nu} = 0 \quad ;$$

c) axial gauges with the LM prescription

$$C_{\mu\nu} = n^*_\mu n_\nu \quad , \quad a = C_\mu = C'_{\mu\nu} = a' = 0 \quad ;$$

d) planar gauge with the LM prescription

$$C_{\mu\nu} = n^*_\mu n_\nu \quad , \quad a = 0 \quad , \quad a' = 1 \quad , \quad n^2 C_\mu = n^*_\mu \quad , \quad n^2 C'_{\mu\nu} = g_{\mu\nu} \quad .$$

A complete list will be given in a forthcoming paper [4].

One could also incorporate the Chan-Halpern flow gauges [15] by adding a term $Q(C'_\mu A^\mu \xi)$ but, if we want the ghost decoupling, the ghost poles should be e.g. of the form $[k_0 \pm i\alpha|\vec{k}|^2]^{-1}$. This shows that the evolution of unphysical degrees of freedom is governed by a parabolic operator. Therefore the theory is potentially pathological, so that this kind of flow gauges should be avoided.

The general expression (15) of the Lagrangian is useful for carrying out the quantization program once for all and getting some other general results. For the discussion of renormalization, it is however appropriate to restrict oneself to special cases. Among them, the ghost free ones are characterized by $C_{\mu\nu} = n^*_\mu n_\nu$ while the physical gauges are those for which $\Box_C$ and $\Box_{C'}$ are elliptic operators.

## 7 SUMMARY AND CONCLUSIONS

All the non-covariant linear gauges can be formulated as class III gauges just like the covariant ones. This yields a consistent, ambiguity free, universal theory which satisfies all the requirements of usual field theory, except for covariance in the unphysical sector. One may even apply to it the formal path-integral methods. If the $\Box_C$ operator is hyperbolic, both physical and unphysical poles are regularized by the causal $i\varepsilon$ prescription. If $\Box_C$ is elliptic, the propagator shows only physical poles. A parabolic $\Box_C$ must be avoided because it is pathological. There are singular $(C_{00} = 0)$ and preferred $(C_{00} \neq 0 \ , \ C_{0k} + C_{k0} = 0)$ frames. In principle, the quantization should not be carried out in a singular frame.

## REFERENCES

[1] A. BURNEL, Phys.Rev.**D26**,442(1982).
[2] A. BURNEL, Phys.Rev.**D36**,1852(1987).
[3] A. BURNEL, Phys.Rev.**D40**,1221(1989).
[4] A. BURNEL, Phys.Lett.**227B**,431(1989).

[5] A. BURNEL, R. KOBES, G. KUNSTATTER and K.W. MAK, in preparation.
[6] R. SOLDATI, this Proceedings.
[7] A. BASSETTO, this Proceedings.
[8] J.C. TAYLOR, this Proceedings.
[9] J. FRENKEL and J.C. TAYLOR, Nucl.Phys.**B109**,439(1976).
[10] S.L. NYEO, Phys.Rev.**D36**,2512(1987).
[11] G. KUNSTATTER, this Proceedings.
[12] I. LAZZIZZERA, this Proceedings.
[13] A.C. KALLONIATIS, this Proceedings.
[14] H. HÜFFEL, P.V. LANDSHOFF and J.C. TAYLOR, Phys.Lett.**217B**,147(1989).
[15] H.S. CHAN and M.B. HALPERN, Phys.Rev.**D35**,540(1986).
[16] S. CARACCIOLO, G. CURCI and P. MENOTTI, Phys.Lett. 113B,311(1982)

# Canonical Quantization of Gauge Theories in Axial Gauges

Kurt Haller

Department of Physics, University of Connecticut,
Storrs, CT 06296-3046, USA

In this review I will discuss an approach to axial gauges based on the canonical quantization of gauge fields. This work takes as its point of departure a Lagrangian in which each field has a canonically conjugate partner so that its commutation rules with all other fields is determined. Primary constraints, which stem from those fields in the Lagrangian whose time derivatives do not also appear in it, are avoided in this approach. As a result, the operator constraints that define the gauge sometimes are replaced by their time derivatives; and the constraints themselves are implemented "weakly" by defining a physical subspace in which the constraint applies, and to which the dynamical time evolution of state vectors remains confined. Another essential feature is the use of ghost states to represent the excitations that appear in longitudinal and timelike components of gauge fields. Zero norm ghost states permit a realistic and satisfactory representation of these components. The quanta of longitudinal fields (such as the electric field in QED, for example) carry no energy or momentum and have no probability of being detected. Ghost excitations pro vide a suitable and natural formalism for incorporating these properties. The ghost excitations useful in gauge theories consist of two species here represented by annihilation operators $a_Q(\mathbf{k})$ and $a_R(\mathbf{k})$ and their respective creation operators $a_Q^*(\mathbf{k})|p\rangle$ and $a_R^*(\mathbf{k})$ (in QCD they carry an additional Lie group index, which we generally suppress). The algebra of the ghost sector requires a small generalization of the algebra of positive metric Hilbert spaces. The norms of any states of the form $a_Q^*(\mathbf{k})|p\rangle$ or $a_R^*(\mathbf{k})|p\rangle$, where $|p\rangle$ is any normalizable ghost-free state, vanish; but the inner product of $a_Q^*(\mathbf{k})|p\rangle$ and $a_R^*(\mathbf{k})|p\rangle$ does not vanish. $a_Q(\mathbf{k})$ and $a_Q^*(\mathbf{k})$ commute, as do $a_R(\mathbf{k})$ and $a_R^*(\mathbf{k})$. But $[a_Q(\mathbf{k}), a_R^*(\mathbf{k}')] = [a_R(\mathbf{k}), a_Q^*(\mathbf{k}')] = \delta_{\mathbf{k},\mathbf{k}'}$ and the unit operator in the one particle sector is given by

$$1 = \sum_{\mathbf{k}} \left(a_Q^*(\mathbf{k})|0\rangle\langle 0|a_R(\mathbf{k}) + a_R^*(\mathbf{k})|0\rangle\langle 0|a_Q(\mathbf{k})\right). \tag{1}$$

A feature of axial gauge theories that has attracted particular attention at this workshop is the ambiguity in the singular propagator (using the convention $g_{\mu\nu}=\delta_{\mu\nu}$)

$$D_{\mu\nu}(k)= \frac{\delta_{ab}}{(k^2- i\varepsilon)}\left[\delta_{\mu\nu} - \frac{(k_\mu n_\nu+k_\nu n_\mu)}{(k\cdot n)} + \frac{n^2 k_\mu k_\nu}{(k\cdot n)^2}\right]. \tag{2}$$

Originally this propagator was evaluated using the principal value (PV) prescription, but later work raised serious questions about that choice; in particular, in the light cone gauge the PV prescription has been shown to lead to uncontrollable infinities, and the Mandelstam-Leibbrandt (ML) prescription has replaced it.[1-3] Efforts have been made to learn whether one or another prescription for the singular propagator can be shown to be derivable from canonical quantization.[4-8] Different approaches to canonical quantization in axial gauges have led to propagators with different prescriptions. But these various canonical quantization procedures have been so dissimilar from each other in so many ways, that the characteristic that correlates with the propagator prescription has not been easy to identify. I will here use the temporal gauge as a model to cast some new light on this question. The temporal gauge has the advantage of being a relatively uncomplicated axial gauge. Prior work on the canonical quantization has led to a propagator which allows for a possible violation of time-translation invariance, but which also includes the PV version of the propagator as a special case.[7] I will discuss an alternate treatment of the temporal gauge which is based on the identical Lagrangian as that earlier work, the same equations of motion, and the same weak constraints. The only difference between the two will be in the explicit dependence of the longitudinal and timelike fields on the ghost excitations. And that difference alone will be shown to make a change in the propagator equivalent to a change in the prescription with which the propagator is evaluated. I remind the reader that the Lagrangian for this theory is

$$\mathcal{L} = -(1/4)f_{ij}f_{ij} + (1/2)f_{i0}f_{i0} + j_iA_i - j_0A_0 - \partial_0A_0G - \bar{\psi}\,(m + \gamma\cdot\partial)\psi \tag{3}$$

where $f_{ij} = \partial_jA_i - \partial_iA_j - 2eA_i\times A_j$ and $f_{i0} = \partial_0A_i + \partial_iA_0 + 2eA_i\times A_0$ for Yang-Mills theory and QCD (I will refer to both these theories, collectively, as QCD hereafter), and $f_{ij} = \partial_jA_i - \partial_iA_j$ and $f_{i0} = \partial_0A_i + \partial_iA_0$ for QED; $j_\mu = ie\bar{\psi}\gamma_\mu\tau\psi$ for QCD and $j_\mu = ie\bar{\psi}\gamma_\mu\psi$ for QED. The equations of motion include Ampere's Law, $\partial_0f_{i0} - \partial_jf_{ij} - J_i = 0$; and also $\partial_if_{i0} + J_0 = (\partial_0 - 2eA_0\times)G$ for QCD and $\partial_if_{i0} + J_0 = \partial_0G$ for QED. The latter two equations would be statements of Gauss's Law, were their respective right hand sides replaced by 0. However the remaining equations of motion fall short of entitling us to assume that much. They are, for QCD, $\partial_0(\partial_0 - 2eA_0\times)G = 0$ and, for QED, $\partial_0\partial_0G = 0$; in addition, there is the equation of motion $\partial_0A_0 = 0$ in QCD and QED. Whenever it occurs, $J_\mu$ refers to the conserved current. In the case of QED $J_\mu = j_\mu$ ; in QCD $J_i = j_i + 2eA_0\times f_{i0} + 2eA_j\times f_{ij}$ and $J_0 = j_0 + 2eA_i\times f_{i0} - 2eA_0\times G$. And $f_{i0}$, the negative electric field, is the momentum conjugate to $A_i$, and $-G$ is the momentum conjugate to $A_0$, so that $[A_i(\mathbf{x}), f_{j0}(\mathbf{y})] = i\delta_{i,j}\delta(\mathbf{x}-\mathbf{y})$, and $[A_0(\mathbf{x}), G(\mathbf{y})] = -i\delta(\mathbf{x}-\mathbf{y})$, while all other commutators of these fields among themselves vanish. Anticommutation rules for the spinor field are as in ref. (7).

Before discussing the implementation of the constraints, we proceed to the main point of this work. The canonical formalism leads to the Hamiltonian

$$H = \int \mathcal{H} d\mathbf{x} \tag{4}$$

where the Hamiltonian density $\mathcal{H}$ is given by $\mathcal{H} = \mathcal{H}_0 + \mathcal{H}_1$, and where

$$\mathcal{H}_0 = (1/2)f_{i0}f_{i0} + (1/4)(\partial_j A_i - \partial_i A_j)(\partial_j A_i - \partial_i A_j) + A_0 \partial_i f_{i0} + \mathcal{H}(\text{fermions}) \tag{5}$$

and

$$\mathcal{H}_1 = j_0 \cdot A_0 - j_i \cdot A_i - 2eA_0 \cdot (A_i \times E_i) + 2e\partial_j A_i \cdot (A_j \times A_i) + e^2 (A_j \times A_i) \cdot (A_j \times A_i) \tag{6}$$

for QCD, and

$$\mathcal{H}_1 = j_0 A_0 - j_i A_i \tag{7}$$

for QED. Up to this point,all the results reviewed have been identical to the material in the original discussion in ref. (7). Only at this point do we make the first and essential departure from that earlier formulation. Whereas we adhere to the standard representation of the transverse parts of $A_i$ and $E_i = -f_{i0}$, we adapt a procedure used by Bassetto *et al*,[5] by Lazzizzera,[6] and by Landshoff and Taylor,[8] and use the following representations of the longitudinal components:

$$A_i^L(\mathbf{x}) = \sum_{\mathbf{k}} \frac{k_i}{k^{3/2}} \Theta(k_3)\left( a_R(\mathbf{k})\, e^{i\mathbf{k}\cdot\mathbf{x}} + a_R^*(\mathbf{k})\, e^{-i\mathbf{k}\cdot\mathbf{x}}\right), \tag{8}$$

$$E_i^L(\mathbf{x}) = i\sum_{\mathbf{k}} \frac{k_i}{\sqrt{k}} \Theta(k_3)\left( a_Q(\mathbf{k})\, e^{i\mathbf{k}\cdot\mathbf{x}} - a_Q^*(\mathbf{k})\, e^{-i\mathbf{k}\cdot\mathbf{x}}\right), \tag{9}$$

and for the timelike $A_0$ and its canonically conjugate G,

$$A_0(\mathbf{x}) = \sum_{\mathbf{k}} \frac{-i\alpha}{\sqrt{k}} \Theta(-k_3)\left( a_Q(\mathbf{k})\, e^{i\mathbf{k}\cdot\mathbf{x}} - a_Q^*(\mathbf{k})\, e^{-i\mathbf{k}\cdot\mathbf{x}}\right), \tag{10}$$

and

$$G(\mathbf{x}) = \sum_{\mathbf{k}} \frac{\sqrt{k}}{\alpha} \Theta(-k_3)\left( a_R(\mathbf{k})\, e^{i\mathbf{k}\cdot\mathbf{x}} + a_R^*(\mathbf{k})\, e^{-i\mathbf{k}\cdot\mathbf{x}}\right). \tag{11}$$

These representations of the fields differ from the ones used in ref. (7) in the following way: In this alternate representation, the longitudinal components of $A_i(\mathbf{x})$ and $E_i(\mathbf{x})$ are superpositions of only those ghost excitations for which $k_3 > 0$; and, to preserve commutation rules, $A_0(\mathbf{x})$ and $G(\mathbf{x})$ are required to be superpositions of only those ghost excitations for which $k_3 < 0$. In ref. (7), $A_i(\mathbf{x})$, $E_i(\mathbf{x})$, $A_0(\mathbf{x})$ and $G(\mathbf{x})$ have representations similar to the ones used here, except that all of them extend over all values of $k_3$. Eqs. (8) - (11) satisfy the equal time commutation rules imposed by the canonical formulation precisely as do the corresponding representations used in ref. (7). The field equations therefore are completely unaffected by the substitution of this alternate representation of the gauge fields. When $H_0$ is expressed in terms of the gluon (or photon) excitations, we find that eqs. (8) - (11) lead to

$$H_0 = \sum_{\mathbf{k}} k\left[ a_n^\dagger(\mathbf{k})\, a_n(\mathbf{k}) + a_Q^*(\mathbf{k}) a_Q(\mathbf{k}) \Theta(k_3) - i(\alpha/2)\left(a_Q(\mathbf{k}) a_Q(-\mathbf{k}) - a_Q^*(\mathbf{k}) a_Q^*(-\mathbf{k})\right)\right] + H_0(f) \tag{12}$$

where $H_0(f)$ represents the Hamiltonian for non-interacting fermions. When we use $H_0$ in the expression for the interaction picture field $A_i(\mathbf{x},t)$, given by $A_i(\mathbf{x},t)=\exp[iH_0t]\, A_i(\mathbf{x})\, \exp[-iH_0t]$, we obtain, for the longitudinal component of $A_i(\mathbf{x},t)$,

$$A_i^L(\mathbf{x},t) = \sum_{\mathbf{k}} \frac{k_i}{2k^{3/2}} \Big\{ \Theta(k_3)\Big[\big(a_R(\mathbf{k})\, e^{i\mathbf{k}\cdot\mathbf{x}} + a_R^*(\mathbf{k})\, e^{-i\mathbf{k}\cdot\mathbf{x}}\big) - ikt\big(a_Q(\mathbf{k})\, e^{i\mathbf{k}\cdot\mathbf{x}} - a_Q^*(\mathbf{k})\, e^{-i\mathbf{k}\cdot\mathbf{x}}\big)\Big] +$$

$$\alpha kt\Theta(-k_3)\big(a_Q(\mathbf{k})\, e^{i\mathbf{k}\cdot\mathbf{x}} + a_Q^*(\mathbf{k})\, e^{-i\mathbf{k}\cdot\mathbf{x}}\big) \Big\}. \qquad (13)$$

From eq. (13) it follows that the longitudinal propagator, $D^L_{i,j}(\mathbf{x}_1,t_1;\, \mathbf{x}_2,t_2) = \langle 0|T\{A_i(\mathbf{x}_1,t_1)\, A_j(\mathbf{x}_2,t_2)\}|0\rangle$, is given by

$$D^L_{ij}(\mathbf{x}_1,t_1;\, \mathbf{x}_2,t_2) = -i\int \frac{d\mathbf{k}}{(2\pi)^3}\, e^{i\mathbf{k}\cdot(\mathbf{x}_1-\mathbf{x}_2)} \left(\frac{k_i k_j}{k^2}\right)\big[\Theta(k_3)(t_1-t_2)\Theta(t_1-t_2) + \Theta(-k_3)(t_2-t_1)\Theta(t_2-t_1)\big]. \qquad (14)$$

To obtain this expression from

$$D_{i,j}(\mathbf{x},t) = -i\int \frac{d\mathbf{k}\,dk_0}{(2\pi)^4} \frac{e^{i(\mathbf{k}\cdot\mathbf{x}-k_0x_0)}}{k^2-i\varepsilon}\left(\delta_{i,j} - \frac{k_i k_j}{k_0^2}\right) \qquad (15)$$

we would have to use the prescription

$$\frac{1}{k_0^2} = \frac{1}{(k_0 + i\varepsilon\,\mathrm{sign}(k_3))^2} \qquad (16)$$

in the integration instead of the PV prescription.

It is significant that this canonical formulation, in which $A_0$ and G are operator valued fields and gauge fixing is weakly implemented in a physical subspace, gives rise to a propagator very similar to one obtained by Lazzizzera, who uses $A_0=0$ identically, in the context of a very different approach to the canonical formulation of the temporal gauge.[6] I am not arguing here that this form of the propagator is preferable to the one given in ref. (7), or to somewhat similar ones used by Cheng and Tsai[4] and by Caracciolo *et al*,[9] or to the ones proposed by Gaigg *et al* [10] or by Landshoff.[11] The point here is merely to use this alternate formulation as a "toy" model, to demonstrate that changing the representation of the gauge fields in terms of ghost operators is precisely what corresponds to a change of prescription in the singular integration in the axial gauge propagator. The argument consists of showing that when the representation of the gauge fields in terms of ghost excitations is changed, but everything else in the theory remains unaltered, the propagator prescription is affected.

To guarantee the viability of the theory it is also necessary to include interactions and to implement the constraints which guarantee the validity of Gauss's Law and fix the gauge to $A_0=0$. The alternate representations of the gauge fields used here do not require any change in the strategy for achieving these objectives. But, with this alternate representation, the forms of the operators that impose these constraints change when we express them in terms of ghost excitations. Explicit calculations show that $[a_Q(\mathbf{k}),H_1]=-\Theta(k_3)k^{-(3/2)}[H,J_0(\mathbf{k})]$ so that $[H,\{a_Q(\mathbf{k})+k^{-(3/2)}J_0(\mathbf{k})\}]=0$ when $k_3>0$, but $[H,a_Q(\mathbf{k})]=0$ when $k_3<0$. We therefore can define the operator

$$\Omega(\mathbf{k}) = \Theta(k_3)\left(a_Q(\mathbf{k}) + \frac{J_0(\mathbf{k})}{k^{3/2}}\right)+\Theta(-k_3)a_Q(\mathbf{k}) \tag{17}$$

so that $[H,\Omega(\mathbf{k})]=0$ in the case of QCD as well as QED. Eqs. (9) and (10) lead to

$$\partial_i E_i(\mathbf{x})-J_0(\mathbf{x})=-\sum_{\mathbf{k}} k^{3/2}\,\Theta(k_3)\left(\Omega(\mathbf{k})\,e^{i\mathbf{k}\cdot\mathbf{x}} + \Omega^*(\mathbf{k})\,e^{-i\mathbf{k}\cdot\mathbf{x}}\right) \tag{18}$$

and to

$$A_0(\mathbf{x})=\sum_{\mathbf{k}} \frac{-i\alpha}{\sqrt{k}}\,\Theta(-k_3)\left(\Omega(\mathbf{k})\,e^{i\mathbf{k}\cdot\mathbf{x}} - \Omega^*(\mathbf{k})\,e^{-i\mathbf{k}\cdot\mathbf{x}}\right). \tag{19}$$

Within the subspace defined by $\Omega(\mathbf{k})|v\rangle=0$, Gauss's Law and the gauge-fixing condition $A_0=0$ are implemented; and since the time evolution operator exp[-iHt] commutes with $\Omega(\mathbf{k})$, state vectors within that subspace are permanently contained within it. In all these respects this formulation of the theory behaves precisely like the one discussed in ref. (7), although the latter led to a propagator evaluated according to a quite different prescription. Finally we consider the special case of QED in which $J_0=j_0$. In that case we can unitarily transform the entire formulation with a similarity transformation in which the unitarily transformed form of $\Omega(\mathbf{k})$ is $a_Q(\mathbf{k})$. The similarity transformation that effects this transformation is $e^D$, where D, in this alternate formulation, is given by

$$D = \sum_{\mathbf{k}} \frac{\Theta(k_3)}{k^{3/2}}\left(a_R{}^*(\mathbf{k})j_0(\mathbf{k}) - a_R(\mathbf{k})j_0(-\mathbf{k})\right). \tag{20}$$

$e^D He^{-D}$, the unitarily transformed Hamiltonian for QED in this gauge, takes the form

$$e^D He^{-D}= H_C+H_Q \tag{21}$$

where $H_C$ is the Hamiltonian for QED in the Coulomb gauge, and $H_Q$ is given by

$$H_Q=\sum_{\mathbf{k}} \Big\{ k\big[\Theta(k_3)\, a_Q{}^*(\mathbf{k})a_Q(\mathbf{k})-i\alpha\Theta(-k_3)\left( a_Q(\mathbf{k})a_Q(-\mathbf{k})- a_Q{}^*(\mathbf{k})a_Q{}^*(-\mathbf{k})\right)\big]$$

$$-\frac{1}{\sqrt{k}}\big[j_0(-\mathbf{k})a_Q(\mathbf{k})\left(\Theta(k_3)+i\alpha\Theta(-k_3)\right)+j_0(\mathbf{k})a_Q{}^*(\mathbf{k})\left(\Theta(k_3)-i\alpha\Theta(-k_3)\right)\big] \Big\}. \tag{22}$$

Eq. (22) demonstrates clearly that the time evolution operators $\exp[-i(e^D He^{-D})t]$ and $\exp[-iH_C t]$ time translate state vectors in identical fashion in the subspace given by the unitarily transformed constraint $a_Q(\mathbf{k})|n\rangle=0$.

The fact that in this alternate representation, in the case of QED, the transformed Hamiltonian $e^D He^{-D}$ has the identical effect in the physical subspace as does the Hamiltonian for the Coulomb gauge, provides instructive insights into the ambiguity in propagator prescriptions. In ref. (7) we found a similar identity between $e^D He^{-D}$ and the Coulomb gauge Hamiltonian, even though, in that latter case, longitudinal and timelike components of gauge fields as well as D extended over all values of $k_3$. How are we to understand the fact that the two formulations describe the same physical theory, but nevertheless lead to propagators with different prescriptions? The following observations are offered to clarify this question: When constraints are fully implemented, the Hamiltonians for QED in various different gauges are entirely equivalent to QED in the Coulomb gauge, and hence also to each other. Technically, this fact is most easily demonstrated by unitarily transforming Hamiltonians and state vectors in the physical subspace so that $\Omega(\mathbf{k})$ is transformed to $a_Q(\mathbf{k})$, and H to $e^D He^{-D}$, which has the form given in eqs. (20) -(22). The detailed form of $H_Q$ depends on the gauge, but in any gauge $H_Q$ always consists of terms in which $a_Q(\mathbf{k})$ or $a_Q{}^*(\mathbf{k})$ appear, but never $a_R(\mathbf{k})$ or $a_R{}^*(\mathbf{k})$. When this form of the theory is used, the dynamical consequences of $e^D He^{-D}$ and the Coulomb gauge Hamiltonian are indistinguishable. However, Feynman rules for the perturbative S-matrix are based not on the transformed Hamiltonian, $e^D He^{-D}$, but on the untransformed H; and the S-matrix is developed in a formulation in which Gauss's Law and gauge constraints are not, in fact, implemented. An analysis summarized in ref. (7) shows that the S-matrix is still correct when these constraints are not respected. But this immunity has subtle limitations. Renormalization constants *are* affected by the failure to implement constraints, and it is not trivial that the renormalized S-matrix is identical in different gauges.[12] It is in this form of the theory that different gauges, and different options within each gauge, develop different propagators. In axial gauges there are different prescriptions for singular integrals, and in covariant gauges there is the choice among the Feynman, the Landau, and the Fried-Yennie options for the Lorentz gauge. There can be many different propagators, as there can be many different forms of H and D for which $e^D He^{-D}$ has the general form described in eq. (21). Some of these propagators may well be too singular to be part of a renormalizable theory. For example, serious questions have been raised even about the renormalizibility of the Coulomb gauge in QED.[13] As has been pointed out in ref. (3), and in the work reviewed therein, some of the axial gauge propagators have this defect.

Since QED and QCD have the same propagators, the part of the preceding discussion that pertains to the dependence of the S-matrix on the form of the propagator applies to QCD as well as to QED. But the demonstration of the equivalence of the interacting theory in all gauges, which can be given for QED by a similarity transformation, does not apply to QCD. A result, relevant to this topic, that can be established for this, as well as other formulations of QCD in the temporal gauge, is that the S-matrix is unitary without Faddeev-Popov ghosts. We will turn to this question in the next paragraph.

The S-matrix for QCD in this theory saturates unitarity in the space of states that contains quarks and transverse gluons only. In QED it is trivial that the S-matrix is unitary in the physical subspace, since it is completely equivalent to QED in the Coulomb gauge. For QCD it is necessary to show that there are no S-matrix elements to final states that include combinations of ghosts, unless those states have zero norm, so that no probability "leaks" to the unobservable part of space occupied by ghost particles. The proof consists of a demonstration that if there is even one single Q-type ghost in a final state, the S-matrix to that state vanishes. Since ghost states have zero norm unless they contain at least one mixed pair (Q and R), that argument is sufficient to complete the proof. We represent the S matrix as

$$S_{f,i} = \delta_{f,i} - 2\pi i\delta(E_f - E_i)T_{f,i}\,, \tag{23}$$

where $T_{f,i} = \langle f | H_1 + H_1 (E_i - H + i\varepsilon)^{-1} H_1 | i \rangle$, and we choose a final state $|f\rangle$ of the form $a_Q{}^*(\mathbf{k})|f'\rangle$, where $|f'\rangle$ designates a state that may contain quarks, transverse gluons, or either variety of gluon ghost, R-type as well as Q-type. This makes the transition amplitude $T_{f,i}$ for transitions from states $|i\rangle$ to $|f\rangle$

$$T_{f,i} = \langle f' | a_Q(\mathbf{k})\{H_1 + H_1 (E_i - H + i\varepsilon)^{-1} H_1\} | i \rangle\,. \tag{24}$$

The identity

$$a_Q(\mathbf{k})\,(E - H + i\varepsilon)^{-1} = (E - H + i\varepsilon)^{-1}\, a_Q(\mathbf{k}) + (E - H + i\varepsilon)^{-1}[\, a_Q(\mathbf{k}), H_1](E - H + i\varepsilon)^{-1} \tag{25}$$

leads to

$$T_{f,i} = \langle \psi_{f'}{}^{(-)}{}_{(E_f)} | [\, a_Q(\mathbf{k}), H_1]\ | \psi_i{}^{(+)}{}_{(E_i)} \rangle + (E_f - E_i)\, \mathcal{F}(f'; i), \tag{26}$$

and $\mathcal{F}(f'; i) = \langle f' | H_1 (E_i - H + i\varepsilon)^{-1} (E_f - H + i\varepsilon)^{-1} [\, a_Q(\mathbf{k}), H_1] | \psi_i{}^{(+)}{}_{(E_i)} \rangle$; and where $\Psi_i{}^{(\pm)}{}_{(E)}$ represents the scattering state given by

$$\Psi_n{}^{(\pm)}{}_{(E_n)} = \{1 + (E_n - H \pm i\varepsilon)^{-1} H_1\} | n \rangle. \tag{27}$$

Here $|n\rangle$ obeys $(H_0 - E_n)|n\rangle = 0$ and the $(\pm)$ designation refers to outgoing (+) and incoming (–) boundary conditions respectively. Since

$$[a_Q(\mathbf{k}), H_1] = -\Theta(k_3) k^{-(3/2)}\, [H, J_0\,(\mathbf{k})], \tag{28}$$

and since $(H_0 - E_n)\, a_Q{}^*(\mathbf{k})|n\rangle = 0$ when $(H_0 - E_n)|n\rangle = 0$, we obtain the result that

$$T_{f,i} = -\Theta(k_3) k^{-(3/2)} \{\langle \psi_{f'}{}^{(-)}{}_{(E)} | J_0(\mathbf{k})\ | \psi_i{}^{(+)}{}_{(E_i)} \rangle\, [E_f - E_i] + i\varepsilon \langle f' | H_1 (E_f - H + i\varepsilon)^{-1} J_0(\mathbf{k}) | \psi_i{}^{(+)}{}_{(E_i)} \rangle$$
$$- i\varepsilon \langle \psi_{f'}{}^{(-)}{}_{(E_f)} | J_0(\mathbf{k}) (E_i - H + i\varepsilon)^{-1} H_1 | i \rangle \} + (E_f - E_i)\, \mathcal{F}(f'; i)\,. \tag{29}$$

The proof that unitarity saturates in the ghost-free sector of this theory consists of the observation that the only contributions to the S-matrix element $S_{f,i}$ between the states $|i\rangle$ and $|f\rangle = a_Q^*(\mathbf{k})|f'\rangle$ that survive "on-shell", when $E_f = E_i$, are the ones in eq. (29) that are proportional to $i\varepsilon$; and these vanish in the limit $\varepsilon \to 0$, except when the $\langle f'|H_1(E_f - H + i\varepsilon)^{-1} J_0(\mathbf{k})|\psi_i{}^{(+)}{}_{(E_i)}\rangle$ or $\langle \psi_{f'}{}^{(-)}{}_{(E_f)}|J_0(\mathbf{k})(E_i - H + i\varepsilon)^{-1} H_1|i\rangle$ develop $(i\varepsilon)^{-1}$ singularities; and that happens only in self-energy corrections to external lines. As we discussed previously, these contributions do not affect scattering cross sections and, in particular, can not keep S-matrix elements to final states with Q-ghost populations from vanishing.

Although this review has largely dealt with two different representations of gauge fields in the temporal gauge, it is likely that there are even other representations that lead to propagators in this gauge with yet different prescriptions. And similar correspondences between propagator prescriptions and representations of gauge fields can be observed in other axial gauges.[14]

The author acknowledges interesting and informative discussions with Profs. A. Bassetto, H. Cheng, P. V. Landshoff, I. Lazzizzera, G. Leibbrandt, E. C. Tsai, and Mr. G. Pollak.

## REFERENCES

1) S. Mandelstam, Nucl. Phys. **B213**, 149 (1983).
2) G. Leibbrandt, Phys. Rev. D **29**, 1699 (1984).
3) A complete review of this problem appears in G. Leibbrandt, Rev. Mod. Phys., **59**, 1067 (1987).
4) H. Cheng and. E. C. Tsai, Phys. Rev. D **36**, 3196 (1987).
5) A. Bassetto, M. Dalbosco, I. Lazzizzera and R. Soldati, Phys. Rev. D **31**, 2012 (1985).
6) I. Lazzizzera, Phys. Lett. **210B**, 188 (1988).
7) K. Haller, Phys. Rev. D **36**, 1830; 1839 (1987).
8) P. V. Landshoff and J. C. Taylor, *The Spacelike Axial Gauge*, DAMTP preprint 89/21.
9) S. Caracciolo, G. Curci, and P. Menotti, Phys. Lett. **113B**, 311(1982).
10) P. Gaigg, M. Kreuzer, and G. Pollak, Phys. Rev. D **38**, 2559 (1988).
11) P.V. Landshoff, Phys. Lett. **169B**, 69 (1986).
12) I. Bialynicki-Birula, Phys. Rev. D **2**, 2877 (1970).
13) F. Strocchi and A. S. Wightman, J. Math. Phys. **15**, 2198 (1974).
14) K. Haller, *Gauge Theory in the Light Cone Gauge*, Univ. of Connecticut preprint (revised).

# Gauge Invariance and BRS Quantum Field Theory

Er-Cheng Tsai

*Department of Physics, National Taiwan University, Taipei, Taiwan R.O.C.*

**Abstract**

**We quantize the BRS gauge field theory canonically and prove that it is equivalent to the theory in the physical gauge without resorting to perturbation theory. The relationship between the quantum states in the BRS theory and those in the physical theory is explicitly given. We also show that one may obtain matrix elements of gauge-invariant operators in the physical theory by calculating corresponding ones in the BRS theory.**

## 1 What is Gauge Invariance?

In classical mechanics, the meaning of gauge invariance is fairly simple. It means that there are some arbitrary functions $\xi_i(x,t)$ of space-time such that physical quantities are independent of $\xi_i(x,t)$, although the field variables which describe the physical quantities may depend on $\xi_i(x,t)$. But it is not trivial to carry this gauge symmetry over to quantum mechanics. To have gauge invariant quantum field theories, we must define a space for the states, a norm and the expressions for physical operators in different gauges. Furthermore, the matrix elements for physical operators calculated in different gauges with different inner products should be equal. For nonabelian gauge field theories, we have found that it is possible to meet theses requirements to formulate gauge invariant quantized theories. Even the BRS theories[1] which involve unphysical ghost fields and do not have a classical analogy can be quantized and proven to be equivalent to the quantized field theories without ghosts [2,3,4].

The BRS perturbation rules for the Feynman gauge are explicitly covariant and perhaps the most convenient rules to use in perturbative calculations. The Feynman gauge is not, strictly speaking, a gauge in the classical sense. For, unlike other gauges, the Feynman gauge does not impose a condition on the components of $A_\mu$ and does not even exist in the classical theory. Kugo and Ojima[5] verified that, with respect to some restricted (physical) subspace, the S-matrix is still unitary. However, their work still leaves some important questions unanswered. For example, the BRS Lagrangian involves unphysical ghost fields, while the gauge invariant Lagrangian does not. The equations of motion in these Lagrangian theories are different. In particular, one may ask: Can one obtain any physical quantity in one theory by calculating the corresponding one in the other? Are these two quantum theories equivalent to each other?

## 2 The Correspondence Formula

We have found that given a physical wavefunction $\psi_w$ in the temporal [2] gauge, it is useful to construct a wavefunction $\psi_{eff}$ in the BRS Lagrangian theory by setting

$$\psi_{eff} = e^{\Theta} \psi_w. \tag{1}$$

This does not mean that $e^{\Theta}$ is the operator which transforms the quantum states of the BRS Lagrangrian theory. The reason is that

$$H_{eff}\psi_{eff} \neq e^{\Theta} H_w \psi_w, \tag{2}$$

where $H_{eff}$ is the Hamiltonian in the BRS Lagrangrian theory and $H_w$ is the Hamiltonian in the temporal gauge. One of the implication of (2) is that, if $\psi_w$ is an eigenstate of $H_w$, $\psi_{eff}$ is not necessarily an eigenstate of $H_{eff}$. Instead we have, as will be show later,

$$H_{eff}\psi_{eff} = e^{\Theta} H_w \psi_w + Q\phi, \tag{3}$$

where $Q$ is the BRS charge generator and $\phi$ is some wavefunction the precise form of which can be easily calculated but is unimportant. Mathematically, (3) implies that $H_{eff}$ cannot be defined if we restrict ourselves to the Hilbert space of wavefunctions of the form (1). Physically, it implies that a quantum state initially in the Hilbert space does not remain in this space as time evolves. Thus we need to expand this Hilbert space into the space of wavefunctions of the form of

$$\psi_{eff} = e^{\Theta} \psi_w + Q\phi. \tag{4}$$

This expanded space is closed under time evolution, but is not isomorphic to the Hilbert space of the physical theory. Indeed, this expanded space is not even a Hilbert space, as the inner product is positive semi-definite, not positive definite.

This is not to say that the BRS Lagrangian theory is not useful, especially if one is content to restrict oneself to the consideration of certain matrix elements.

Quite the opposite, the BRS Lagrangian theory is perhaps more useful than it has been generally given credit for. It can be used not only for the calculation of the S-matrix, which is the transition amplitude of infinite time duration, but also for the calculation of the transition amplitude of any finite time duration. As a matter of fact, one can obtain the matrix elements of a gauge-invariant operator $O$ in the physical theory by calculating the corresponding ones in the BRS Lagrangian theory, with the wavefunctions in these two theories related by (1)

$$\langle\psi_{eff}|O_{eff}|\psi_{eff}\rangle = \langle\psi_w|O_w|\psi_w\rangle_w. \tag{5}$$

We call such formulae equating matrix elements in these two theories correspondence formulae. A correspondence formula is exact, valid for all values of the coupling constant. Indeed, no reference to perturbation theory is necessary. This formulation is readily applicable to any specific gauge field theory, in particular, to the Weinberg-Salam-theory. (With the correspondence formula, the proof of unitarity of the Feynman-gauge Feynman rules in such a theory is rendered unnecessary.)

### 3 The Derivation

Let's derive the correspondence formula which equates the finite time transition amplitude in the temporal gauge to that in the Feynman gauge. We shall consider a pure Yang-Mills theory with the Lagrangrian

$$L = -\frac{1}{4}F^a_{\mu\nu}F^{\mu\nu a}, \tag{6}$$

where

$$F^a_{\mu\nu} = \partial_\mu A^a_\nu - \partial_\nu A^a_\mu - gf^{abc}A^b_\mu A^c_\nu.$$

The Lagrangian (6) is gauge invariant and, as it is independent of $\frac{dA^a_0}{dt}$, the dynamical variable conjugate to $A^a_0$ cannot be defined. A way to get around this is to choose the temporal gauge $A^a_0 = 0$. The usual procedure of canonical quantization can then be applied. In particular the Hamiltonian density, denoted by $H_w$, is

$$H_w = \int d^3x \frac{1}{2} \left[ \vec{\pi}^a \cdot \vec{\pi}^a + \vec{B}^a \cdot \vec{B}^a \right]. \tag{7}$$

The Gauss law is missing, and one imposes it as a supplementary condition on the quantum state:

$$G^a|\psi_w\rangle = (\vec{D} \cdot \vec{\pi})^a |\psi_w\rangle = 0. \tag{8}$$

Another way to quantize this nonabelian gauge field theory is to add to the Lagrangian (6) a term $-\frac{1}{2}(\partial_\mu A^{\mu a})^2$ which contains $\frac{dA_0^a}{dt}$, but is not gauge invariant.

$$L_F = -\frac{1}{4} F^a_{\mu\nu} F^{\mu\nu a} - \frac{1}{2}(\partial_\mu A^{\mu a})^2. \tag{9}$$

The BRS effective Lagrangian is obtained from $L_F$ by adding a term with Grassman fields $\xi$ and $\eta$ (ghost fields) [3] to make it invariant under the BRS transformation.

$$\delta A^{\mu a} = (D^\mu \xi)^a = (\partial^\mu \xi)^a - g f^{abc} A^{\mu b} \xi^c, \tag{10}$$

$$\delta \xi^a = \frac{1}{2} g f^{abc} \xi^b \xi^c, \tag{11}$$

$$\delta \eta^a = i \pi_0^a, \tag{12}$$

The Lagrangian

$$L_{eff} = -\frac{1}{4} F^a_{\mu\nu} F^{\mu\nu a} - \frac{1}{2}(\partial_\mu A^{\mu a})^2 - i(\partial^\mu \eta)^a (D_\mu \xi)^a \tag{13}$$

is invariant under the BRS transformation.

The BRS Lagrangian $L_{eff}$ yields the conserved quantity

$$Q = \int d^3x \left[ (\vec{D}\vec{\pi})^a \cdot \xi^a + i \pi_\eta^a \pi_0^a + \frac{1}{2} g f^{abc} \pi_\xi^a \xi^b \xi^c \right], \tag{14}$$

where

$$\pi_0^a = -(\partial_\mu A^{\mu a}),$$

$$\pi_\eta^a = -i(D^0 \xi)^a,$$

$$\pi_\xi^a = i\frac{d\eta^a}{dt},$$

with

$$\left[\pi_\eta^a(\vec{x}), \pi_\eta^b(\vec{y})\right]_+ = \left[\pi_\xi^a(\vec{x}), \pi_\xi^b(\vec{y})\right]_+ = -i\delta^{ab}\delta^{(3)}(\vec{x}-\vec{y}). \tag{15}$$

(Note (15) implies that $\pi_\eta$ and $\pi_\xi$ are anti-hermitian operators.) $Q$ is a Grassman operator satisfying

$$Q^2 = 0. \tag{16}$$

The variation of a field under a BRS transformation can now be alternatively expressed as the commutator (for a field with bose statistics) or anti-commutator (for a Grassman field) of this field with $Q$. For example, it is easy to verify from (14) that

$$\delta A^{\mu a} = i\,[Q, A^{\mu a}] = (D^\mu \xi)^a, \tag{17}$$

$$\delta \xi^a = i\,[Q, \xi^a]_+ = \frac{1}{2} g f^{abc}\xi^b\xi^c, \tag{18}$$

$$\delta \eta^a = i\,[Q, \eta^a]_+ = i\pi_0^a, \tag{19}$$

and

$$\delta\pi_\xi^a = i\left[Q, \pi_\xi^a\right]_+ = (\vec{D}\cdot\vec{\pi})^a - g f^{abc}\pi_\xi^b\xi^c. \tag{20}$$

With this definition, $\delta^2$ of any operator vanishes

$$\delta^2 = 0.$$

The effective Hamiltonian $H_{eff}$ corresponding to (13) is

$$H_{eff} = H_w + i\int d^3x\,[Q, \Delta]_+, \tag{21}$$

where $H_w$ is given by (7) and

$$\Delta = A_0^a\pi_\xi^a + i(\vec{\nabla}\cdot\vec{A^a})\eta^a + \frac{1}{2}i\pi_0^a\eta^a. \tag{22}$$

The particular form for $\Delta$ appears immaterial. The important point in (21) is that $H_{eff}$ differs fro $H_w$ by a BRS variation.

Let $|\psi_w\rangle$ be a physical quantum state in the physical theory and hence satisfies (8), and let us denote

$$\psi_w(\vec{A}) \equiv \langle \vec{A}|\psi_w\rangle, \tag{23}$$

where $|\vec{A}\rangle$ is an eigenstate of the field operator. The wavefunction $\psi_{eff}$ in the BRS theory is constructed as

$$\psi_{eff}(\vec{A},\xi,\eta) \equiv e^{\Theta}\psi_w(\vec{A}), \tag{24}$$

where

$$\Theta = \Theta_1 + \Theta_2, \tag{25}$$

$$\Theta_1 = i\int d^3x\eta^a\delta I^a(\vec{A}), \tag{26}$$

and

$$\Theta_2 = \frac{1}{2}\int d^3x(A_0^a - iI^a)\sqrt{-\nabla^2}(A_0^a - iI^a), \tag{27}$$

where $I^a$ may be any real valued function of $\vec{A}$. This wavefunction satisfies

$$Q\psi_{eff} = 0. \tag{28}$$

For the Feynman gauge, $I^a$ is chosen as

$$I^a = \frac{1}{\sqrt{-\nabla^2}}(\vec{\nabla}\cdot\vec{A})^a$$

and

$$\Theta_1 = -i\int d^3x\eta^a\frac{1}{\sqrt{-\nabla^2}}(\vec{\nabla}\cdot\vec{D}\xi)^a \tag{29}$$

and

$$\Theta_2 = \frac{1}{2}\int d^3x(\sqrt{-\nabla^2}A_0^a - i\vec{\nabla}\cdot\vec{A}^a)\frac{1}{\sqrt{-\nabla^2}}(\sqrt{-\nabla^2}A_0^a - i\vec{\nabla}\cdot\vec{A}^a). \tag{30}$$

Equations (21) and (28) give

$$H_{eff}\psi_{eff} = H_w\psi_{eff} + iQ\Delta\psi_{eff}. \tag{31}$$

By using that $[Q, H_w] = 0$, it is easy to prove

$$e^{-iH_{eff}t}\psi_{eff} = e^{-iH_w t}\psi_{eff} + iQ\phi. \tag{32}$$

and the matrix elements are equal

$$\langle\psi^{(1)}_{eff}|e^{-iH_{eff}t}|\psi^{(2)}_{eff}\rangle = \langle\psi^{(1)}_{eff}|e^{-iH_w t}|\psi^{(2)}_{eff}\rangle. \tag{33}$$

Equation (33) has not yet in the desired form, as $H_w$ operates on $e^{\Theta}\psi_w$ and not directly on $\psi_w$, as it is the case in a physical theory. It can be shown that

$$e^{-iH_w t}e^{\Theta}\psi_w = e^{\Theta}e^{-iH_w t}\psi_w + iQ\phi'. \tag{34}$$

Thus we have

$$e^{-iH_{eff}t}e^{\Theta}\psi_w = e^{\Theta}e^{-iH_w t}\psi_w + iQ\overline{\phi}. \tag{35}$$

Equation (35) shows that, while a wavefunction of the form $e^{\Theta}\psi_w$ does not remain in this form as it evolves in time, it differs from this form only by a term of the form $Q\phi$. Indeed, a wavefunction of the form of

$$e^{\Theta}\psi_w + iQ\phi. \tag{36}$$

remains to be in this form at all times. We shall call the quantum theory with the Hamiltonian $\psi_{eff}$ in the space of (36) the effective theory. The left side of (33) is a transition amplitude in the effective theory. Substituting (35) into (33), we get

$$\langle\psi^{(1)}_{eff}|e^{-iH_{eff}t}|\psi^{(2)}_{eff}\rangle = \langle\psi^{(1)}_{w}|e^{\Theta^{\dagger}+\Theta}e^{-iH_w t}|\psi^{(2)}_{w}\rangle. \tag{37}$$

Referring to (26,27), we have

$$\Theta^{\dagger} + \Theta = 2\Theta_1 + \Omega_3 + \Omega_4, \tag{38}$$

where

$$\Omega_3 \equiv \int d^3x A^a_0\sqrt{-\nabla^2}A^a_0, \tag{39}$$

and

$$\Omega_4 \equiv - \int d^3x I^a \sqrt{-\nabla^2} I^a. \tag{40}$$

The only factor on the right side of (37) which is dependent on $A_0, \eta, \xi$ is $exp(2\Theta_1 + \Omega_3)$, as $\Omega_4, H_w$ and $\psi_w$ are independent of these variables. Thus the functional integration over $A_0, \eta, \xi$ which one performs in taking the matrix element of (37) can be explicitly carried out. The function $\Omega_3$ is negative definite if $A_0$ is imaginary. Thus we integrate over the imaginary values of $A_0$, which is just the indefinite metric. The integral is a constant, which can be set to unity if the quantum states are properly normalized. The functional integrations over $\eta$ and $\xi$ can also be carried out giving a determinant which can be interpreted as the contributions of ghosts. The right side can then be reduced to the transition amplitude of the physical theory. Thus we have

$$\langle \psi_{eff}^{(1)} | e^{-iH_{eff}t} | \psi_{eff}^{(2)} \rangle = \langle \psi_w^{(1)} | e^{-iH_w t} | \psi_w^{(2)} \rangle_w . \tag{41}$$

where the subscript outside of the parenthesis indicates that the matrix element in the r.h.s. of (41) is that of the physical theory, i.e. it involves functional integrations over $\vec{A}$ but not over $A_0, \eta, \xi$ , with a weight-factor originating from gauge fixing.[2]

Similarly, we may derive a correspondence formulae for other gauge invariant operators. Examples of such operators are elements of the Poincaré group: time translations (already discussed), spatial translations, rotations and Lorentz boosts. There also exits a correspondence formula for the Wilson loop [7].

## 4 Conclusions

Equation (41) is an exact formula, and is valid between states $|\psi_w\rangle$ satisfying the Gauss law. If we choose $|\psi_{w1}\rangle$ and $|\psi_{w2}\rangle$ to be vacuum states, and if we make a perturbative expansion of (41) and use the adiabatic hypothesis, the l.h.s. (r.h.s.) of (41) is the perturbation series for the S-matrix in the Feynman (temporal ) gauge. The propagator for the longitudinal gluon on the r.h.s. of

(41) is

$$D_{LL}(x,x') = -i\delta^{(3)}(\vec{x}-\vec{x}')t_>, \tag{42}$$

where $t_>$ is the larger of $x_0$ and $x_0'$. Also, for the transverse gluon, we have

$$D_{ij}^{T}(x,x') = i\int \frac{d^4k}{(2\pi)^4}\frac{1}{k^2+i\epsilon'}\left(\delta_{ij}-\frac{k_ik_j}{\vec{k}^2}\right). \tag{43}$$

In addition

$$D_{0\mu}(x,x') = D_{\mu 0}(x,x') = 0. \tag{44}$$

This is the propagator given by eq.(8) of reference [6] with $\alpha = -1$. The other choice of $\alpha = 1$ in reference [6], also giving the correct answer, can be obtained from our formalism by reversing the roles of $\psi_{w1}^*$ and $\psi_{w2}$. Note that, in this prescription, we pay the price of translational invariance for the propagator but gain the advantage that $D_{0\mu}$ is strictly zero.

Finally, we mention that the correspondence formula (41) can be extended to the covariant $\alpha$ gauge, the spatial axial gauge, and the Coulomb gauge.

Our result also solves the puzzle whether the Feynman rules for the quantum Wilson loop in the axial gauge are consistent with those in other gauges, to all orders of the coupling constant and for all contours in the space-time region. In particular, we show that one must include the contribution of ghost loops if the principal value prescription is used for the axial gauge propagator [7].

# References

[1] C. Becchi, A. Rouet and R. Stora, Phys. Lett. **52B**, 344 (1974);

[2] H. Cheng and E.C. Tsai, Canonical Quantization of Nonabelian Gauge Field Theories and Feynman Rules, Chinese Journal of Physics **V25**, 95,1987.

[3] H. Cheng and E.C. Tsai, Phys. Lett. **176B**, 130 (1986).

[4] H. Cheng and E.C. Tsai, Phys. Rev. **D40**, 1246 (1989).

[5] T. Kugo and I. Ojima, Phys. Lett. **73B**, 459 (1978);

[6] S. Caracciolo, G. Curci and P. Menotti, Phys. Lett. **113B**, 311 (1982).

[7] H. Cheng and E.C. Tsai, Phys. Rev. **D36**, 3196 (1987).

# Some Aspects of the Fock-Schwinger and Poincaré Gauges

Carlos A. P. Galvão and Patricio Gaete

Centro Brasilieiro de Pesquisas Físicas - DCP,
Rua Dr. Xavier Sigaud 150, Rio de Janeiro, Brasil.

Let us consider the gauge-invariant description of the electromagnetic interaction between the Maxwell field and a charged matter field $\phi(x)$, as given by the Lagrangian

$$L = -\frac{1}{4} F_{\mu\nu}F^{\mu\nu} + L_M(\phi, D\phi) \quad (I.1)$$

where $F_{\mu\nu} = \partial_\mu A_\nu - \partial_\nu A_\mu$ is the field-strength tensor, $A_\mu(x)$ are the gauge potentials, and $L_M$ is the Lagrangian for the matter field, constructed with the appropriate covariant derivatives D. We shall consider only Abelian gauge fields. Our analysis and results can be extended to non--Abelian gauge fields as well.

Gauge invariance is required as a fundamental symmetry principle of Nature and, in the case we are concerned with, it means that the theory can be equivalently described by the fields $A_\mu$, $\phi$, or by their gauge transformed counterparts,

$$A_\mu^+ = A_\mu + \partial_\mu \Delta(x) \quad , \quad \phi = e^{ie\Delta(x)} \quad . \quad (I.2)$$

The price paid by requiring that (I.1) be invariant under (I.2) is that of using more variables for the $A_\mu(x)$ field than is actually necessary to describe the physical phenomena. When applying the theory to some specific situation, one must eliminate those spurious variables or,in other words, one must break the gauge freedom of the theory. This is where the gauge conditions come into play, since the procedure to break the gauge invariance is that of imposing extra conditions $f_n(A) = 0$ on the theory.

Obviously, there are some guidelines for choosing gauge conditions. In principle, one is guided by the simplicity of the resulting theory. One can also take into account the fact that the gauge conditions should express some specific physical peculiarity of the system, even though this is not always the case. (For an excellent review of the many most used covariant and non-covariant gauge conditions see the paper by Leibbrandt, Ref. [1].)

A long time ago, V. Fock[2] and J. Schwinger[3] used the covariant

*Talk presented by C.A.P. Galvao

gauge condition

$$(x^{\mu}-\xi^{\mu})A_{\mu}(x) = 0 \quad . \tag{I.3}$$

Here $\xi = (\xi^{\mu})$ is an arbitrary point in space-time, which we shall take as $\xi = 0$ since there is no loss of generality. We shall refer to (I.3) as the "FS gauge". Our concern in this talk shall be with the noncovariant version of (I.3) but let us make some comments on the FS gauge to start with, since it has some interesting consequence.

First of all, (I.3) fulfills the important properties of being attainable (in the sense that it can be attained from any field configuration by a suitable gauge transformation) and it fixes the gauge (if two potentials satisfy (I.3) they differ by an homogeneous function of zeroth degree, which can be taken as a constant, so that they are related by the identity gauge transformations. It clearly breaks the Poincaré invariance of the theory and as a consequence it renders a theory which is free from Faddeev-Popov ghosts and leads to a particularly simple structure for the Green's functionals. But it does introduce technical complications, for instance, in the computation of Feynman diagrams. Attention to these facts have been called by Kummer and Weiser[4], who among other things developed a consistent perturbation theory in the FS gauge.

The most remarkable property of the FS gauge is that it enables one to express the potentials in terms of the field strength-tensor as

$$A_{\mu}(x) = \int_0^1 d\lambda \lambda x^{\nu} F_{\mu\nu}(\lambda x) \quad . \tag{I.4}$$

The revival of the interest in the FS gauge is mostly due to this fact, (which is also valid for non-Abelian gauge fields also). This property has been used in the study of nonperturbative effects in QCD of strong interactions[5]. It also provides, in principle, a way to formulate a gauge theory in terms of the field-strength tensor instead of the potentials[6]. (In Ref. [7] we list some more papers where the FS gauge has been used. The list is, of course, incomplete.)

One interesting point that has been stressed in [8] is that equation (I.4) follows from the application of the Poincaré Lemma to the field strength two-form simultaneously with covariant Maxwell's equations and no use of any gauge condition. The FS gauge then follows as a consequence of (I.4). In the same paper it is also shown that the use of the Poincaré Lemma in conjunction with the noncovariant form of Maxwell's equations lead to the following expressions for the vector potential $\vec{A}$ and scalar potential $A_0$:

$$\vec{A}(t,\vec{z}) = - \int_0^1 d\lambda \; \lambda \; \vec{z}\times\vec{B}(t,\lambda\vec{z}) \quad , \tag{I.5a}$$

$$A_0(t,\vec{z}) = + \int_0^1 d\lambda\ \vec{z}.\vec{E}(t,\lambda\vec{z}) \quad , \tag{I.5b}$$

where $\vec{E}$ and $\vec{B}$ are the electric and magnetic field vectors. These expressions allows one to obtain the multipole expansions for the potentials in a straightforward manner. They read:

$$\vec{A}(t,\vec{z}) = -\vec{z}\times\left[\sum_{n=0}^{\infty} \frac{n+1}{(n+2)!} (\vec{z}.\nabla_y)^n\ \vec{B}(t,\vec{y})\right]_{\vec{y}=0} \quad , \tag{I.6a}$$

$$A_0(t,\vec{z}) = + \vec{z}.\left[\sum_{n=0}^{\infty} \frac{1}{(n+1)!} (\vec{z}.\nabla_y)^n\ \vec{E}(t,\vec{y})\right]_{\vec{y}=0} \quad , \tag{I.6b}$$

where we used the notation $\nabla_y \equiv \{\partial/\partial y_i\}$ .

Expressions (I.5a,b) and (I.6a,b) are very useful in the context of nonrelativistic quantum mechanics where it is desirable to use the fields themselves instead of the potentials. They have been largely used in connection with problems on quantum optics and molecular electrodynamics. See [9] and references there-in.

In terms of the field-strength tensor expressions (I.5a,b) read

$$A_j(t,\vec{z}) = - \int_0^1 d\lambda\ \lambda\ z_i\ F_{ij}(t,\ \vec{z}) \quad , \tag{I.7a}$$

$$A_0(t,\vec{z}) = - \int_0^1 d\lambda\ z_i\ F_{oi}(t,\lambda\vec{z}) \quad . \tag{I.7b}$$

Now, from equation (I.7a) it is clear that the vector potential satisfy the condition

$$\vec{z}.\vec{A}(t,\vec{z}) = 0 \quad , \tag{I.8}$$

which is the noncovariant version of the FS gauge. The argument can now be reversed: by imposing (I.8) one can easily show that (I.7a) holds.

Suppose now that one wants to implement (I.8) as a gauge condition in the Hamiltonian description of the electromagnetic field. Since two first-class constraints show up in that description, two gauge conditions are necessary to fix the gauge. As we will show in the next section, the other gauge condition compatible with (I.8) and the constraints is

$$A_0(t,\vec{z}) = - \int_0^1 d\lambda\ \vec{z}.\vec{\pi}(t,\lambda\vec{z}) \quad , \tag{I.9}$$

where $\vec{\pi} = \vec{E}$ is the momentum associated with the vector potential. Equation (I.9) is precisely the same as (I.7b).

The gauge conditions (I.8) and (I.9) define the Poincaré gauge for the electromagnetic field. One can easily prove that they are attainable from any field configuration and that once they are imposed the gauge is fixed up to an unimportant homogeneous function of zeroth degree.

## II - QED IN THE POINCARÉ GAUGE

The Lagrangian for the Maxwell and Dirac fields in interaction is

$$L = -\frac{1}{4} F_{\mu\nu}F^{\mu\nu} + i\overline{\psi}(\not{D}-m)\psi \quad . \tag{II.1}$$

$\psi = \psi(x)$ is a four component spinor, $\overline{\psi} = \gamma^0\psi^+$, $\not{D} \equiv \gamma^\mu D_\mu \equiv \gamma^\mu(\partial_\mu - ieA_\mu)$ and, according to our conventions $(\gamma^0)^2 = -1$, $(\gamma_k)^2 = 1$.

Following the standard procedure of passing to the Hamiltonian formulation one finds the canonical Hamiltonian

$$H_c = \int d^3x \, [\, \tfrac{1}{2}\pi_i\pi_i + \tfrac{1}{4} F_{ij}F_{ij} + \pi_\psi\gamma^0(\gamma_j\partial_j - ie\gamma_jA_j + im)\psi + $$
$$+ \pi_i\partial_iA_0 - e\overline{\psi}\gamma^0A_0\psi] \tag{II.2}$$

and the first class constraints

$$\Omega_1(x) = \pi_0(x) \approx 0 \quad , \quad \Omega_2(x) = \partial_i\pi_i - e\overline{\psi}\gamma^0\psi \approx 0 \quad , \tag{II.3}$$

where $\pi_\alpha = \partial L/\partial\dot{A}^\alpha = -F_{o\alpha}$, $\pi_\psi = \partial L/\partial\dot{\psi} = i\overline{\psi}\gamma^0$ are the momenta.

Imposing the gauge condition

$$\Omega_4(x) = X_iA_i \approx 0 \quad , \tag{II.4}$$

and demanding that it must be preserved by the time evolution of the system one finds the differential equation

$$\Omega_3(x) = (\vec{X}.\vec{\nabla})A_0 + \vec{X}.\vec{\pi} \approx 0 \quad , \tag{II.5}$$

whose solution can be expressed as

$$\overline{\Omega}_3(x) = A_0 + \vec{X}. \int_0^1 d\alpha \; \vec{\pi}(t,\alpha\vec{x}) \approx 0 \tag{II.6}$$

(See equation (I.9).) Of course, it is matter of convenience to use (II.5) or (II.6) as a gauge condition since both are compatible with (II.2) and (II.3). Using (II.5), (II.4) and (II.3) we find the following nonvanishing Dirac brackets:

$$\{A_0(x);A_k(Z)\}^* = Z_kC^{-1}_{13}(Z,X) - \frac{\partial}{\partial X_k} C^{-1}_{12}(Z,X) \quad , \tag{II.7a}$$

$$\{A_i(x),\pi_k(Z)\}^* = \delta_{ik}\delta^{(3)}(X-Z) + Z_k\frac{\partial}{\partial X_i} C^{-1}_{24}(X,Z) \quad , \tag{II.7b}$$

$$\{A^T_i(x),\pi^T_k(Z)\}^* = \delta_{ik}\delta^{(3)}(\vec{X}-\vec{Z}) - \frac{\partial}{\partial X_i}\frac{\partial}{\partial Z_k}\frac{1}{4\pi\,|\vec{X}-\vec{Z}|} \quad , \tag{II.7c}$$

$$\{A_0(x),A^T_k(Z)\}^* = Z_k\, C^{-1}_{13}\,(Z,X) + \frac{\partial}{\partial Z_k}\frac{1}{4\pi|\vec{X}-\vec{Z}|} \quad , \tag{II.7d}$$

$$\{\psi(x),\pi_\psi(Z)\}^* = -\delta^{(3)}(\vec{X}-\vec{Z}) \quad , \tag{II.7e}$$

$$\{\psi(x),A_0(Z)\}^* = -ie\psi(x)C_{21}^{-1}(X,Z) \quad , \tag{II.7f}$$

$$\{\psi(x),\pi_k(Z)\}^* = -ie\psi(x)C_{24}^{-1}(X,Z)Z_k \quad , \tag{II.7g}$$

$$\{\pi_\psi(x),A_0(Z)\}^* = -ie\pi_\psi(x)C_{21}^{-1}(X,Z) \quad , \tag{II.7h}$$

$$\{\pi_\psi(x),\pi_k(Z)\}^* = ie\pi_\psi(x)C_{24}^{-1}(X,Z)Z_k \quad , \tag{II.7i}$$

$$\{\psi(x),\pi_k{}^T(Z)\}^* = -ie\left(Z_k\, C_{24}^{-1}(X,Z) + \frac{\partial}{\partial Z_k}\frac{1}{4\pi|\vec{X}-\vec{Z}|}\right)\psi(x) \quad , \tag{II.7j}$$

$$\{\pi_\psi(x),\pi_k^T(Z)\}^* = ie\left(Z_K\, C_{24}^{-1}(X,Z) + \frac{\partial}{\partial Z_k}\frac{1}{4\pi|\vec{X}-\vec{Z}|}\right)\pi_\psi(x) \quad , \tag{II.7k}$$

where

$$C_{12}^{-1}(X,Z) = -\int_0^1 d\beta\, \beta^2 \int_0^1 d\alpha\, \alpha^2\, X.X\delta^{(3)}(\alpha\beta\vec{X}-\vec{Z}) \quad , \tag{II.8}$$

$$C_{13}^{-1}(X,Z) = -C_{24}^{-1}(X,Z) = \int_0^1 d\alpha\, \alpha^2\, \delta^{(3)}(\alpha\vec{X}-\vec{Z}) \quad . \tag{II.9}$$

We now observe that performing the usual procedure of canonical quantization by converting the Dirac brackets into equal-time commutation or anticommutation relations according to the prescription $\{A,B\}^* \to 1/ih[A,B]_\pm$ the brackets (II.7) do not lead to any ambiguity ordering problems. (This fact remains true for non-Abelian gauge fields [10].) In spite of the non-canonical character of the Dirac brackets (II.7) the Hamiltonian operator in the interaction picture can be evaluated and the result is

$$H = -ie\int d^3x\pi_\psi\gamma^0\gamma_i A_i\psi - \frac{e^2}{2}\int d^3xd^3y\pi_\psi(y)\pi_\psi(x)\psi(x)C_{12}^{-1}(y,x) . \tag{II.10}$$

Expression (II.10) has the same formal structure as the one obtained by Kiefer and Rothe [11] for the Coulomb gauge.

Now, from equations (II.7) and (II.10) one sees that the essential features of the Coulomb gauge are somehow contained in the Poincaré gauge. In fact one can modify the Poincaré gauge conditions in such a way as to reproduce the same results as in the Coulomb gauge. The key point is to recall that the Coulomb gauge condition $\partial_i A_i = 0$ completely eliminates the nonphysical longitudinal component $\vec{A}^L$ of the vector potential from the theory, while condition (II.3) does not. However, if instead of (II.3) one imposes

$$\aleph_4(x) = \vec{X}.\vec{A}^L \approx 0 \quad , \tag{II.11}$$

where $A_i^L = -\partial_i \nabla^{-2} \partial_j A_j$ are the longitudinal components of the vector potential, then the resulting description of QED is essentially the same as that in the Coulomb gauge [1,2].

Acknowledgments: CAPG wishes to express his acknowledgments to the Fundação de Amparo à Pesquisa do Estado do Rio de Janeiro (FAPERJ) for financial support.

REFERENCES

[1] G. Leibbrandt, Rev. Mod. Phys. 59(4), 1067 (1987).

[2] V.A. Fock, Sov. Phys. 12, 404 (1937).

[3] J. Schwinger, Phys. Rev. 82, 684 (1952).

[4] W. Kummer, J. Weiser, Z. Phys. C31, 105 (1986).

[5] V.A. Novikov, M.A. Shifman, A.J. Vainshtain, V.J. Zakharov, Fortschr. Phys. 32, 586 (1984).

[6] L. Durand, E. Mendel, Phys. Rev. D26, 1368 (1982), and Phys. Rev. D30, 1754 (1984).

[7] J. Zuk, Phys. Rev. D34, 1791 (1986).
C. Cronström, Phys. Lett. 90B, 267 (1980).
M.A. Shifman, Nucl. Phys. B173, 13 (1980).
E.V. Shuryak, A.J. Vainshtein, Nucl. Phys. B201, 141 (1982).

[8] W.C. Brittin, W. Rodman Smuthe, W. Wyss, Am. J. Phys. 50(8), 693 (1982).

[9] Donald B. Kobe, Am. J. Phys. 50(2), 128 (1982).

[10] Carlos A.P. Galvão, B.M. Pimentel, "Non Abelian Gauge Fields in the Poincaré Gauge", preprint, CBPF-NF-006/88.

[11] C. Kiefer, K.D. Rothe, N. Cimento 83A, 140 (1984).

[12] Carlos A.P. Galvão, Patricio Gaete, B.M. Pimentel, "Quantum Electrodynamics in the Poincaré Gauge". Instituto de Física Teórica, preprint IFT/P-01/89. Submitted for publication.

# Introduction

After an overview of the consequences of introducing a fixed direction in Minkowski space in noncovariant gauges in general (Bassetto), and after a discussion of special problems of the temporal gauge (Landshoff), one of the inventors of the Leibbrandt-Mandelstam (L.M.) prescription which brought a breakthrough in the mathematical treatment of noncovariant gauges describes first practical attempts to go beyond one-loop computations in Feynman-diagrams (Leibbrandt). The LM-prescription leads to nonstandard renormalization off the light-cone. Nevertheless, in the first one loop computation of divergent parts in a physical S-matrix element the disturbing nonpolynomial terms are shown to cancel (Pollak). Any prescription must be tested in observable "quantities". The only known example where a test at the two-loop level was performed is the computation of the Wilson-loop. As shown in the review of Soldati despite a careful definition of test-function spaces, not all prescriptions are equally useful. The same aspect, but from a different angle, is analyzed by Lee.

The Coulomb-gauge is another example of a noncovariant gauge. Here the problem is not related to the ghost-pole $\boldsymbol{k}^2 = 0$ in momentum integrals $\int d^4k$, but to $k_0$-integrations in higher loop orders (Taylor).

The remaining contributed papers deal with consequences for the spurious poles from relations between different noncovariant and covariant gauges and with attempts at alternative prescriptions.

# II

# Spurious Poles and Problems with Feynman-Integrals

# Good and Bad News Concerning Axial Gauges

A. Bassetto

Dipartimento di Fisica "G.Galilei", Via Marzolo 8, Padova, Italy
Istituto Nazionale di Fisica Nucleare, Sezione di Padova, Italy

Axial gauges, characterized by the presence of a constant gauge vector $n_\mu$, became popular at the end of the seventies, when perturbative calculations were successfully performed concerning deep inelastic structure functions and hadron fragmentation distributions in $e^+e^-$ annihilation, starting from Quantum Chromodynamics (QCD) [1], [2].

Their main virtue, besides the triviality of the Faddeev-Popov determinant, is their natural partonic interpretation: there is indeed a close correspondence between independent fields and "physical" degrees of freedom. This is the root of the dominance of dressed ladder diagrams when resumming the collinear singularities of the theory in inclusive distributions [1], [2].

Quite apart from perturbative QCD, the tight relation between field components and quanta was crucial in the formulation of some supersymmetric theories (typically SUSY Yang-Mills N=4) [3], [4], [5]) or even string theories [6].

However, it was also soon realized that the singularities in the vector propagator, peculiar of those gauges (often called "spurious singularities", as they are indeed a gauge artifact), although generally harmless in tree diagram calculations, were instead the source of troublesome expressions in loop calculations [7], [8].

They call indeed for a prescription to be handled in Feynman diagrams, and then some questions naturally arise:

i) are there prescriptions which are completely consistent ?

ii) are different prescriptions related through residual gauge transformations ?

iii) which is the structure of the Hilbert space of the various cases ?

In order to answer those questions we find it particularly suitable to characterize the various gauge choices according to their (apparent) violation of the Lorentz covariance.

Covariant gauges (<u>0-degree violation</u>) have been extensively studied and will not be discussed here.

The axial gauge fixing $n^\mu A_\mu = 0, A_\mu$ being the vector potential, is an example of a condition which entails a 1-degree violation.

Obviously, three cases are possible: the space-like ($n^2 < 0$), the time-like ($n^2 > 0$) and the light-like ($n^2 = 0$) ones.

In the space-like case canonical quantization is possible by making use of the Dirac procedure for constrained systems [9]. It is not restrictive to choose a frame of reference such that $n_\mu = (0, 0, 0, 1)$. By requiring the finiteness of the total energy-momentum of the vector field one realizes that the vector potentials cannot vanish uniformly at spatial infinity [10]. Instead we get

$$\lim_{x_3 \to +\infty} A_0(t, \bar{x}) - \lim_{x_3 \to +\infty} A_0(t, \bar{x}) = -\int_{-\infty}^{+\infty} F_{03}(t, x_\perp, \zeta) d\zeta, \tag{1}$$

$x_\perp$ denoting the transverse components of $x_\mu$ and $F_{\mu\nu}$ the field-strenght tensor. More generally one can prove that

$$\lim_{x_3 \to +\infty} A_\alpha(t, \bar{x}) \neq \lim_{x_3 \to -\infty} A_\alpha(t, \bar{x}), \quad \alpha = 1, 2. \tag{2}$$

As a consequence the potentials in this (and other similar) gauge(s) are intrinsically non compactifiable.

These boundary conditions are obviously related, via Fourier transform, to the prescription for handling the spurious singularity of the propagator [11]. Antisymmetric boundary conditions $\lim_{x_3 \to +\infty} = -\lim_{x_3 \to -\infty}$ lead to the Cauchy principal value (CPV) prescription for the singularity $(nk)^{-1}$. They can be changed by performing a residual gauge transformation [10]

$$[\mathcal{G}(t, [u]), A_\alpha(t, x_\perp, x_3)] = i D_\alpha u(x_\perp), \tag{3}$$

$D_\mu$ being the covariant derivative and $\mathcal{G}$ the generator of the residual gauge transformations

$$\mathcal{G}(t, [u]) = -\int d^2 x_\perp u(x_\perp) \int_{-\infty}^{+\infty} d\zeta D^\alpha F_{0\alpha}, \tag{4}$$

(for simplicity we are considering only the pure Yang-Mills case).

The finiteness of the total energy-momentum entails its weak vanishing [10].

Concerning the double pole $(nk)^{-2}$ affecting the vector propagator, one cannot square the CPV prescription, as one would obtain a meaningless expression from a mathematical viewpoint. Usually one treats it as minus the derivative of CPV. In so doing, however, the positivity of the sum over the polarization vectors at the Feynman poles is lost. This difficulty was also pointed out in ref. [12]. No choice exists satisfying simultaneously the two conflicting requirements

i) hermiticity of $A_\mu$,

ii) possibility of handling products of such a prescription in a consistent way from a mathematical viewpoint.

If, for instance, we impose the vanishing of the potentials in the limit $x_3 \to -\infty$, we obtain, in the free field case,

$$\begin{aligned}\left[\tilde{A}_\alpha^{(-)}(\bar{k}), \tilde{A}_\beta^{(+)}(\bar{k}')\right] &= \left[\delta_{\alpha\gamma} + \frac{k_\alpha k_\gamma}{k_\perp^2}\left(\frac{|\bar{k}|}{k_3 - i\varepsilon} - 1\right)\right] \times \\ &\left[\delta_\beta^\gamma + \frac{k'_\beta k'^\gamma}{k'^2_\perp}\left(\frac{|\bar{k}'|}{k'_3 + i\varepsilon} - 1\right)\right]\delta(\bar{k} - \bar{k}'),\end{aligned} \tag{5}$$

a meaningless expression.

We conclude that the axial gauge, in which all redundant degrees of freedom have been eliminated by solving the constraints, is indeed too singular. As a test, the exponentiation $0(g^4)$ of the Wilson loop does not occur [13] (see Soldati's contribution to this workshop).

The temporal case ($n^2 > 0$) cannot be canonically quantized by repeating this procedure. As a matter of fact $A_0$ is a Lagrange multiplier, which, in the Dirac procedure, cannot be used in a gauge fixing, but should be rather determined after solving the constraints. Moreover, in setting $A_0 = 0$, Gauss' law is usually lost as an operator equation. At best one can envisage procedures in which it is recovered in a weak sense.

It is impossible here to review the huge literature existing on the temporal gauge. Suffices it to say that a calculation $0(g^4)$ of the Wilson loop using the standard propagator with the CPV prescription

$$\mathcal{D}_{\mu\nu}(k) = -\frac{1}{k^2 + i\varepsilon}\left(g_{\mu\nu} - \frac{n_\mu k_\nu + n_\nu k_\mu}{nk} + \frac{k_\mu k_\nu}{(nk)^2}n^2\right) \tag{6}$$

does not exponentiate correctly [14]; for instance non translationally invariant terms might be necessary.

A one parameter family of possible propagators (and related vertices) has been obtained [15] in the interesting original approach of ref. [16], based on a "finite time" Feynman kernel. Unfortunately in this treatment polynomiality of the interaction is lost and the resulting Feynman rules are rather unwieldy.

The light-cone gauge obviously avoids the double pole in the free propagator ($n^2 = 0$), whereas the polarization sum turns out to be positive definite on the Feynman poles, no matter what prescription. Canonical quantization can be performed by assuming the validity of a "null plane" commutator algebra. In light-cone coordinates the evolution is supposed to occur along $x^+ = \dfrac{x^0 + x^3}{\sqrt{2}}$ and then we essentially recover an axial formulation [9]. In particular the spurious singularity is again to be interpreted as CPV.

However the light-cone gauge, quantized in this way, has problems on its own, when performing loop calculations. Indeed, it is clear that in higher order loops multiple spurious poles in the same variable occur, calling for a regularizing prescription. It is also known since a long time that Feynman diagrams computed in the spacelike axial gauge exhibit bad divergences in the limit $n^2 \to 0$ [17], making naive power counting totally unreliable.

An example is provided by the integral

$$I = \int \frac{d^4k}{k^2(k+q)^2} P\left(\frac{1}{nk}\right) \tag{7}$$

which is finite for spacelike $n_\mu$, but behaves like $(log|n^2|)^2$ when $n^2 \to 0$. The same integral, if computed at $n^2 = 0$ in $2\omega$ space-time dimensions, gives rise to a double pole at $\omega = 2$.

This is a dramatic result, for instance in the N=4 supersymmetric Yang-Mills theory, as pointed out in ref. [8], since this integral enters the one loop calculation of the renormalization group *β-function* through the vector contribution to the vector selfenergy and gives a singular term which does not cancel against the corresponding fermion and scalar contributions, thereby jeopardizing the finiteness of the model. Note, that we are considering a gauge invariant quantity, which should be a priori insensitive to gauge artifacts. These considerations make clear that our discussion about "spurious" singularities is not of an academic nature and cast some doubts about conclusions derived using light cone gauge without taking these subtleties carefully into account.

In conclusion, homogeneous axial gauges with a prescription in the propagator entailing a 1-degree violation of the Lorentz symmetry, do not provide us with a reliable framework for performing perturbative calculations.

Another possibility of avoiding the double spurious pole in the vector propagator is given by the planar gauge choice [1]. In the space-like case ($n_\mu = (0,0,0,1)$), the equation concerning $A_3$ reads [10]

$$D_3 \partial^3 A_3 = 0 \tag{8}$$

By imposing suitable boundary conditions at spatial infinity, $D_3$ can be (at least perturbatively) inverted and, thereby, $A_3$ obeys a free-field equation.

Actually it contributes with a negative sign to the total Hamiltonian [10]. In other words, it is a ghost, which, however, being free, can be consistently extruded from the Hilbert subspace $\mathcal{H}$ spanned by the "physical" states

$$A_3^{(-)}|\Phi> = 0 \quad \forall|\Phi> \in \mathcal{H}. \tag{9}$$

Canonical quantization can be performed using Dirac brackets and one obtains the free propagator in the form $\mathcal{D}_{\mu\nu}(k) = -\frac{1}{k^2+i\varepsilon}\left(g_{\mu\nu} - \frac{n_\mu k_\nu + n_\nu k_\mu}{nk}\right)$. The double pole of eq. (6) is just cancelled by the ghost contribution. The spurious simple pole can be interpreted as CPV (antisymmetric boundary conditions on the vector potentials).

We have a quantization scheme relying on redundant unphysical degrees of freedom. Moreover, problems occur beyond the tree level. As a matter of fact higher order poles appear in loop calculations, which require a suitable systematic regularization procedure.

A possibility, suggested in ref. [18], is the combined use of the characteristic function $\varrho(x)$ which adiabatically switches on and off the interaction (and becomes equal to $g$ only at the very end of the calculation) and of a space $\tau$ of test functions which suitably vanish in correspondence with the spurious singularities involved. One first defines recursively the Green's functions as functionals over $\tau$, thereby getting rid of possible ill-defined quantities, and then continues the result to the entire $L_2$ ($\tau$ is dense over $L_2$ in the $L_2$ topology) in order to obtain the ingredients to be inserted in the next order loop, after subtraction of the divergences (see, to this regard, Soldati's contribution to this workshop).

This seemingly esoteric procedure simply translates in a clean mathematical language, empirical rules people usually adopt when using such a gauge.

Still there is no guarantee that the result obtained in this way is physically sensible, as in the regularization procedure important singularities might be lost. In ref. [18] it has been shown up to order $g^4$ that, following those recipes, the Wilson loop exponentiation occurs, at variance with the homogeneous axial case [13]. We believe the ghost $nA$ contribution makes the difference.

We would like to comment upon the more crucial role the adiabatic switching function plays in the context of noncovariant gauge theories. Always in a perturbative framework, the root of the interaction picture is the possibility of identifying at a given time (say $t = 0$) "free" and "interacting" fields. If we denote by $F_{\mu\nu}^{(0)}$ and $F_{\mu\nu}$ the corresponding field-strength tensors and by $H^{(0)}$ and $H$ the related Hamiltonians, the request of having $H^{(0)}$ finite at a given time imposes the constraint (in a weak sense)

$$\int_{-\infty}^{+\infty} d\zeta \partial_\alpha F_{0\alpha}^{(0)}(t, x_\perp, \zeta) = 0, \tag{10}$$

whereas the finiteness of $H$ at the same time requires (always weakly)

$$D_3^{-1} D_\alpha F_{0\alpha} \Big|_{x_3=-\infty}^{x_3=+\infty} = 0 \tag{11}$$

(we recall that $F_{0\alpha}(F_{0\alpha}^{(0)})$ are independent canonical variables). Eqs. (10) and (11) generally are conflicting requirements if we want to identify $F_{0\alpha}$ and $F_{0\alpha}^{(0)}$ at a given time.

This paradoxon is solved by making use of the switching function $\varrho(x)$. As a matter of fact the total Hamiltonian density now reads

$$\mathbf{H}(x) = \mathbf{H}^{(0)}(x) + \varrho(x)\mathbf{H}_{int}(x) \tag{12}$$

and, thanks to $\varrho(x)$, $H$ turns out to be automatically finite once $H^{(0)}$ does. Of course the adiabatic limit and the assumption concerning the completeness of the asymptotic states are much more delicate in this context.

The renormalization in this framework has been studied and solved (at least formally) in refs. [19], [20]. It relies on the fortunate circumstance that, with only one gauge vector, the normal power counting criterion for convergence works and the counterterms are local [17]. These results have been confirmed in ref. [21], in the light of the regularization procedure discussed above.

The planar temporal case ($n^2 > 0$) in our opinion suffers from difficulties analogous to the ones of the corresponding homogeneous case, as far as the canonical quantization is concerned. For this reason it will not be discussed here.

In conclusion, the only gauge choice with one gauge vector, which does not exhibit inconsistencies so far, is the space-like planar case. We stress that it makes essential use of unphysical redundant degrees of freedom and requires a very delicate treatment from a mathematical point of view.

In the light cone case ($n^2 = 0$) Mandelstam [5] and Leibbrandt [22] eventually proposed the following prescription (ML)

$$\frac{1}{[nk]} \equiv \frac{1}{nk + i\varepsilon \; sign\hat{n}k} = \frac{\hat{n}k}{nk\hat{n}k + i\varepsilon}, \tag{13}$$

(the last equality holds in the sense of distributions), with a second vector $\hat{n}_\mu(\hat{n}^2 = 0)$ which, in a suitable reference frame, takes the form $(n_0, -\bar{n})$.

In order to understand how this second vector enters the theory (thereby giving rise to a 2-degree apparent violation of the Lorentz covariance), we briefly review the canonical derivation of eq. (13) performed in [23].

Starting from the Lagrangian density

$$\mathcal{L} = -\frac{1}{4} F_{\mu\nu} F^{\mu\nu} - \lambda n^\mu A_\mu \tag{14}$$

(fermions are disregarded for simplicity), where $\lambda$ is a Lagrange multiplier, one easily derives the equations of motion

$$\begin{aligned} D_\nu F^{\mu\nu} &= -n^\mu \lambda, \\ nA &= 0, \end{aligned} \tag{15}$$

and, as a consequence,

$$(n\partial)\lambda = 0. \tag{16}$$

In the null-plane formalism, eq. (16) is interpreted as a constraint, leading, under suitable boundary conditions, to the solution $\lambda = 0$ and, thereby, to the strong fulfilment of the Gauss law and to the CPV prescription for the $(nk)^{-1}$ singularity we have discussed before.

But eq. (16) actually is an evolution equation, which naturally possesses non-vanishing solutions. When $\lambda \neq 0$, the Gauss operator does not vanish anymore and we are led again to a formulation with redundant degrees of freedom.

Canonical quantization is performed by imposing the following equal time commutation relations

$$[A_i^a(t,\bar{x}), F_{0j}^b(t,\bar{y})] = i\delta^{ab}\delta_{ij}\delta(\bar{x}-\bar{y}); \tag{17}$$

$i,j = 1,2,3$ and $a,b$ are indices of the adjoint representation of the algebra of SU(N). In the radiation case ($g = 0$), the potentials can be written as a sum of a "physical" part $T_\mu(\bar{k})\delta(k^2)$, such that

$$nT = kT = 0 \tag{18}$$

and of an "unphysical" one

$$\Gamma_\mu(k) = n_\mu \frac{\delta(nk)}{k_\perp^{1/2}}\,\ell(\bar{k}) + \frac{ik_\mu}{k_\perp^{3/2}}\,\delta(nk)f(\bar{k}), \tag{19}$$

where $\ell(\bar{k}) = k_\perp^{-3/2}\tilde{\lambda}(\bar{k})$.

$T_\mu$ can be decomposed in frequencies and quantized in the usual way. The operators $\ell$ and $f$ can also be decomposed in frequencies and the algebra (17) entails the crossed algebra

$$[\ell_\mp^a(\bar{k}), f_\pm^b(\bar{q})] = \pm\delta^{ab}\delta(\bar{k}-\bar{q}), \tag{20}$$

all the other commutators vanishing.

From those algebras the free propagator is univocally obtained in the form

$$\mathcal{D}_{\mu\nu}^{ab}(x) = \frac{i\delta^{ab}}{(2\pi)^4}\int \frac{d^4k e^{ikx}}{k^2+i\varepsilon}\Big(-g_{\mu\nu} + \frac{n_\mu k_\nu + n_\nu k_\mu}{[nk]}\Big), \tag{21}$$

where $[nk]^{-1}$ has precisely the ML expression (13).

Among the virtues of the ML prescription, there is the possibility of performing Wick's rotation without extra terms. This, in turn, allows for a power counting criterion for convergence of Feynman integrals, in which, however, the components of any integration vector that are orthogonal to $n_\mu$ are to be treated separately [24]. It is remarkable that, with this prescription, the integral in eq. (7) turns out to be convergent in agreement with the result obtained in other gauges [8].

Several comments concerning this formulation relying on two gauge vectors are in order.

First of all, there is a further (apparent) violation of Poincaré covariance; moreover, Gauss' law does not hold as an operator equation. As a matter of fact the Gauss' expression is equal to the Lagrange multiplier $\lambda$, which behaves like a genuine field propagating along a generating line of the light-cone [23]. Eq. (13) can be essentially interpreted as its propagator. This field gives rise to negative norm states.

Fortunately it propagates according to a free equation (see eq. (16)), even in the interacting theory. As a consequence, a decomposition in frequencies, stable under time evolution, is possible and one can thereby select in the Fock space of states, which is of indefinite norm, a subspace $\mathcal{H}$ of semi-definite norm in which Poincaré-covariance, Gauss' law and unitarity of the formal perturbative S-matrix are recovered, by means of the condition.

$$\lambda^{(-)}|\Phi>=0, \quad \forall|\Phi>\in \mathcal{H}. \tag{22}$$

We remark that, in particular, the Lorentz covariance is not recovered via the Dirac bracket formalism, but rather as a weak condition in a suitable space of states. The price this formulation pays for having good properties from a mathematical viewpoint is the presence of unphysical degrees of freedom.

The renormalization procedure in this framework deserves a special discussion. The presence of a second gauge vector gives rise to a plethora of tensorial structures which can be used a priori to build possible counterterms; besides, the one particle irreducible (1PI) vertices exhibit, already at the level of primitive divergences, singularities at $\omega = 2$, the residues of which are nonpolynomial in the external momenta. To remove them, nonlocal counterterms are required, further increasing the number of possibilities. As a final pathology, 1PI vertices stay divergent, no matter the number of external lines.

All those difficulties have been overcome in ref. [24] with the use of several ingredients. First of all a careful analysis of the allowed non local singular terms was performed, taking into account a different power counting between longitudinal and transverse components, homogeneity properties with respect to the vectors $n_\mu$ and $\hat{n}_\mu$, the appropriate Lee-Ward identities and a kind of continuity of the ML prescription with respect to prescriptions with only one gauge vector [25].

As a matter of fact it is known [17], [21] that, with only one gauge vector, the counterterms are local. The same reasoning could be extended to the light-cone case with the CPV prescription, were it possible to avoid the pinches between Feynman and spurious poles. If they are disregarded, one gets a result which is clearly unphysical, but local. This is exactly what happens with the ML prescription, setting $\hat{n}_\mu = n_\mu$ afterwards (eq. (13) becomes indeed CPV under such a replacement).

Other a priori possible independent singular local structures were discarded, as they would conflict, if present, with the Lorentz invariance of the "physical" S-matrix elements [19]. Eventually only one nonlocal structure $\Omega$ survives, defined as

$$nD\Omega = \frac{nF\hat{n}}{n\hat{n}}, \tag{23}$$

$D$ being the covariant derivative and $F$ the field-strength tensor (obviously $nD$ is inverted according to the ML prescription).

The "canonical" transformation, relating "bare" and "renormalized" vector potentials ($A_\mu$) and Dirac spinors ($\psi$) is given by

$$A_\mu^{(0)} = Z_3^{1/2}\left[A_\mu - \left(1 - \tilde{Z}_3^{-1}\right)n_\mu\Omega\right],$$

$$\psi^{(0)} = \left(Z_2\tilde{Z}_2\right)^{1/2}\left[1 - \left(1 - \tilde{Z}_2^{-1}\right)\frac{\hat{n}n}{2n\hat{n}}\right]\psi, \tag{24}$$

$$g_0 = Z_3^{-1/2}g,$$

and involves four independent renormalization constants [24].

This result can be extended to SUSY Yang-Mills $N = 4$ [26]. In this case one can show that $Z_2 = Z_3 = 1$ at any loop order, whereas $\tilde{Z}_2$ and $\tilde{Z}_3$ keep divergent. In other words, only the "transverse" Green functions of SUSY Yang-Mills $N = 4$ are finite. The "nontransverse" Green's functions are ultraviolet divergent; nevertheless, the S-matrix elements, corresponding to those divergences, vanish, in agreement with the partial results obtained in the Fermi-Feynman gauge [27].

In conclusion the light-cone gauge is acceptable, provided the ML prescription (eq. (13)) is adopted. In this case a ghost is present and nonlocal infinities occur in the loop expansion. Nevertheless, the theory is perturbatively renormalizable and sensible, when restricted to a suitable space of "physical" states.

The one vector and two vector formulations do not differ by residual gauge transformations, but in their very content of degrees of freedom. They are in our opinion different theories.

At this point one might wonder whether a two vector formulation is possible also when $n^2$ and $\hat{n}^2$ are different from zero. Of course the ensuing theories are much more involved in comparison with the light-cone choice and hence the interest is, to some extent, of academic nature. Nevertheless, they might be useful to explore continuity properties in $n_\mu$ and $\hat{n}_\mu$ and to make comparisons with the corresponding one vector formulations.

Two options are possible, the axial [28], [29], [30] and the planar one, with free propagators given respectively by eqs. (6) and (21), in which the singularity $(nk)^{-1}$ is interpreted according to the prescription (13) (to this regard we recall that the ML distribution can be squared).

Still, in spite of this good feature, we are afraid that the double spurious pole might again be troublesome, for a different reason. As a matter of fact to square eq. (13), which behaves like a Feynman propagator with regard to its pole, gives rise to terms of a "higher derivative" type. It is known since a long time [31] that multi-pole ghosts are produced by this pathology. Therefore it would be mandatory to check that they are harmless in this instance with respect to unitarity. Moreover, this property should survive renormalization.

As a consequence, we shall limit ourselves in the following to discuss the planar case [32]. This is most easily done, starting from the Lagrangian

$$\mathcal{L} = -\frac{1}{4}F_{\mu\nu}F^{\mu\nu} + \frac{1}{2n^2}\partial_\mu\phi\partial^\mu\phi + \lambda(\phi - nA), \tag{25}$$

$\phi$ being an auxiliary field and $\lambda$ a Lagrange multiplier (again fermions are disregarded). From eq. (25) we derive the Euler-Lagrange equations

$$\begin{aligned} D^\nu F_{\mu\nu} &= -n_\mu\lambda, \\ \partial^2\phi &= n^2\lambda, \\ \phi &= nA, \end{aligned} \tag{26}$$

and, as a consequence,

$$nD\lambda = 0. \tag{27}$$

In the space-like case eq. (27), under suitable boundary conditions, gives $\lambda = 0$ and we are back to the previous one-vector formulation (eq. (8)). But in the spacelike case[1] we can also search for nonvanishing solutions of eq. (27), using two gauge vectors.

This is most easily accomplished by choosing a Lorentz frame in which $n_\mu = (n_0, 0, 0, n_3)$ and $\hat{n}_\mu = (n_0, 0, 0, -n_3)$, $|n_3| > |n_0|$.

One can set up a quantization procedure starting from equal time canonical commutators which turn out to involve also the quantum $\lambda$, by essentially repeating the procedure of ref. [23]. One gets the free field propagator in the form of eq. (21), where the singularity $(nk)^{-1}$ is now to be interpreted according to the ML prescription (13). Of course unphysical degrees of freedom are present; besides the quantum $\lambda$, there is also the component $nA$, which does not obey a free field equation anymore (see eqs. (26)). They can and do produce negative norm states in the Fock space.

As a consequence the first problem to be solved is to understand the structure of the subspace of the "physical" states. This has been done in ref. [32], following the BRST procedure [33].

The Lagrangian (25) together with the additional term

$$\mathcal{L}_g = i\eta nD\omega, \tag{28}$$

[1] In the time-like case the quantum associated to $\lambda$ would "propagate" as a tachyon. In this situation it is not clear whether the argument, used in ref. [23] to define the perturbative S-matrix as the weak limit between localizable states of the evolution operator, is tenable.

$\eta$ and $\omega$ being Faddeev-Popov (FP) ghosts, is invariant under the transformation

$$\begin{aligned} \delta A_\mu &= \zeta D_\mu \omega\,, \\ \delta\omega &= -\frac{1}{2}\zeta g\omega \times \omega\,, \\ \delta\eta &= -i\zeta\lambda\,, \\ \delta\lambda &= \delta\phi = 0\,, \end{aligned} \tag{29}$$

$\zeta$ being a constant Grassmann parameter. The related nilpotent conserved charge is

$$Q_B = n_0 \int d^3\bar{x}\Big[\omega\cdot(\lambda + \frac{1}{2}ig\omega\times\eta)\Big]. \tag{30}$$

The scale transformation for the FP ghosts is generated by the conserved charge

$$Q_G = -in_0 \int d^3\bar{x}\omega\cdot\eta. \tag{31}$$

A first subspace $\mathcal{V}$ of the Fock space can therefore be defined by imposing the conditions

$$Q_B|\Phi>=0, \quad Q_G|\Phi>=0, \quad \forall|\Phi>\in\mathcal{V}, \tag{32}$$

which guarantee the decoupling of the FP ghosts and of the quantum $\lambda$ from the physical sector of the asymptotic states. As a matter of fact eqs. (32) imply in the bosonic sector

$$\lambda_{as}^{(-)}|\Phi>=0, \quad \forall|\Phi>\in\mathcal{V}. \tag{33}$$

Still the subspace $\mathcal{V}$ contains quanta belonging to the unphysical field $\phi = nA$ which does not obey a free field equation (see eqs. (26)). Fortunately one can show that

$$\partial^2\phi = n^2[\eta, Q_B]. \tag{34}$$

As a consequence the restriction $\phi_R$ of the field $\phi$ in $\mathcal{V}$ obeys a free-field equation. Then we can eventually single out a positive semidefinite norm space $\mathcal{H}\subset\mathcal{V}$, by the condition

$$\phi_R^{(-)}|\Phi>=0, \quad \forall|\Phi>\in\mathcal{H}. \tag{35}$$

This condition is obviously the analogue of eq. (9) in the one vector case.

Perturbative one loop calculations have started using the propagator (21) with the ML prescription. The one loop vector self energy has been computed in refs. [34], [35], [36], whereas in refs. [37], [38] the fermion self energy and the fermion-fermion- vector vertex have been obtained at one loop order together with a check of the related Lee-Ward identity. In ref. [38] the four fermion vertex has also been obtained and the gauge invariance of the related S-matrix elements has been checked.

These results exhibit remarkable continuity properties in the limit $n^2 \to 0$, where they smoothly reproduce the light cone situation (with the ML prescription). In the limit $n_0 \to 0$ they should reproduce the spacelike planar results with the CPV prescription, provided the limit is understood in the sense of the theory of distributions, namely introducing all the necessary regularizations in the intermediate steps [25].

Unfortunately renormalization seems to be a formidable task. Singularities at $\omega = 2$ with nonpolynomial residues in the external momenta now affect also Green's functions which are expected to diverge no matter the number of external lines, whereas the circumstance that $n^2$ and $\hat{n}^2$ are different from zero, increases further the number of a priori possible tensorial structures to be used as counterterms. It is likely that several nonlocal independent structures are needed; in such a case it would be a nontrivial task to relate a given divergent term to the structure which is designed to re-absorb it.

From the computational side, already at one loop level the three vector vertex seems to require the use of algebraic manipulation programs.

Therefore, the two vector planar formulation we have just described seems to encounter serious difficulties both of a conceptual and of a technical nature. A solution, if any, does not seem within one's reach in a near future.

## References

[1] Yu. L. Dokshitzer, D.I. Dyakonov and S.I. Troyan, Phys. Reports **58** (1980) 270 and references therein.

[2] A. Bassetto, M. Ciafaloni and G. Marchesini, Phys. Reports **100** (1983) 201 and references therein.

[3] M. Namazie, A. Salam and J. Strathdee, Phys. Rev. **D28** (1983) 269.

[4] L. Brink, O. Lindgren and B.E.W. Nilsson, Nucl. Phys. **B212** (1983) 401.

[5] S. Mandelstam, Nucl. Phys. **B213** (1983) 149.

[6] see e.g. M.B. Green, J.H. Schwarz and E. Witten, Superstring Theory - Cambridge Univ. Press 1987, and relevant references therein.

[7] G. Curci, W. Furmanski and R. Petronzio, Nucl. Phys. **B175** (1980) 27.

[8] D.M. Capper, J.J. Dulwich and M.J. Litvak, Nucl. Phys. **B241** (1984) 463.

[9] for a general treatment, see: A. Hanson, T. Regge and C. Teitelboim, Constrained Hamiltonian Systems. Accademia Nazionale dei Lincei - Roma 1976 - and references therein.

[10] A. Bassetto, I. Lazzizzera and R. Soldati, Nucl. Phys. **B236** (1984) 319.

[11] A. Bassetto, I. Lazzizzera and R. Soldati, Phys. Lett. **107B** (1981) 278.

[12] N. Nakanishi, Phys. Lett. **131B** (1983) 381.

[13] G. Nardelli and R. Soldati, DFUB, Bologna preprint 7 (1989).

[14] S. Caracciolo, G. Curci and P. Menotti, Phys. Lett. **113B** (1982) 311.

[15] J.P. Leroy, J. Micheli and G.C. Rossi, Nucl. Phys. **B232** (1984) 511.

[16] G.C. Rossi and M. Testa, Nucl. Phys. **B163** (1980) 109, **B176** (1980) 477.

[17] W. Kummer, Acta Phys. Austr. **41** (1975) 315.

[18] A. Bassetto and R. Soldati, Nucl. Phys. **B276** (1986) 517.

[19 A. Andrasi and J.C. Taylor, Nucl. Phys. **B192** (1981) 283.

[20] A.I. Mil'shtein and V.S. Fadin, Yad. Fiz.**34** (1981) 140; Engl. transl: Sov. J. Nucl. Phys. **34** (1981) 779.

[21] A. Bassetto, G. Nardelli and R. Soldati, Mod. Phys. Lett. **A3** (1988) 1663.

[22] G. Leibbrandt, Phys. Rev. **D29** (1984) 1699.

[23] A. Bassetto, M. Dalbosco, I. Lazzizzera and R. Soldati, Phys. Rev. **D31** (1985) 2012.

[24] A. Bassetto, M. Dalbosco and R. Soldati, Phys. Rev. **D36** (1987) 3138.

[25] A. Bassetto and R. Soldati, Phys. Rev. **D37** (1988) 3065.

[26] A. Bassetto and M. Dalbosco, Mod. Phys. Lett. **A1** (1988) 65.

[27] O.V. Tarasov and A.A. Vladimirov, Phys. Lett. **96B** (1980) 94;
M.T. Grisaru, M. Rocek and W. Siegel, Phys. Rev. Lett. **45** (1980) 1063;
W. Caswell and D. Zanon, Phys. Lett. **100B** (1981) 152.

[28] P. Gaigg, M. Kreuzer, M. Schweda and O. Piguet, J. Math. Phys. **28** (1987) 2781;
P. Gaigg and M. Kreuzer, Phys. Lett. **205B** (1988) 530.

[29] G. Leibbrandt, Nucl. Phys. **B310** (1988) 405.

[30] I. Lazzizzera, Phys. Lett. **210B** 188 and preprint U.T.F. 146 (1988) Trento (Nuovo Cimento to appear).

[31] see for instance: N. Nakanishi, Prog. Theor. Phys. Suppl. **51** (1972) 1.

[32] A. Bassetto, I. Lazzizzera, G. Nardelli and R. Soldati, Phys. Lett. **228B** (1989) 235.

[33] C. Becchi, A. Rouet and R. Stora, Phys. Lett. **52B** (1974) 344; Commun. Math. Phys. **42** (1975) 127; Ann. Phys. **98** (1976) 287;
I.V. Tyutin, Lebedev preprint FIAN n. 39 (1975) unpublished.

[34] G. Nardelli and R. Soldati, Phys. Lett. **206B** (1988) 495.

[35] P. Gaigg, M. Kreuzer and G. Pollak, Phys. Rev. **D38** (1988) 2559.

[36] G. Leibbrandt and S.L. Nyeo, Mod. Phys. Lett. **A3** (1988), 1085:

[37] G. Nardelli, Phys. Rev. **D40** (1989) 2662.

[38] G. Pollak, Phys. Rev.**D40** (1989) 2057.

# The Temporal Gauge: Where Do We Stand?

Peter V. Landshoff

Department of Applied Mathematics and Theoretical Physics,
University of Cambridge,
Silver Street, Cambridge CB3 9EW, England

**Abstract:** A review is given of recent attempts to establish a useful prescription for perturbation-theory calculations in the temporal gauge $A_0 = 0$, at both zero and nonzero temperature. These attempts are not yet completely successful.

## 1 Introduction

Axial gauges are often used, both for theoretical analyses and for numerical calculations. This is because they are more physical than covariant gauges, in that their fields contain fewer unphysical degrees of freedom. A consequence is that using them often simplifies calculations, and also individual Feynman diagrams may be more directly identified with physical processes. (For example, the perturbation-theory analysis of the Altarelli-Parisi equation usually uses an axial gauge.)

Because axial gauges are so familiar, it is surprising that they are still not completely understood. This is in spite of a great deal of recent work on the subject. I am going to begin by giving a general review of the present position as I see it, and then in later sections I will go back and discuss some of the details.

An axial gauge is defined by imposing on the field $A_\mu(x)$ the condition $n.A = 0$, where $n$ is some fixed vector. I shall explicitly consider the case of the temporal axial gauge $A_0 = 0$, though everything I say is equally applicable to spacelike axial gauges [2], such as $A_3 = 0$. Lightcone gauges, for which $n^2 = 0$, are somewhat different [3], and I shall not discuss them.

We are still not completely sure that axial gauges exist, in a *useful* sense. By this, I mean three things:

1. The field $A_0$ really is zero, so that the elements $D_{00}$, $D_{0i}$ and $D_{i0}$ of the propagator $D_{\mu\nu}$ all vanish.

2. The propagator that connects points at times $t_1$ and $t_2$ depends only on $t_1 - t_2$ and not on $t_1 + t_2$. This translation invariance is needed if there is to be a simple momentum-space perturbation theory.

3. Faddeev-Popov ghosts should decouple from the theory. After all, the motivation for using an axial gauge is to remove unphysical degrees of freedom. Much of recent work does not succeed in meeting all three of these conditions. I think that the conditions are important, and so I shall discuss here mainly those papers that do try to satisfy them.

In a simple Lagrangian approach to the gauge $n.A = 0$, one obtains the propagator by including a gauge-fixing term $\frac{1}{2\lambda}(n.A)^2$, inverting the coefficient of the term quadratic in $A$, and then letting $\lambda \to 0$. This produces

$$D_{\mu\nu}(k) = \frac{1}{k^2}\left[-g_{\mu\nu} + \frac{k_\mu n_\nu + n_\mu k_\nu}{n.k} - n^2\frac{k_\mu k_\nu}{(n.k)^2}\right] . \tag{1.1}$$

In the temporal-gauge case, where $n = (1,0,0,0)$, this becomes

$$D_{00} = D_{0i} = D_{i0} = 0$$

$$D_{ij} = \frac{1}{k^2}\left[\delta_{ij} - \frac{k_i k_j}{k_0^2}\right] . \tag{1.2}$$

In applications, this propagator has to be integrated over the components of $k$, in particular over $k_0$. The question that has to be answered is how this integration deals with the double pole at $k_0 = 0$.

If one has a prescription for handling this double pole, one should test it by using it to calculate some gauge invariant quantity and comparing the answer with that which is obtained with the Feynman gauge. It has been found [1], [2] that a Wilson loop is a good choice for this gauge-invariant quantity. Taking the loop to be rectangular in shape and with one side (of length $T$) in the time direction and much longer than the other (whose length is $L$), gives a fairly simple but powerful diagnostic test of whether one's choice of prescription for the double pole does at least have a possibility of being correct. So far, the calculations of the Wilson loop have been taken up to order $g^4$ in perturbation theory. In order to be judged successful, they have to reproduce the exponentiation that is found in the Feynman gauge: for $T \gg L$

$$W(T,L) \sim \exp(-iTV(L)) = 1 - iTV(L) - \tfrac{1}{2}T^2V^2 \dots , \tag{1.3}$$

(where $V(L)$ is actually the quark-antiquark potential).

The traditional belief was that the correct prescription was the principal-value prescription

$$\frac{1}{k_0^2} \to \frac{1}{2}\left[\frac{1}{(k_0+i\epsilon)^2} + \frac{1}{(k_0-i\epsilon)^2}\right] , \tag{1.4}$$

where $\epsilon \to 0$ at the end of a calculation. But the Wilson-loop test has shown that this prescription is *wrong* for nonabelian gauge theories: it does not give exponentiation.

An alternative is the $\alpha$-prescription [2]

$$\frac{1}{k_0^2} \to \frac{1}{k_0^2+\alpha^2} , \tag{1.5}$$

where $\alpha \to 0$ at the end of a calculation. Individual graphs diverge like negative powers of $\alpha$, but these divergences cancel when the graphs are added together, and the correct answer is obtained for the Wilson loop. This is a nontrivial test, and the optimistic attitude is to believe that it indicates that the prescription is valid generally, but it is not yet known whether this is the case.

A third prescription is what I will call the Vienna prescription [4], which is a generalisation to the temporal gauge of a prescription first applied by Mandelstam and Leibbrandt [3] to the light-cone gauge. This prescription, in its simplest form, picks out one of the components of the 3-vector $\mathbf{k}$, say $k_1$, and then reads

$$\frac{1}{k_0^2} \to \left(\frac{k_1}{k_0 k_1 \pm i\eta}\right)^2 , \tag{1.6}$$

where $\eta \to 0$ at the end of a calculation. The Vienna prescription gives the right answer for the Wilson loop [5]. It *probably* also gives the right answer for all other calculations, though this has not yet been proved completely.

The obstacle that stands in the way of completing the proof of the $\alpha$ or the Vienna prescription is the existence of pinching poles. The $\alpha$ prescription certainly has pinching poles: the denominator in (1.5) has poles at $k_0 = \pm i\alpha$, which pinch the real-axis integration contour when $\alpha \to 0$. It is the existence of these pinching poles that produces the negative powers of $\alpha$ in individual graphs, and it also makes it uncertain [6] whether the prescription is generally valid with $D_{00} = D_{0i} = D_{i0} = 0$. As far as is known [5], [7], no pinching poles occur in applications of the Vienna prescription. If one could show that this is always true, one would know that the prescription is correct.

However, this would not be the end of the problem. The Vienna prescription leads to really nasty integrals and so in practice it is unpleasant to use [8]. So even if it is correct, there is a strong incentive to continue to investigate other prescriptions, such as the $\alpha$ prescription, where the integrals are nicer.

In order to derive the Vienna prescription, one fixes the gauge so that [9]

$$A_0 = \zeta \frac{1}{\partial_1} \frac{1}{\partial^2} \boldsymbol{\partial}.\mathbf{A} \ . \tag{1.7}$$

It does not matter that this relation is nonlocal. It is easy to impose canonical quantisation, for example in the Hamiltonian formalism [9], [10], [11]. The formalism contains unphysical particles, which correspond to the longitudinal component of the gauge field, and one has to impose a constraint that these are not present in physical states:

$$\langle \text{phys} | \boldsymbol{\partial}.\boldsymbol{\pi} | \text{phys} \rangle = 0 \ , \tag{1.8}$$

where $\boldsymbol{\pi}$ is the momentum field canonically conjugate to $\mathbf{A}$. (This equation is just Gauss's law.) When $\zeta \to 0$, it would appear from (7) that one retrieves $A_0 = 0$, as desired. But this limit may not be straightforward, because of possible pinching poles. (In order to retrieve precisely the prescription (1.6), one has to let $\zeta \to 0$ through imaginary values, $\zeta = \pm i\eta$ with $\eta$ real.)

The $A_0 = 0$ gauge would seem [12] to be particularly suitable for calculations at finite temperature $T$. The presence of the heat bath then anyway breaks the Lorentz symmetry, so that covariant gauges lose some of their usual advantage, and the temporal gauge preserves the rotational symmetry that is still present. As usual, the thermal average of a physical quantity $Q$ is given by

$$\langle Q \rangle = \sum_i \langle i | Q e^{-H/T} | i \rangle \ , \tag{1.9}$$

but, unlike in the scalar-field case, there is a restriction on the summation: it extends only over physical states $|i\rangle$ that satisfy the constraint (1.8). To implement this, one needs the Hamiltonian approach to field quantisation. James [13] has recently shown that it can be used straightforwardly to obtain the real-time finite-temperature field theory.

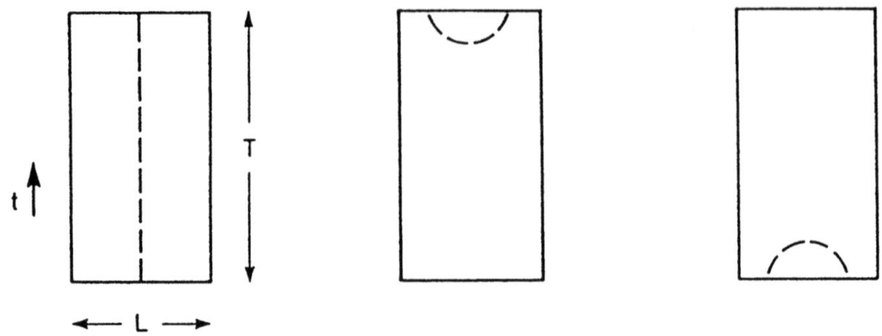


Figure 1. Lowest order diagrams for the Wilson loop in the gauge $A_0 = 0$

If one wants an imaginary-time theory, things are much more difficult. This is for two reasons. Firstly, in the imaginary-time approach there is a summation over discrete values of $k_0$, instead of integration. The summation includes the value $k_0 = 0$, so the presence of the double pole at this point is particularly awkward. Secondly, the usual derivation of the imaginary-time formalism depends crucially on the periodicity of the propagator in imaginary time (it takes equal values at time $t$ and $t - i/T$). However, the longitudinal-field propagator turns out not to have this periodicity [13]. It is not known what one can do about this.

## 2 Wilson Loop Calculations

The calculation of the Wilson loop that I described in the last section is rather delicate, which is why it provides a good diagnostic test. I shall now fill in some of the details about this calculation.

As is well known, the definition of the Wilson loop in $SU(N)$ gauge theory is

$$W = \frac{1}{N}\langle 0|\mathrm{tr}Pe^{ig\oint A_\mu dx^\mu}|0\rangle = \frac{1}{N} - \frac{g^2}{2N}\langle 0|\mathrm{tr}\oint\oint dx_\mu dy_\nu A^\mu(x)A^\nu(y)|0\rangle + \ldots . \quad (2.1)$$

Here, $P$ denotes ordering of the colour matrices round the integration contour and, in the case of a calculation in Minkowski space, time ordering. Take the loop to be rectangular, with sides of lengths $T$ and $L$ and with the $T$-side parallel to the time direction. In the gauge $A_0 = 0$, with $D_{00} = D_{0i} = D_{i0} = 0$, three diagrams contribute in order $g^2$; these

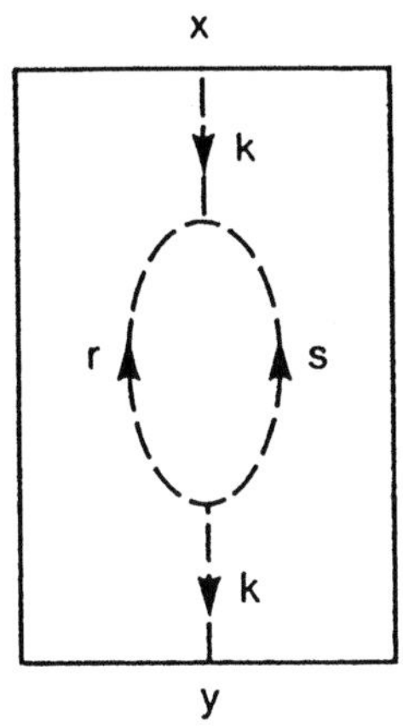


Figure 2. A contribution to the Wilson loop in order $g^4$

are shown in figure 1. One finds that the sum of the diagrams is independent of the prescription (though this is not true of the separate diagrams). All the prescriptions, including the principal-value one, give for $T \gg L$ the answer $-iTV_0(L)$, where

$$V_0(L) = \frac{-g^2 c_F}{4\pi^\omega}\Gamma(\omega-1)L^{2-2\omega} , \tag{2.2}$$

where $\omega = \frac{1}{2}(n-1)$, with $n$ the dimension of space-time. In order $g^4$ there are many diagrams. The most delicate one is that shown in figure 2, which contributes

$$-g^2 c_A c_F \int \frac{d^n k}{(2\pi)^n} \int_0^L \int_L^0 e^{ik_0T - ik_1(x_1-y_1)} \Delta_{11}(k) = 2g^2 c_A c_F \int \frac{d^n k}{(2\pi)^n} \frac{e^{ik_0T}}{k_1^2} \Delta_{11}(k) , \tag{2.3}$$

where $\Delta_{ij}$ is the propagator with the one-loop insertion. A part of $\Delta_{ij}$ is

$$\Delta_{ij}(k) = \frac{k_i k_j}{(k^2)^2}\left(\frac{1}{k_0^2}\right)^2 \frac{g^2}{(2\pi)^n} \int d^n r d^n s \delta^{(n)}(r+s+k) \frac{1}{r^2 s^2}[1 - \frac{\mathbf{s}^2}{r_0^2}][(\mathbf{r}.\mathbf{k})^2 - \mathbf{r}^2\mathbf{s}^2] . \tag{2.4}$$

The reason why the diagram is delicate is obvious: it requires one to square the distribution $\frac{1}{k_0^2}$. With the principal-value prescription (1.4) one must do the integrations before letting $\epsilon \to 0$; otherwise the prescription has no well-defined meaning. One finds that the loop integration over $r$ or $s$ vanishes like $\epsilon$, and the $k_0$ integration diverges like $\frac{1}{\epsilon}$, so that one ends up with a finite answer. Other prescriptions give different answers for this integral.

The conclusion from the Wilson-loop tests is, then, that things become sensitive in next-to-lowest order calculations. However, this is true only for nonabelian gauge theory: it seems that in the abelian case even the principal-value prescription works [1], [2].

## 3 The $\alpha$-Prescription

In response to the discovery of Carracciolo et al. [1] that the principal-value prescription gives the wrong answer for the Wilson loop, I invented the $\alpha$ prescription and showed that it gives the right answer for this calculation [2]. This prescription is

$$D_{00} = D_{0i} = D_{i0} = 0$$

$$D_{ij} = \frac{1}{k^2 + i\epsilon}\left[\delta_{ij} - \frac{k_i k_j}{k_0^2 + \alpha^2}\right] . \tag{3.1}$$

In my first attempt at the calculation, I set $\alpha = i\epsilon$ and let $\epsilon \to 0$ after the integrations, but this gave the wrong answer. The reason for this is that it is equivalent to initially giving the gauge particle a mass (here complex), and letting this go to zero at the end of the calculation. This has long been known to be an incorrect procedure [14]. Instead, therefore, after the integrations I first let $\epsilon \to 0$ and then afterwards $\alpha \to 0$, and found that this gave the right answer. In fact this is more satisfactory, because it means that the procedures in Euclidean and Minkowski space are equivalent: in Euclidean space one does not needs $\epsilon$ at all and so sets it to zero from the start.

In the calculation, one encounters the integral

$$\begin{aligned} I(T) &= \frac{1}{2\pi}\int dk_0 \frac{e^{ik_0 T}}{k_0^2 + \alpha^2} \\ &= \frac{1}{2\alpha} - \tfrac{1}{2}|T| + O(\alpha) , \end{aligned} \tag{3.2}$$

and also its square

$$[I(T)]^2 = \frac{1}{4\alpha^2} - \frac{2|T|}{\alpha} + \tfrac{1}{2}|T|^2 + O(\alpha) . \tag{3.3}$$

The terms involving inverse powers of $\alpha$ cancel when one adds together the various diagrams. Notice, however, that the finite term in $[I(T)]^2$ is not the square of that in $I(T)$: this is the consequence of letting $\alpha \to 0$ only after the integrations.

There have been attempts [6] to derive the $\alpha$ prescription, or something like it, by using a gauge-fixing condition such as

$$A_0 = \alpha \frac{1}{\partial^2} \partial . \mathbf{A} \ . \tag{3.4}$$

The result is very close to (4.1), but it does differ from it in a significant way. For example, it gives

$$D_{0i}(k) = \frac{1}{k^2} \frac{i\alpha k_i}{k_0^2 + \alpha^2} \ . \tag{3.5}$$

When $\alpha \to 0$ this does not vanish: its limit is

$$\frac{k_i}{\mathbf{k}^2} \pi \delta(k_0)$$

Similarly, $D_{00} \neq 0$, and also $D_{ij}$ differs just a little from what is given in (3.1). The result is that there are many extra diagrams, in which gluons attach to the vertical sides of the Wilson loop.

It may be verified explicitly [15] that the sum of all these extra diagrams, together with some ghost terms, vanishes. It is not known whether this is an accident, or a general property. I should like to believe that is will always happen, so that one may use (3.1) and forget the extra complications that result from the analysis of [6] – otherwise the prescription is of little use!

## 4 The Vienna Prescription

The Vienna prescription reads

$$D_{00} = D_{0i} = D_{i0} = 0$$

$$D_{ij} = \frac{1}{k^2} \left[ \delta_{ij} - \frac{k_i k_j}{(k_0 \pm i\eta/k_1)^2)} \right] \ . \tag{4.1}$$

As I have already said, this prescription gives the right answer for the Wilson loop, and very likely is generally valid, but it results in nasty integrals.

To obtain the prescription, one imposes the gauge condition (1.7). With $\zeta$ real, and a suitable prescription for the inverse differential operators, this condition preserves the hermiticity of the fields, which is important for the quantisation procedure [9]. (It is convenient to think of the nonlocal condition (1.7) as implemented in momentum space; the exact way in which one treats the $k_1 = 0$ singularity does not matter in the end, for example one may use a principal-value prescription.)

To find the propagator, consider the asymptotic fields, with the gauge coupling switched off. Use the gauge condition (1.7) to eliminate $A_0$. Then, from the resulting Lagrangian one can easily calculate $\boldsymbol{\pi}$ in terms of $\dot{\mathbf{A}}$ and $\mathbf{A}$ and so write down the canonical commutation relations. One can also calculate the Hamiltonian and obtain the equations of motion from it, and solve them. The general solution is

$$\boldsymbol{\partial}.\boldsymbol{\pi}(t,\mathbf{x}) = \int \frac{d^3k}{(2\pi)^3}\theta(k_1)\{e^{-(i\zeta t/k_3 - i\mathbf{k}.\mathbf{x})}p(\mathbf{k}) + e^{i\zeta t/k_3 - i\mathbf{k}.\mathbf{x}}p^\dagger(\mathbf{k})\}\,,$$

$$\boldsymbol{\partial}.\mathbf{A}(t,\mathbf{x}) = t\boldsymbol{\partial}.\boldsymbol{\pi}(t,\mathbf{x}) + i\int \frac{d^3k}{(2\pi)^3}\theta(k_1)\{e^{-(i\zeta t/k_3 - i\mathbf{k}.\mathbf{x})}q(\mathbf{k}) - e^{i\zeta t/k_3 - i\mathbf{k}.\mathbf{x}}q^\dagger(\mathbf{k})\}\,, \quad (4.2)$$

with the commutation relations

$$[p(\mathbf{k}), p^\dagger(\mathbf{k}')] = [q(\mathbf{k}), q^\dagger(\mathbf{k}')] = [p(\mathbf{k}), q(\mathbf{k}')] = 0\,,$$

$$[q(\mathbf{k}), p^\dagger(\mathbf{k}')] = \mathbf{k}^2(2\pi)^3\delta^{(3)}(\mathbf{k}-\mathbf{k}')\,. \quad (4.3)$$

Eliminating $A_0$ from the Lagrangian causes the loss of the field equation $\frac{\delta L}{\delta A_0} = 0$, which is just Gauss's law. This is imposed as the constraint (1.8) on the physical states. There are two obvious alternative ways to implement this:

$$\text{either}\;\; p(\mathbf{k})|\text{phys}\rangle = 0 \quad\text{or}\quad p^\dagger(\mathbf{k})|\text{phys}\rangle = 0\,. \quad (4.4a)$$

The contribution of the longitudinal field to $H$ is

$$H^L = \int \frac{d^3k}{(2\pi)^3}\theta(k_3)\frac{1}{\mathbf{k}^2}\{p^\dagger(\mathbf{k})p(\mathbf{k}) + \frac{\eta}{k_3}(p^\dagger(\mathbf{k})q(\mathbf{k}) + p(\mathbf{k})q^\dagger(\mathbf{k}))\}\,, \quad (4.5)$$

so that if the vacuum is to be an eigenstate of the Hamiltonian we should impose

$$\text{either}\;\; q(\mathbf{k})|0\rangle = 0 \quad\text{or}\quad q^\dagger(\mathbf{k})|0\rangle = 0\,, \quad (4.4b)$$

according to which choice we make in (5.4a). It is then straightforward to calculate that with the first choices in (5.4)

$$i\langle 0|TA_i^L(x_1)A_j^L(x_2)]|0\rangle = -\int \frac{d^4k}{(2\pi)^4}\frac{k_ik_j}{\mathbf{k}^2}\frac{e^{-ik.x}}{(k_0 - (\zeta - i0)/k_1)^2}\,. \quad (4.6)$$

The second choices in (4.4) give just the complex-conjugate result.

Physical quantities should be independent of $\zeta$. So we can make an analytic continuation in $\zeta$ without changing anything, provided only that we ensure that during the continuation the pole in the integrand does not cross the integration contour. In particular, we may continue to pure-imaginary values, $\zeta \to \mp i\eta$, where the sign depends on which choices we make in (4.4). Thus we retrieve either of the prescriptions (1.6).

A complete set of physical states is those states that contain only transverse gauge particles. One can check that the interaction does not result in unphysical states. To do this, note that the completeness relation reads

$$1 = \sum |\ \rangle\langle\ | + \int \frac{d^3k}{(2\pi)^3}\theta(k_1)\frac{1}{\mathbf{k}^2}\left\{q^\dagger(\mathbf{k})\sum |\ \rangle\langle\ | p(\mathbf{k}) + p^\dagger(\mathbf{k})\sum |\ \rangle\langle\ | q(\mathbf{k})\right\} + \ldots , \tag{4.7}$$

where $\sum |\ \rangle\langle\ |$ denotes a sum over states $|\ \rangle$ containing only transverse particles. Hence the probability of the interaction creating one longitudinal particle is an integral of

$$\langle\ | S^\dagger q^\dagger |\ \rangle\langle\ | pS |\ \rangle . \tag{4.8}$$

It is straightforward to show [11], [16] that $pS|\ \rangle = 0$, so that this probability vanishes. Similar arguments apply to the creation of more than one longitudinal particle.

## 5 Langevin-Equation Approach

In quantum field theory one has to calculate the path integral

$$\frac{1}{Z}\int \mathcal{D}A\, A(x_1)A(x_2)\ldots A(x_n)e^{-S[A]} . \tag{5.1}$$

The Langevin-equation approach to this problem [17] involves the following procedure:

1. Introduce a fifth dimension $\tau$ and make the field $A$ become a function of $\tau$ as well as the ordinary space-time $x$. Introduce also a gaussian white noise $\eta(x,\tau)$.

2. Solve the Langevin equation
$$\frac{\partial A}{\partial \tau} = -\frac{\delta S}{\delta A} + \eta . \tag{5.2}$$

3. Form the product
$$A(x_1,\tau)A(x_2,\tau)\ldots A(x_n,\tau) \tag{5.3}$$
and average it over all choices of the white-noise function $\eta$.

Then when $\tau \to \infty$ it is claimed that one obtains just the desired path integral (6.1). This is easy to prove when $A$ is a scalar field [18]. When it is a gauge field, there is reason to believe that the procedure works also, and that there is no need explicitly to introduce ghosts or to fix the gauge [17].

In the gauge-field case, there are 4 Langevin equations (5.2), one for each component of the field. Lim suggested [19] that one might fix the gauge in the path integral (5.1) by inserting a $\delta$-functional $\delta[A_0]$ in the integrand, so as to remove $A_0$ from the formalism. Then there would be only three Langevin equations, one for each component of $\mathbf{A}$. This leads to a well-defined propagator (1.2), with the prescription

$$\frac{1}{k_0^2} \to \frac{1-e^{-2k_0^2\tau}}{k_0^2}\ , \tag{5.4}$$

with $\tau \to \infty$ at the end of any calculation that uses the propagator. For finite $\tau$ there is no pole, so that this is a well-defined prescription.

I applied the Wilson-loop test to this prescription and found that it fails [20]. The reason is almost certainly because it is incorrect to fix the gauge by simply inserting $\delta[A_0]$ in the path integral, because then one loses one of the four field equations (Gauss's law). However, one can put this right [20]. To obtain the propagator, consider the asymptotic field, that is switch off the gauge-field coupling. Suppose $A_0(x,\tau) = 0$. Then the zeroth component of the free-field Langevin equation (6.2) tells us that

$$A_0(x,\tau+\delta\tau) = \delta\tau\left\{-\tfrac{\partial}{\partial t}\partial.\mathbf{A}(x,\tau) + \eta_0(x,\tau)\right\}\ . \tag{5.5}$$

Evidently $A_0(x,\tau+\delta\tau)$ is not zero, but we can make it zero if we now make a gauge transformation, which reads

$$A_\mu(x,\tau+\delta\tau) \to A_\mu + \partial_\mu\Omega\ , \tag{5.6}$$

with

$$\Omega = \partial.\mathbf{A} - \int_{t_0}^{t} dt'\ \eta_0(t',\mathbf{x},\tau)\ . \tag{5.7}$$

The resulting propagator has been derived before [22]. It is not translation-invariant; hence it does not have a simple momentum-space representation. Instead of a Fourier transform of $\frac{1}{k_0^2}$, giving a dependence on $t_1 - t_2$, where $t_1$ and $t_2$ are the times at the ends of the propagator, there is the integral

$$\frac{1}{2\pi}\int \frac{dp_0}{p_0^2}\left[e^{ip_0t_1} - e^{ip_0t_0}\right]\left[e^{-ip_0t_2} - e^{-ip_0t_0}\right]\ . \tag{5.8}$$

When $t_0 \to \pm\infty$, this becomes

$$-\tfrac{1}{2}|t_2 - t_1| \mp \tfrac{1}{2}(t_1 + t_2) \pm t_0\ , \tag{5.9}$$

which is the form suggested in [1].

Because this propagator is not translation invariant, so that there is no simple momentum-space representation, it is not really very *useful*. Also, when the interaction is switched on, it seems likely that one will find a rather complicated vertex in Feynman graphs [21].

## 6 Summary

More work is needed. Firstly, the proof of the Vienna prescription remains to be completed. Secondly, other prescriptions for which the integrals are simpler need exploring further, such as the $\alpha$ prescription. Thirdly, it would be useful to set up an imaginary-time finite-temperature gauge field theory, if this is possible.

## References


[1] S. Carracciolo, G. Curci and P. Menotti, Phys. Lett. 113B (1982) 311.

[2] P. V. Landshoff, Phys. Lett. **169B** (1986) 69.

[3] G. Leibbrandt, Rev. Mod. Phys. **59** (1987) 1067.

[4] P. Gaigg and M. Kreuzer, Phys. Lett. **205B** (1988) 530.

[5] H. Hüffel, P. V. Landshoff and J. C. Taylor, Phys. Lett. **217** (1989) 147.

[6] F. Steiner, Phys. Lett. **173B** (1986) 321;
H. Cheng and E. C. Tsai, Phys. **D36** (1987) 3196;
H. S. Chan and M. B. Halpern, Phys. Rev. **D33** (1988) 540.

[7] A. Andrasi and J. C. Taylor, Nucl. Phys. **B310** (1988) 222 .

[8] E. Bagan and C. P. Martin, Cambridge preprint DAMTP 89-12;
P. Gaigg, M. Kreuzer and G. Pollak, Phys. Rev. **D38** (1988) 2559.

[9] P. V. Landshoff, Phys. Lett. **227B** (1989) 427.

[10] I. Lazzizzera, Phys. Lett. **210B** (1988) 188.

[11] K. Haller, Phys. Rev. **D36** (1987) 1830 and 1839;
A. Burnel, Phys. Rev. **D40** (1989) 1221.

[12] U. Heinz, K. Kajantie and T. Toimela, Phys. Lett. **183B** (1987) 96.

[13] K. A. James, Cambridge preprint DAMTP 89-20.

[14] H. van Dam and M. Veltman, Nucl. Phys. **B22** (1970) 397.

[15] H. Cheng and E. C. Tsai, *private communication.*

[16] I. Lazzizzera, *private communication.*

[17] G. Parisi and Wu Yongshi, Sci. Sin. **24** (1981) 483;
P. H. Damgaard and H. Hüffel, Phys. Rep. **152** (1987) 227.

[18] H. Hüffel and P. V. Landshoff, Nucl. Phys. **B260** (1985) 545.

[19] S. C. Lim, Phys. Rev. **D32** (1985) 1384.

[20] P. V. Landshoff, Phys. Rev. **D35** (1987) 766 .

[21] R. Horgan and P. V. Landshoff, *in preparation.*

[22] M. Lavelle, M. Schaden and A. Vladikas, Phys. Lett. **203B** (1988) 441;
A. C. Kalloniatis, *these proceedings.*

# Unified-Gauge Formalism at Two Loops

George Leibbrandt

Department of Mathematics and Statistics, University of Guelph,

Guelph, Ontario, Canada

## 1 Introduction

In massless Yang-Mills theory, with Lagrangian density

$$L_{YM} = -\frac{1}{4}(F^a_{\mu\nu})^2 - \frac{1}{2\alpha}(n\cdot A^a)^2, \quad \alpha\to 0, \tag{1.1}$$

the axial-type gauges (pure axial, light-cone, temporal) are defined by [1-4]

$$n_\mu A^a_\mu = 0, \quad \mu = 0,1,2,3, \quad n^2 = n_0^2 - \vec{n}^2, \tag{1.2}$$

where $n_\mu = (n_0,\vec{n})$ is a fixed vector and $A^a_\mu$ a massless Yang-Mills field, $a = 1,2,\ldots,N^2-1$ for SU(N); $\alpha$ is the gauge parameter. We use dimensional regularization and a (+,-,-,-) metric. The propagator reads (for $\alpha = 0$)

$$G^{ab}_{\mu\nu} = \frac{-i\,\delta^{ab}}{(q^2+i\epsilon)}\left[\delta_{\mu\nu} - \frac{(q_\mu n_\nu + q_\nu n_\mu)}{q\cdot n} + \frac{n^2 q_\mu q_\nu}{(q\cdot n)^2}\right], \quad \epsilon > 0, \tag{1.3}$$

and the four-gluon vertex of zero-loop order

$$V^{abcd}_{\mu\lambda\sigma\rho} = -ig^2\Big[f^{abe}f^{cde}(\delta_{\mu\sigma}\delta_{\lambda\rho} - \delta_{\mu\rho}\delta_{\lambda\sigma}) + f^{ace}f^{bde}(\delta_{\mu\lambda}\delta_{\sigma\rho} - \delta_{\mu\rho}\delta_{\lambda\sigma})$$
$$+ f^{ade}f^{cbe}(\delta_{\mu\sigma}\delta_{\lambda\rho} - \delta_{\mu\lambda}\delta_{\sigma\rho})\Big]\,. \tag{1.4}$$

The poles of $(q\cdot n)^{-\beta}$, $\beta = 1,2$, were handled originally [5] by the principal-value (PV) prescription [3,4,6-9],

$$PV\,\frac{1}{q\cdot n} = \frac{1}{2}\lim_{\mu\to 0}\left(\frac{1}{q\cdot n + i\mu} + \frac{1}{q\cdot n - i\mu}\right), \quad \mu > 0,$$

but more recently by the unifying-gauge prescription for axial-type gauges, namely [10-13]

$$\frac{1}{q\cdot n} = \lim_{\epsilon\to 0}\frac{q\cdot n^*}{q\cdot n\, q\cdot n^* + i\epsilon}, \quad \epsilon > 0. \tag{1.5}$$

Formula (1.5) is a generalization of the prescription for the light-cone gauge developed independently by MANDELSTAM [14] and the author [15]. The noncovariant vector $n_\mu$ and its dual, $n^*_\mu$, have the following structure:

$$n_\mu = \begin{cases} n^{\ell c}_\mu = (|\vec{n}|,\vec{n}), & n_0^2 = \vec{n}^2, \\ n^{ax}_\mu = (|\vec{n}_\perp|,\vec{n}), & n_0^2 < \vec{n}^2, \\ n^{temp}_\mu = (n_0,\vec{n}_\perp, -i|\vec{n}_\perp|), & n_0^2 > \vec{n}^2; \end{cases} \tag{1.6a}$$

$$n^*_\mu = \begin{cases} (n^{\ell c}_\mu)^* = (|\vec{n}|,-\vec{n}), \\ (n^{ax}_\mu)^* = (|\vec{n}_\perp|,-\vec{n}), \\ (n^{temp}_\mu)^* = (n_0,\vec{n}_\perp,\ i|\vec{n}_\perp|). \end{cases} \tag{1.6b}$$

The unifying prescription (1.5) is valid for the pure axial, light-cone, temporal and planar gauge and seems to give consistent results to one loop [16-27].

The purpose of the present calculation is to test the unifying prescription (1.5) in detail for the two-loop self-energy diagram, shown in Fig. 1. A partial test had already been carried out in 1989 by HÜFFEL, LANDSHOFF and TAYLOR [28] on the Wilson loop.

Of special importance and concern are the following points:

(i) Can all two-loop integrals be calculated explicitly? In particular, can we obtain the coefficients of the single and double poles? What about the finite parts?

(ii) Are any of the integrations ambiguous?

(iii) Are there any unexpected technical problems?

(iv) Do new integrals arise at the two-loop level which are absent at the one-loop level?

Before describing the highlights of this calculation, I should like to caution that the computation for the sunset diagram Fig. 1 is not quite complete. Furthermore, I decided to calculate everything by hand, because I wanted to be "in control of the situation", ready to develop new formulae if and when the need arose. It turned out to be a wise decision.

## 2 Structure of Sunset Graph

The selfenergy graph in Fig. 1, sometimes called the sunset diagram, is the most complicated of the two-loop diagrams. Its amplitude reads

$$\Pi^{af}_{\mu\nu}(p) = \frac{1}{3!}\iint \frac{dqdk}{(2\pi)^{4\omega}} V^{abcd}_{\mu\lambda\sigma\rho}\, G^{dg}_{\rho\alpha}(q)\, G^{ch}_{\sigma\beta}(p-k-q)\, G^{be}_{\lambda\xi}(k)\, V^{ghef}_{\alpha\beta\xi\nu}\,,$$
$$dq = d^{2\omega}q,\ dk = d^{2\omega}k; \tag{2.1}$$

$G^{ab}_{\mu\nu}$ and $V^{abcd}_{\alpha\beta\mu\nu}$ are given in eqs. (1.3) and (1.4), respectively. Let us write

$$\Pi^{af}_{\mu\nu}(p) = \delta^{af}[\Pi_{\mu\nu}(n^2=0) + \Pi_{\mu\nu}(n^2\neq 0)], \tag{2.2}$$

where the "light-cone" part reads [29]

$$\Pi_{\mu\nu}(n^2=0) = 3\lambda(N)(2\omega-3)[A(p)\delta_{\mu\nu} - B(p)(k_\mu n_\nu + k_\nu n_\mu) - C(p)n_\mu n_\nu],$$
$$\lambda(N) = \frac{-i\ g^4 N^2}{2(2\pi)^{4\omega}}, \tag{2.3}$$

while

$$\Pi_{\mu\nu}(n^2\neq 0) = \lambda(N)n^2[\Pi(1)\delta_{\mu\nu} + \Pi(2)k_\mu k_\nu + \Pi(3)n_\mu n_\nu$$
$$+ \Pi(4)(k_\mu n_\nu + k_\nu n_\mu) + \Pi(5)(q_\mu k_\nu + q_\nu k_\mu)], \tag{2.4}$$

with N = 3 for SU(3). The coefficients A,B,C, are the same as in ref.[29]:

$$A(p) = \iint dq\ dk[q^2k^2Q^2]^{-1} = \iint dR,\ dR = \frac{dq\ dk}{q^2k^2Q^2},\ Q_\mu = (p-k-q)_\mu, \tag{2.5a}$$

$$B(p) = \iint dR(k\cdot n)^{-1}, \tag{2.5b}$$

$$C(p) = \iint dR\ k\cdot q[q\cdot nk\cdot n]^{-1}. \tag{2.5c}$$

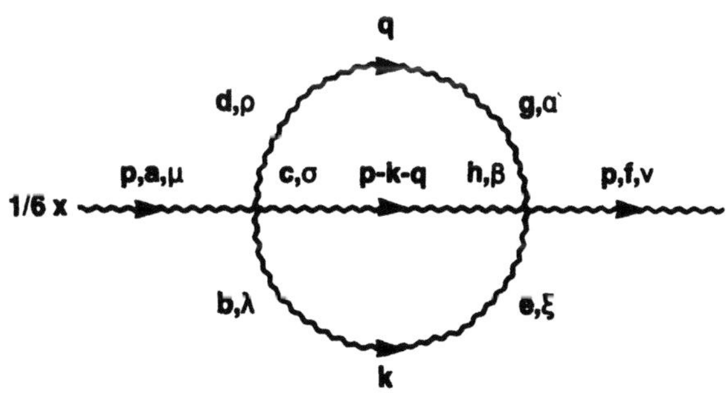


Fig. 1. Two-loop Yang-Mills self-energy.

It is convenient to re-write the five terms in eq. (2.4) as follows:

$$\Pi(1)\delta^{\mu\nu} = 4\Pi_1{}^{\mu\nu} - 8\Pi_2{}^{\mu\nu} + 2n^2\Pi_3{}^{\mu\nu},$$

$$\Pi(2)k^\mu k^\nu = -2\Pi_4{}^{\mu\nu} + 4n^2\Pi_5{}^{\mu\nu} - 4n^2\Pi_6{}^{\mu\nu} + 3n^4\Pi_7{}^{\mu\nu},$$

$$\Pi(3)n^\mu n^\nu = -\Pi_8{}^{\mu\nu},$$

$$\Pi(4)(k^\mu n^\nu + k^\nu n^\mu) = 4\Pi_9{}^{\mu\nu} + 4\Pi_{10}{}^{\mu\nu} + 2n^2\Pi_{11}{}^{\mu\nu} - 4\Pi_{12}{}^{\mu\nu} - 2n^2\Pi_{13}{}^{\mu\nu} + \mu \leftrightarrow \nu,$$

$$\Pi(5)(q^\mu k^\nu + q^\nu k^\mu) = 3\Pi_{14}{}^{\mu\nu} - 3n^2\Pi_{15}{}^{\mu\nu} + 2n^2\Pi_{16}{}^{\mu\nu} - \tfrac{3}{2}n^4\Pi_{17}{}^{\mu\nu} + \mu \leftrightarrow \nu\ . \tag{2.6}$$

The double integrals $\Pi_j{}^{\mu\nu}$, j = 1,...,17, are listed in Appendix A. For example,

$$\Pi_1{}^{\mu\nu} = \delta^{\mu\nu} \iint dR\ k^2(k\cdot n)^{-2}, \tag{2.7}$$

$$\Pi_2{}^{\mu\nu} = \delta^{\mu\nu} \iint dR\ q\cdot k(q\cdot n k\cdot n)^{-1}, \tag{2.8}$$

$$\Pi_3{}^{\mu\nu} = \delta^{\mu\nu} \iint dR(q\cdot k)^2[(q\cdot n)^2(k\cdot n)^2)]^{-1}. \tag{2.9}$$

All integrals contain overlapping divergences, the more difficult ones being $\Pi_3{}^{\mu\nu}$, $\Pi_7{}^{\mu\nu}$, $\Pi_{11}{}^{\mu\nu}$, $\Pi_{13}{}^{\mu\nu}$, $\Pi_{16}{}^{\mu\nu}$ and $\Pi_{17}{}^{\mu\nu}$.

## 3 Initial Steps in Computation

(a) The first step is to apply the decomposition formula [30]

$$\frac{1}{q\cdot n(p-q)\cdot n} = \frac{1}{p\cdot n}\left[\frac{1}{q\cdot n} + \frac{1}{(p-q)\cdot n}\right],\quad p\cdot n \neq 0, \tag{3.1}$$

to all integrals containing multiple, noncovariant factors in the denominator.

(b) It is convenient to defer any vector or tensor integrals, whenever possible, to the <u>second</u> integration. Take, for instance, the term

$$I^{\mu\nu} = \int dq \int \frac{dk}{D}\, k^\mu k^\nu (q\cdot k)^2, \qquad D = D\ (p,q,k,n). \tag{3.2}$$

Rather than perform the k-integration in (3.2), we first let $q_\mu \leftrightarrow k_\mu$,

$$I^{\mu\nu} \Rightarrow \int dq\ q^\mu q^\nu \int \frac{dk}{D}(q\cdot k)^2 = I_0^{\mu\nu}, \tag{3.3}$$

and only then integrate over dk in $I_0^{\mu\nu}$, by applying the procedure in part (c) below.

(c) To evaluate

$$\int \frac{dk}{D}(q\cdot k)^2 = q_\sigma q_\lambda \int \frac{dk}{D} k^\sigma k^\lambda, \tag{3.4}$$

or any integral having the general form

$$\int \frac{dk}{D}(k_\mu, k_\mu k_\nu, k_\mu k_\nu k_\sigma, \ldots), \tag{3.5}$$

we apply the tensor method which is the only viable technique for two-loop integrals, in my opinion. The tensor method is trivial for covariant gauges but requires "additional work" for noncovariant-gauge integrals. Two examples follow.
(i) For the integral $I_\mu = \int dk\, k_\mu/D$ we make either the ansatz [11]

$$I_\mu = A\, p_\mu + B\, n_\mu + C\, F_\mu, \tag{3.6a}$$

where $F_\mu$ is related to the dual vector $n_\mu^*$, or the ansatz

$$I_\mu = A' p_\mu + B' n_\mu + C' n_\mu^*. \tag{3.6b}$$

Since $F_\mu$ is a null vector, the structure (3.6a) is computationally more convenient than (3.6b). In the temporal gauge, for instance,

$$F_\mu^{temp} = (\vec{n}_\perp/\sigma,\ n_4/\sigma, i),\ \sigma^2 = \vec{n}_\perp^2 + n_4^2.$$

Multiplication of (3.6a) successively by $p_\mu$, $n_\mu$, $F_\mu$ yields three equations which can be solved uniquely for A,B,C. We note that integration of $I_\mu$ in component form is not a viable alternative, simply because it becomes too messy, especially by the time the second integration over dq is completed.

(ii) Next we demonstrate the tensor method for the integral

$$K_{\mu\nu} = \int \frac{dk\ k_\mu k_\nu}{k^2(p-k)^2(k\cdot n)^2}\ . \tag{3.7}$$

The proper ansatz turns out to be

$$\begin{aligned} K_{\mu\nu} &= [A\delta_{\mu\nu} + B(n_\mu F_\nu + n_\nu F_\mu) + C\, F_\mu F_\nu] \\ &\quad + \{Dn_\mu n_\nu + E(p_\mu n_\nu + p_\nu n_\mu) + G\, p_\mu p_\nu + H(p_\mu F_\nu + p_\nu F_\mu)\}, \\ &= K_{\mu\nu}^{pole} + \{\text{finite terms}\}, \end{aligned} \tag{3.8}$$

where $K_{\mu\nu}^{pole}$ and {finite terms } = $f_{\mu\nu}$ contain the pole part and finite portion of $K_{\mu\nu}$, respectively. The coefficients A,B,...,H, are found in the usual way.

Successful application of the tensor method to the "first" integration over dk yields about a dozen distinct types of divergent integrals. We shall show in Section 4 how to represent the poles of these integrals in terms of the two scalar integrals X and Y.

## 4 Reduction of One-Loop Integrals

The general aim is to express the pole parts of all necessary one-loop integrals in terms of two scalar integrals Y and X:

$$Y(p) = \int \frac{dq}{q^2(p-q)^2} = \frac{i\pi^2\Gamma(2-\omega)[\Gamma(\omega-1)]^2(p^2)^{\omega-2}}{\Gamma(2\omega-2)}, \tag{4.1}$$

$$X(p) = \int dq[(p-q)^2(q\cdot n)^2]^{-1}. \tag{4.2}$$

We shall demonstrate the procedure for the integral $K_{\mu\nu}$ given in eqs.(3.7) - (3.8):

$$K_{\mu\nu} = [A\delta_{\mu\nu} + B(n_\mu F_\nu + n_\nu F_\mu) + C\, F_\mu F_\nu] + f_{\mu\nu}. \tag{4.3}$$

We know from explicit computation [11] that only the first three coefficients, in square brackets, are proporational to simple poles. In other words, to obtain the divergent part of $K_{\mu\nu}$ it suffices to determine A,B,C. Multiplication of (4.3) in turn by $\delta^{\mu\nu}$, $n^\mu n^\nu$, $n^\mu p^\nu$, $p^\mu p^\nu$ yields four equations:

$$[2\omega A + 2Bn\cdot F + 0] + \delta^{\mu\nu}f_{\mu\nu} = \int \frac{dk}{(k-p)^2(k\cdot n)^2} = X\ , \tag{4.4}$$

$$[n^2A + 2n^2n\cdot F\ B + C(n\cdot F)^2] + n^\mu n^\nu f_{\mu\nu} = \int \frac{dk}{k^2(p-k)^2} = Y\ , \tag{4.5}$$

$$[\ldots] + n^\mu p^\nu f_{\mu\nu} = \int \frac{dk\ p\cdot k}{k^2(p-k)^2 k\cdot n} = Z\ , \tag{4.6}$$

$$[\ldots] + p^\mu p^\nu f_{\mu\nu} = \int \frac{dk(p\cdot k)^2}{k^2(p-k)^2(k\cdot n)^2} = W, \tag{4.7}$$

or

$$[2\omega A + 2Bn\cdot F + 0] = X - \delta^{\mu\nu}f_{\mu\nu} = X'\ , \tag{4.8}$$

$$[n^2A + 2n^2n\cdot F\ B + C(n\cdot F)^2] = Y - n^\mu n^\nu f_{\mu\nu} = Y'\ , \tag{4.9}$$

$$[\ldots] = Z - n^\mu p^\nu f_{\mu\nu} = Z'\ , \tag{4.10}$$

$$[\ldots] = W - p^\mu p^\nu f_{\mu\nu} = W'. \tag{4.11}$$

First we solve system (4.8) – (4.10) for A and B, constructing the ratio $(\frac{A}{B})^{(4.8)-(4.10)}$; next we solve system (4.8), (4.10), (4.11) for A and B, constructing $(\frac{A}{B})^{(4.8),(4.10),(4.11)}$, and then finally equate the two ratios. The equality

$$(\tfrac{A}{B})^{(4.8)-(4.10)} = (\tfrac{A}{B})^{(4.8),(4.10),(4.11)} \tag{4.12}$$

leads to <u>two</u> identities:

$$p\cdot F\ Y' - n\cdot F\ Z' = 0, \tag{4.13}$$

$$2(\omega-1)n\cdot F(p\cdot F\ Z' - n\cdot F\ W') + D_0 X' = 0, \tag{4.14}$$

which, in turn, yield the following relations among the four scalar integrals X, Y, Z, W:

$$Z = (p\cdot F/n\cdot F)Y + Z^{\text{finite}}, \tag{4.15}$$

$$W = \frac{1}{(n\cdot F)^2}\left[(p\cdot F)^2\ Y + \frac{D_0}{2(\omega-1)}\ X\right] + W^{\text{finite}}, \tag{4.16}$$

where

$$Z^{\text{finite}} = n_\mu(p_\nu - (p\cdot F/n\cdot F)n_\nu)f^{\mu\nu}, \tag{4.17}$$

$$W^{\text{finite}} = \left[p_\mu p_\nu - (n\cdot F)^{-2}\left((p\cdot F)^2 n_\mu n_\nu + \frac{D_0}{2(\omega-1)}\ \delta_{\mu\nu}\right)\right]f^{\mu\nu}, \tag{4.18}$$

$$D_0 = p^2(n\cdot F)^2 - 2p\cdot n\ n\cdot F\ p\cdot F + n^2(p\cdot F)^2. \tag{4.19}$$

Systematic application of this technique to the other integrals enables us to express the divergent part of all one-loop integrals in terms of X(p) and Y(p). (See Appendix B.) For example, $K_{\mu\nu}$ in eq.(3.7) becomes

$$K_{\mu\nu}(p) = \frac{Y(p)}{(n\cdot F)^2}\ F_\mu F_\nu$$

$$+ \frac{X(p)}{2(\omega-1)}\left[\delta_{\mu\nu} - \frac{(n_\mu F_\nu + n_\nu F_\mu)}{n\cdot F} + \frac{n^2}{(n\cdot F)^2}\ F_\mu F_\nu\right] + k_{\mu\nu}(\text{finite}), \tag{4.20}$$

where $k_{\mu\nu}$(finite) is known.

## 5 Final Integration

We illustrate the second integration, i.e. the integration over dq, by computing the double integrals $\Pi_1^{\mu\nu}$ and $\Pi_3^{\mu\nu}$.

### 5.1 The Integral $\Pi_1^{\mu\nu}$

$$\Pi_1^{\mu\nu} = \delta^{\mu\nu} \iint \frac{dq\ dk}{q^2Q^2(k\cdot n)^2} \overset{q\leftrightarrow k}{=} \delta^{\mu\nu} \int \frac{dq}{(q\cdot n)^2} \int \frac{dk}{k^2Q^2}, \tag{5.1}$$

$$= \delta^{\mu\nu} \int \frac{dq}{(q\cdot n)^2} Y(p-q), \quad Q_\mu = (p-q-k)_\mu. \tag{5.2}$$

Substituting for Y(p-q) from eq. (4.1), we get

$$\Pi_1^{\mu\nu} = \frac{i\ \delta^{\mu\nu}\pi^2\Gamma(2-\omega)[\Gamma(\omega-1)]^2}{\Gamma(2\omega-2)} \int \frac{dq}{[(p-q)^2]^{2-\omega}(q\cdot n)^2} \quad . \tag{5.3}$$

The remaining one-loop integral

$$\int dq\ [[(p-q)^2]^{2-\omega}(q\cdot n)^2]^{-1} = X(2-\omega), \tag{5.4}$$

can be computed explicitly to yield

$$\Pi_1^{\mu\nu} = \delta^{\mu\nu} \frac{i\pi^2\ \Gamma(2-\omega)[\Gamma(\omega-1)]^2}{\Gamma(2\omega-2)} \quad .$$

$$\cdot \frac{i\pi^2}{\rho^4\Gamma(2-\omega)} \left\{ \frac{n^2\Gamma(3-2\omega)}{2} \int_0^1 \frac{d\xi\ \xi^{2\omega-2}}{(1-B\xi)^{3/2}} [g(P)]^{2\omega-3} \right.$$

$$+ \Gamma(4-2\omega)\left\{ (\vec{P}\cdot\vec{N})^2 \int_0^1 \frac{d\xi\ \xi^{2\omega-3}}{(1-B\xi)^{5/2}} + 2i\rho P_4(\vec{P}\cdot\vec{N}) \int_0^1 \frac{d\xi\ \xi^{2\omega-3}}{(1-B\xi)^{3/2}} \right.$$

$$\left.\left. - \rho^2 P_4^2 \int_0^1 \frac{d\xi\ \xi^{2\omega-3}}{(1-B\xi)^{1/2}} \right\} [g(P)]^{2\omega-4} \right\}, \tag{5.5}$$

where

$$g(P) = P_4{}^2 + \frac{(\vec{P}\cdot\vec{N})^2}{\rho^2(1-B\xi)}\,, \quad B = 1 - \sigma^2/\rho^2, \tag{5.6}$$

$$\rho^{temp} = \rho^{ax} = |\vec{n}_\perp|,\ \rho^{\ell c} = |\vec{n}|;\ \vec{P}_\perp = \vec{p}_\perp,\ P_3 = p_4,\ P_4 = p_3, \tag{5.7}$$

$$\sigma^2 = \vec{n}_\perp^{\,2} + n_4^2\quad \vec{N}_\perp = \vec{n}_\perp,\ N_3 = n_4,\ N_4 = n_3.$$

Clearly the double integral $\Pi_1{}^{\mu\nu}$ possesses only a simple pole as $\omega\to 2^+$.

## 5.2 The Integral $\Pi_3{}^{\mu\nu}$

A more challenging overlapping divergence occurs in the integral $\Pi_3{}^{\mu\nu}$,

$$\Pi_3{}^{\mu\nu} = \delta^{\mu\nu}\iint \frac{dq\ dk\ (q\cdot k)^2}{q^2k^2Q^2(q\cdot n)^2(k\cdot n)^2} = \delta^{\mu\nu}\int \frac{dq\ q^\sigma q^\lambda}{q^2(q\cdot n)^2}\int \frac{dk\ k_\sigma k_\lambda}{k^2Q^2(k\cdot n)^2}\,. \tag{5.8}$$

Using formula (4.20) for the k-integral, we obtain

$$\Pi_3{}^{\mu\nu} = \delta^{\mu\nu}\int \frac{dq\ q^\sigma q^\lambda}{q^2(q\cdot n)^2}\left\{\frac{Y(p-q)}{(n\cdot F)^2}F_\sigma F_\lambda \right.$$

$$\left. + \frac{X(p-q)}{2(\omega-1)}\left[\delta_{\sigma\lambda} - \frac{(n_\sigma F_\lambda + n_\lambda F_\sigma)}{n\cdot F} + \frac{n^2}{(n\cdot F)^2}F_\sigma F_\lambda\right] + k_{\sigma\lambda}(\text{finite})\right\}, \tag{5.9}$$

where

$$X = X^{pole} + X^{finite}.$$

To extract the coefficients of the double or single poles from (5.9) we need to compute the following two integrals:

$$\text{(i)}\ K^{\sigma\lambda}(2-\omega) = \int \frac{dq\ q^\sigma q^\lambda}{q^2[(p-q)^2]^{2-\omega}(q\cdot n)^2}\,; \tag{5.10}$$

$$\text{(ii)}\ \chi^{\sigma\lambda} = \int \frac{dq\ q^\sigma q^\lambda}{q^2(q\cdot n)^2}X^{pole}(p-q) = \frac{i\,\pi^2n^2\Gamma(2-\omega)}{2(\rho^2)^\omega}\int_0^1 \frac{d\xi\ \xi^{\omega-1}}{(1-B\xi)^{\omega-1/2}}$$

$$\cdot\int \frac{dq\ q^\sigma q^\lambda}{q^2(q\cdot n)^2[((\vec{P}-\vec{Q})\cdot\vec{N})^2 + M(P_4-Q_4)^2]^{2-\omega}}\,, \tag{5.11}$$

with $M = \rho^2(1-B\xi) = \rho^2 + n^2\xi$; $\rho^{temp} = \rho^{ax} = |\vec{n}_\perp|$, $\rho^{\ell c} = |\vec{n}|$. $K^{\sigma\lambda}(2-\omega)$ and $\chi^{\sigma\lambda}$ may be evaluated with the help of the tensor method and the appropriate identities.

It is also worth noting, in the context of dimensional regularization, that [31]

$$\int \frac{dq}{q^2(q\cdot n)^2} = 0 \qquad (5.12)$$

as expected, but that

$$\lim_{\omega\to 2^+} \int \frac{dq}{q^2[(p-q)^2]^{2-\omega}(q\cdot n)^2} \qquad (5.13)$$

is different from zero and finite.

## 6 Summary

The Yang-Mills selfenergy $\Pi^{af}_{\mu\nu}$ has been calculated in a unified-gauge formalism which includes the axial, temporal and light-cone gauge, and - with an appropriate change of the gauge-fixing term - the planar gauge. The corresponding unified-gauge prescription

$$\frac{1}{q\cdot n} = \lim_{\epsilon\to 0} \frac{q\cdot n^*}{q\cdot n q\cdot n^* + i\epsilon}, \quad \epsilon > 0,$$

works to two loops for both simple and double poles, $(q\cdot n)^{-\beta}$, $\beta = 1,2$.

The main results are:

(a) $$\Pi^{af}_{\mu\nu}(p) = \delta^{af} \sum_{j=0}^{2} \frac{1}{\epsilon^{2-j}} \Big\{ A_j\delta_{\mu\nu} + B_j p_\mu p_\nu + C_j(p_\mu n_\nu + p_\nu n_\mu) + D_j n_\mu n_\nu$$
$$+ E_j(p_\mu n^*_\nu + p_\nu n^*_\mu) + G_j(n_\mu n^*_\nu + n_\nu n^*_\mu) + H_j n^*_\mu n^*_\nu \Big\}, \quad \epsilon = 2-\omega.$$

The coefficients $\{A_0,\ldots,H_0\}$ can be obtained explicitly; the coefficients $\{A_1,\ldots,H_1\}$ and $\{A_2,\ldots,H_2\}$ are derivable but not necessarily in closed form.

(b) All single and double integrals are well defined. Ambiguities do not arise, and integrals of the type

$$\int \frac{dq\ (1, q_\mu, q_\mu q_\nu)}{q^2[(p-q)^2]^{2-\omega} q\cdot n}$$

can be evaluated explicitly.

(c) No additional problems are caused by overlapping divergences.

(d) The basic approach to the evaluation of two-loop noncovariant-gauge integrals is:

(i) to derive identities between certain one-loop integrals; for instance, $Z = (p\cdot F/n\cdot F)Y + Z^{\text{finite}}$, i.e.

$$\int \frac{dq\ p\cdot q}{q^2(p-q)^2 q\cdot n} = \frac{p\cdot F}{n\cdot F}\int \frac{dq}{q^2(p-q)^2} + \text{Finite Part};$$

(ii) to exploit these identities to express the divergent parts of all one-loop integrals in terms of the scalar integrals

$$Y = \int dq[q^2(p-q)^2]^{-1}, \quad \text{and } X = \int dq[(p-q)^2(q\cdot n)^2]^{-1};$$

(iii) to apply the tensor method to the first as well as the second integrals, i.e. to both the k- and q- integrations.

As for future work in this area, the most urgent task is to complete the present calculation. Knowledge of the coefficients $\{A_0,\ldots,H_0\}$ of the double poles should give us further insight into the structure of the counterterms and hence into the renormalization program. An intriguing question in this connection concerns the non-local terms: can they be handled consistently, for instance, in the framework of the Becchi-Rouet-Stora formalism? [32]

## Acknowledgements

It gives me great pleasure to thank Professor John Ellis and the staff of the Theory Division for their hospitality and assistance during my stay at CERN in 1989, where most of the calculations were carried out. I should also like to thank Professors A. Bassetto, J. Ellis, I. Lazzizzera, M. Schweda, R. Soldati and Drs. J. Alberty, S.-L. Nyeo, G. Nardelli, P. Gaigg and G. Pollak for several discussions. Finally, I am grateful to the organizers of the Workshop on Physical and Nonstandard Gauges for their hospitality and financial support. This research was supported in part by the Natural Sciences and Engineering Research Council of Canada under grant No. A8063.

## Appendix A

## The Double Integrals $\Pi_j^{\mu\nu}$

We list the seventeen double integrals $\Pi_j^{\mu\nu}$ referred to in eq. (2.6) in the main text. These integrals constitute the selfenergy part $\Pi_{\mu\nu}(n^2\neq 0)$ defined in eq. (2.4).

$$\Pi_1^{\mu\nu} = \delta^{\mu\nu} \iint dR\; k^2(k\cdot n)^{-2},$$

$$\Pi_2^{\mu\nu} = \delta^{\mu\nu} \iint dR\; q\cdot k(q\cdot n k\cdot n)^{-1},$$

$$\Pi_3^{\mu\nu} = \delta^{\mu\nu} \iint dR (q\cdot k)^2[(q\cdot n)^2(k\cdot n)^2]^{-1},$$

$$\Pi_4^{\mu\nu} = \iint dR\; \frac{1}{(k\cdot n)^2}\, k_\mu k_\nu,$$

$$\Pi_5^{\mu\nu} = \iint dR\; \frac{Q^2}{(k\cdot n)^2(Q\cdot n)^2}\, k_\mu k_\nu,$$

$$\Pi_6^{\mu\nu} = \iint dR\; \frac{q\cdot Q}{q\cdot n(k\cdot n)^2 Q\cdot n}\, k_\mu k_\nu,$$

$$\Pi_7^{\mu\nu} = \iint dR\; \frac{(q\cdot Q)^2}{(q\cdot n)^2(k\cdot n)^2(Q\cdot n)^2}\, k_\mu k_\nu,$$

$$\Pi_8^{\mu\nu} = \iint dR\; \frac{q\cdot(p-q)k\cdot Q}{q\cdot n(k\cdot n)^2 Q\cdot n}\, n_\mu n_\nu,$$

$$\Pi_9^{\mu\nu} = \iint dR\; \frac{q\cdot k}{q\cdot n(k\cdot n)^2}\, k_\mu n_\nu,$$

$$\Pi_{10}^{\mu\nu} = \iint dR\; \frac{q\cdot Q}{q\cdot n k\cdot n Q\cdot n}\, k_\mu n_\nu,$$

$$\Pi_{11}^{\mu\nu} = \iint dR\; \frac{q\cdot Q k\cdot Q}{q\cdot n(k\cdot n)^2(Q\cdot n)^2}\, k_\mu n_\nu,$$

$$\Pi_{12}^{\mu\nu} = \iint dR \, \frac{q^2}{k\cdot n(q\cdot n)^2} k_\mu n_\nu ,$$

$$\Pi_{13}^{\mu\nu} = \iint dR \, \frac{(q\cdot Q)^2}{k\cdot n(q\cdot n)^2(Q\cdot n)^2} k_\mu n_\nu ,$$

$$\Pi_{14}^{\mu\nu} = \iint dR \, \frac{1}{q\cdot n k\cdot n} q_\mu k_\nu ,$$

$$\Pi_{15}^{\mu\nu} = \iint dR \, \frac{q\cdot k}{(q\cdot n)^2(k\cdot n)^2} q_\mu k_\nu ,$$

$$\Pi_{16}^{\mu\nu} = \iint dR \, \frac{k\cdot Q}{q\cdot n(k\cdot n)^2 Q\cdot n} q_\mu k_\nu ,$$

$$\Pi_{17}^{\mu\nu} = \iint dR \, \frac{q\cdot Q k\cdot Q}{(q\cdot n)^2(k\cdot n)^2(Q\cdot n)^2} q_\mu k_\nu .$$

## Appendix B

Application of the technique developed in Section 4 leads to the following one-loop integrals as functions of the scalar integrals Y and X.

$$\int \frac{dk}{k^2(q-k)^2} = Y(q),$$

$$\int \frac{dk}{(q-k)^2(k\cdot n)^2} = X(q),$$

$$\int \frac{dk \; k_\mu}{k^2(q-k)^2 k\cdot n} = \frac{Y(q)}{n\cdot F} F_\mu + \text{F.P.},$$

$$\int \frac{dk \; k_\mu k_\nu}{k^2(q-k)^2(k\cdot n)^2} = \frac{Y(q)}{(n\cdot F)^2} F_\mu F_\nu$$

$$+ \frac{X(q)}{2(\omega-1)} \left[ \delta_{\mu\nu} - \frac{(n_\mu F_\nu + n_\nu F_\mu)}{n\cdot F} + \frac{n^2}{(n\cdot F)^2} F_\mu F_\nu \right] + \text{F.P.},$$

$$\int \frac{dk}{(q-k)^2 k\cdot n} = \frac{2q\cdot F}{n\cdot F} Y(q) + \text{F.P.},$$

$$\int \frac{dk}{k^2(q-k)^2(k\cdot n)^2} = -\frac{X(q)}{q^2} + \text{F.P.},$$

$$\int \frac{dk\ k^2}{(q-k)^2(k\cdot n)^2} = \frac{4(q\cdot F)^2}{(n\cdot F)^2}\ Y(q) + \left[-q^2 + \frac{2D_0}{(\omega-1)(n\cdot F)^2}\right]X(q) + \text{F.P.},$$

$$\int \frac{dk\ k_\mu}{(q-k)^2(k\cdot n)^2} = \left(q_\mu - n_\mu \frac{q\cdot F}{n\cdot F}\right)X(q)$$

$$+ \frac{F_\mu}{(n\cdot F)^2}\left(2q\cdot F\ Y(q) + (n^2 q\cdot F - q\cdot nn\cdot F)X(q)\right) + \text{F.P.},$$

$$\int \frac{dk\ k^2}{(q-k)^2 k\cdot n} = \frac{2q\cdot F}{(\omega-1)(n\cdot F)^3}\ [(5-2\omega)(n\cdot F)^2\ q^2$$

$$-(\omega+1)n\cdot Fq\cdot Fq\cdot n + n^2(q\cdot F)^2]\ Y(q) + \text{F.P.},$$

$$\int \frac{dk\ k_\mu k_\nu}{k^2(q-k)^2 k\cdot n} = \frac{Y(q)}{2(n\cdot F)^2}\left\{\frac{n\cdot Fq\cdot F}{\omega-1}\,\delta_{\mu\nu} - \frac{q\cdot F}{\omega-1}\,(n_\mu F_\nu + n_\nu F_\mu) + \frac{n^2\ q\cdot F}{(\omega-1)n\cdot F}\,F_\mu F_\nu\right.$$

$$\left. + n\cdot F(q_\mu F_\nu + q_\nu F_\mu) - q\cdot n\ F_\mu F_\nu\right\} + \text{F.P.},$$

where $D_0 = n^2(q\cdot F)^2 - 2q\cdot nn\cdot Fq\cdot F + q^2(n\cdot F)^2$, and F.P. means "finite part".

## References


[1] W. Kummer: Acta Phys. Austriaca 14, 149 (1961).

[2] R.L. Arnowitt, S.I. Fickler: Phys. Rev. 127, 1821 (1962).

[3] J. Schwinger: Phys. Rev. 130, 402 (1963).

[4] Y.-P. Yao: J. Math. Phys. 5, 1319 (1964).

[5] G. Leibbrandt: Rev. Mod. Phys. 59, 1067 (1987).

[6] I.M. Gel'fand, G.E. Shilov: Generalized Functions (Academic, New York, 1964), Vol.1.

[7] W. Kummer: Acta Phys. Austriaca 41, 315 (1975).

[8] J. Frenkel, J.C. Taylor: Nucl. Phys. B109, 439 (1976).

[9] W. Konetschny: Phys. Rev. D28, 354 (1983).

[10] G. Leibbrandt: Seminar at the Summer Theoretical Physics Institute in Quantum Field Theory, University of Western Ontario, London, July 28-August 10, 1985.

[11] G. Leibbrandt: Nucl. Phys. B310, 405 (1988).

[12] P. Gaigg, M. Kreuzer, O. Piguet, M. Schweda: J. Math. Phys. 28, 2781 (1987).

[13] P. Gaigg, M. Kreuzer: Phys. Lett. B205, 530 (1988).

[14] S. Mandelstam: Nucl. Phys. B213, 149 (1983).

[15] G. Leibbrandt: Phys. Rev. D29, 1699 (1984).

[16] E. Bagan, C.P. Martin: Phys. Lett. B223, 187 (1989).

[17] A. Bassetto, I. Lazzizzera, G. Nardelli, R. Soldati: Phys. Lett. B228, 235 (1989).

[18] A. Bassetto, R. Soldati: Phys. Rev. D37, 3065 (1988).

[19] A. Burnel: Phys. Rev. D40, 1221 (1989).

[20] P. Gaigg, M. Kreuzer, G. Pollak: Phys. Rev. D38, 2559 (1988).

[21] P.V. Landshoff: Phys. Lett. B227, 427 (1989).

[22] I. Lazzizzera: Phys. Lett. B210, 188 (1988).

[23] G. Leibbrandt, S.-L. Nyeo: Mod. Phys. Lett. A3, 1085 (1988).

[24] G. Leibbrandt: Phys. Rev. D39, 1752 (1989).

[25] G. Nardelli, R. Soldati: Phys. Lett. B206, 495 (1988).

[26] G. Nardelli: "One-loop quark self-energy and quark-quark-gluon vertex in the planar gauge with the Mandelstam-Leibbrandt prescription", Trento preprint #U.T.F. 179, March 1989.

[27] G. Pollak: Phys. Rev. D40, 2027 (1989).

[28] H. Hüffel, P.V. Landshoff, J.C. Taylor: Phys. Lett. B217, 147 (1989).

[29] G. Leibbrandt, S.-L. Nyeo: J. Math. Phys. 27, 627 (1986).

[30] D.M. Capper, G. Leibbrandt: Phys. Rev. D25, 1002 (1982); 1009 (1982).

[31] D.M. Capper, G. Leibbrandt: J. Math. Phys. 15, 82 (1974).

[32] C. Becchi, A. Rouet, R. Stora: Phys. Lett. 52B, 344 (1974); Commun. Math. Phys. 42, 127 (1975).

# Renormalization and the S – Matrix in YM Theory in Axial Gauges

Günter Pollak

Institut für Theoretische Physik, Technische Universität Wien,
Wiedner Hauptstr. 8-10, A – 1040 Wien, AUSTRIA

**Abstract:** We present a method to calculate one-loop Feynman integrals in the generalized Leibbrandt-Mandelstam prescription by choosing an appropriate Lorentz frame. We obtain closed formulas in terms of hypergeometric functions for the basic integrals. Within this prescription, we then consider QCD in a class of axial gauges comprising the homogeneous axial gauge and the planar gauge. We examine the one-loop UV-divergent terms of 1PI vertices some of which are nonlocal. Therefore, we must restrict ourselves to a nonstandard renormalization where only the S-matrix is made finite by a finite number of local counterterms. Calculating quark-quark scattering we show at the one-loop level how the nonlocal UV-divergences cancel in an on-shell S-matrix element and how the gauge-dependent local UV-divergences vanish by a nonmultiplicative renormalization of the external fermion lines.

## 1 Introduction

It is well known that in gauge theories the gauge condition $n_\mu A^\mu = 0$ where $n_\mu$ is a fixed gauge vector [1] leads to poles $1/pn$ in the boson and ghost propagators. These poles must be regularized when one calculates Feynman integrals. For a long time, the principal value (PV) prescription [2]

$$\mathbf{P}\frac{1}{(pn)^\beta} \equiv \lim_{\varepsilon\to+0}\frac{1}{2}\left(\frac{1}{(pn+i\varepsilon)^\beta}+\frac{1}{(pn-i\varepsilon)^\beta}\right), \qquad \beta = 1,2,\dots \tag{1.1}$$

has been used, but it fails for $n^2 = 0$ [3],[4] and turns out to be inconsistent anyway in Wilson loop calculations [5]. To obtain a consistent prescription, it seems to be necessary to include a second "dual" vector $n^*_\mu = (n_0, -\boldsymbol{n})$. Leibbrandt [4] and Mandelstam [6] independently proposed this for the light-cone (LC) gauge $n^2 = 0$. Later, the Leibbrandt-Mandelstam (LM) prescription has been successfully generalized to other choices of $n_\mu$ and $n^*_\mu$ [7],[8],[9],[10]. One of several equivalent versions of this prescription reads

$$\left(\frac{1}{(pn)^\beta}\right)_{LM} = \lim_{\epsilon\to+0} \frac{1}{(pn + i\epsilon\,\mathrm{sgn}(pn^*))^\beta} . \qquad (1.2)$$

This prescription has been shown to emerge naturally from canonical quantization [11]. (Other methods to obtain the LM prescription from canonical quantization in the axial gauge use gauge-fixing terms involving $n^*_\mu$ or $e$ explicitely [12].) Wilson loop calculations [13] yielding consistent results in the temporal gauge $n_\mu = (1,0,0,0)$ give even stronger support to this prescription. Thus, it appears to be the only correct way, up to now, to regularize the poles $1/pn$ which are specific to axial gauges.

In this talk we first present a method to calculate one-loop Feynman integrals in this prescription. $n_\mu$ and $n^*_\mu$ are only restricted by the condition

$$(nn^*)^2 - n^2\,(n^*)^2 > 0. \qquad (1.3)$$

We then discuss the ultraviolet (UV) divergent terms of one-particle-irreducible (1PI) vertices to one-loop order which have the unpleasant feature of being *nonlocal.* (They contain factors $1/pn$ where $p_\mu$ is an external momentum.) Thus the usual renormalization procedure cannot be performed and we must restrict ourselves to a "nonstandard" renormalization where *only the S-matrix* is made finite by a finite number of local counterterms. In an explicit example we show how the nonlocal UV-divergences cancel in an on-shell S-matrix element and how gauge-dependent local UV-divergences vanish by a *nonmultiplicative* renormalization of the external fermion lines.

## 2 One-Loop Integrals in the Generalized LM Prescription

### 2a Convenient Lorentz Frames and the Conditions for Wick Rotation

The poles of $1/(p^2 + i\epsilon)$ terms occurring in ordinary Feynman integrals lie in the second and the fourth quadrant of the complex $p_0$-plane. This makes it possible to deform the path of integration from the real to the imaginary axis without encountering these poles. (Wick rotation). We are going to show [8] that for real vectors $n_\mu$ and $n^*_\mu$ satisfying (1.3) there exists a Lorentz frame in which the poles of the distribution (1.2) lie in the second or the fourth quadrant, provided $n^2 \neq 0$. For $n^2 = 0$, however, $nn^* > 0$ must be satisfied too. To demonstrate this, we first choose an appropriate Lorentz frame (i. e. a convenient choice of integration variables).

**Lemma:** *For $n_\mu$ and $n^*_\mu$ linearly independent, the Lorentz invariant condition (1.3) is equivalent to the existence of a Lorentz frame where these vectors have the form*

$$n_\mu = (n_0, 0, 0, n_3)\ , \qquad n^*_\mu = (n^*_0, 0, 0, n^*_3)\ . \qquad (2.1)$$

*Proof:* It is obvious that (1.3) is necessary for (2.1). To demonstrate that it is sufficient we distinguish two cases.

**Case 1: Neither $n_\mu$ nor $n^*_\mu$ is spacelike.**

In this case (1.3) is always satisfied for linearly independent vectors $n_\mu$ and $n^*_\mu$. To cast them in the form of (1.2), we first rotate in 3-dim. Euclidean space to get

$$n_\mu = n_0\,(1,0,a,b)\,, \qquad n^*_\mu = n^*_0\,(1,0,a,c)\,. \tag{2.2}$$

Note that $b$ and $c$ cannot both be zero, otherwise $n_\mu$ and $n^*_\mu$ would not be linearly independent. As both vectors are not spacelike we have $a^2 < 1$ and by a Lorentz boost we eliminate $a$ in $n_\mu$ and in $n^*_\mu$ at the same time.

**Case 2: One of the vectors $n_\mu$ or $n^*_\mu$ is spacelike.**

Without loss of generality let this vector be $n_\mu$. We boost away $n_0$. Because of (1.3) $n^*_0 \neq 0$ after this boost. Rotating in 3-dim. Euclidean space we find a Lorentz frame where

$$n_\mu = n_3\,(0,0,0,1)\,, \qquad n^*_\mu = n^*_0\,(1,0,a,b)\,. \tag{2.3}$$

Now we have $a^2 < 1$ because of (1.1) which means that a Lorentz boost eliminates $a$ in $n^*_\mu$, leaving $n_\mu$ unchanged this time. □

We now use a Lorentz frame (2.1) to examine the position of the poles of (1.2) in the complex $p_0$-plane. We observe that for

$$n^*_\mu = \lambda(n_0, 0, 0, -n_3), \qquad \lambda > 0. \tag{2.4}$$

the poles lie in the second or the fourth quadrant. This corresponds to the Lorentz invariant conditions (in addition to (1.3))

$$\mathrm{sgn}(n^2) = \mathrm{sgn}\left((n^*)^2\right)\,, \qquad nn^* > 0. \tag{2.5}$$

These conditions are too strong, however, because the position of the poles of (1.2) remains the same if $n^*_\mu$ is replaced by $\bar{n}_\mu = \lambda(n^*_\mu + \alpha n_\mu)$ where $\lambda \in \mathbb{R}^+$ and $\alpha \in \mathbb{R}$. Hence, if the vector $n^*_\mu$ lies in the hatched or the unhatched half plane of figs. 1 it is sufficient that after a suitable Lorentz boost this half plane contains one vector $\bar{n}_\mu$ of the form (2.4). For $n^2 > 0$ or $n^2 < 0$, if $n^*_\mu$ lies in the unhatched half plane we immediately find such a vector $\bar{n}_\mu$ in this half plane. If $n^*_\mu$ lies in the hatched half plane, after a suitable Lorentz boost the dark region of the hatched half plane contains the vector $\bar{n}_\mu$ we are looking for. For a lightlike vector $n_\mu$, however, we need the additional condition $nn^* > 0$, because only the unhatched half plane contains a vector $\bar{n}_\mu$ of the form (2.4) and this situation is invariant under Lorentz transformations.

In other words, $\bar{n}_\mu = \pm n_\mu + \bar{\lambda} n^*_\mu$ for $\bar{\lambda}$ sufficiently small but positive satisfies $\mathrm{sgn}(n^2) = \mathrm{sgn}\left(\bar{n}^2\right)$ and $n\bar{n} > 0$, provided $n^2 \neq 0$. For $n^2 = 0$, $nn^* > 0$ must hold to achieve $n\bar{n} > 0$. $\bar{n}^2 = n^2 = 0$ is then satisfied by $\bar{n}_\mu = n^*_\mu - ((n^*)^2/2nn^*)n_\mu$.

## 2b Calculating the Basic One-Loop Integral

Methods to calculate integrals in the LM prescription (1.2), using dimensional regularization, have been developed by various authors [4],[14],[7],[9],[10]. A quite elegant technique that differs from these approaches is to use the relation

$$\left(\frac{1}{qn}\right)_{LM} = \mathrm{P}\left(\frac{1}{qn}\right) - i\pi\,\mathrm{sgn}(qn^*)\,\delta(qn) \tag{2.6}$$

and to start from the integral

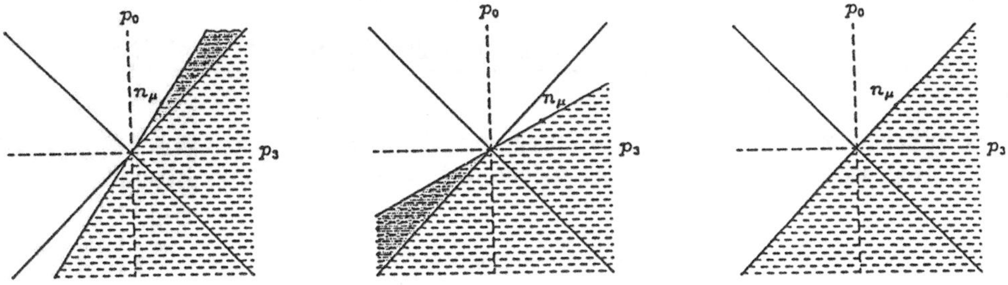


**Fig. 1.** Position of the vector $\tilde{n}_\mu$ for timelike, spacelike and lightlike $n_\mu$

$$J(\alpha,1) := \int d^{2\omega}q \frac{1}{(q^2+2pq-L)^\alpha} \mathrm{sgn}(qn^*)\delta(qn)\,. \tag{2.7}$$

We calculate this integral in the frame (2.1). We expect to get a result in terms of $L$,$p^2$,$pn$,$pn^*$,$n^2$,$(n^*)^2$ and $nn^*$ which take the same value in the original frame. We perform a change of integration variables

$$\begin{aligned} q_0 &= \frac{n_3 qn^* - n_3^* qn}{n_3 n_0^* - n_3^* n_0}, \qquad q_3 = \frac{n_0 qn^* - n_0^* qn}{n_3 n_0^* - n_3^* n_0}, \\ dq_0 dq_3 &= \frac{d(qn)d(qn^*)}{|n_3 n_0^* - n_3^* n_0|} = \frac{d(qn)d(qn^*)}{+\sqrt{(nn^*)^2 - n^2(n^*)^2}} := \frac{d(qn)d(qn^*)}{R}, \end{aligned} \tag{2.8}$$

and rewrite the integrand of (2.7) in terms of $qn$ and $qn^*$.

$$J(\alpha,1) = \int \frac{d(qn^*)}{R} \mathrm{sgn}(qn^*) \int d(qn)\delta(qn) \int d^{2\omega-2}q_\perp$$
$$\left[-(q_\perp+p_\perp)^2 - \frac{n^2}{R^2}\left(qn^* + pn^* - \frac{nn^*}{n^2}(qn+pn)\right)^2 + \frac{(qn+pn)^2}{n^2} - p^2 - L\right]^{-\alpha}. \tag{2.9}$$

We introduce the abbreviations

$$z = L + p^2 - \frac{(pn)^2}{n^2}, \qquad u = \frac{nn^*}{n^2} pn - pn^*\,. \tag{2.10}$$

We perform the $qn$-integration (where a possible $i\varepsilon$-term is absorbed) and change the remaining integration variables to

$$Q_\perp = q_\perp + p_\perp, \qquad x = qn^* - u\,. \tag{2.11}$$

This yields

$$J(\alpha,1) = (-1)^\alpha \frac{1}{R} \int_{-|u|}^{|u|} dx\, \mathrm{sgn}(x+u) \int d^{2\omega-2}Q_\perp \left(Q_\perp^2 + x^2\frac{n^2}{R^2} + z\right)^{-\alpha}. \tag{2.12}$$

The sign function has enabled us to restrict the limits of the $dx$-integration to $|x| \leq |u|$ because the integrand is an even function of $x$. Therefore the condition

$$L + p^2 > \frac{(pn)^2}{n^2} - \frac{n^2}{R^2}u^2 = \frac{-n^2(pn^*)^2 + 2(nn^*)(pn)(pn^*) - (n^*)^2(pn)^2}{R^2} \tag{2.13}$$

is sufficient for a positive definite integrand. If $L + p^2 > 0$ this can always be achieved for small values of $pn$ and $pn^*$. Euclidean $dQ_\perp$-integration then is straightforward

$$J(\alpha, 1) = \pi^{\omega-1}(-1)^\alpha \frac{\Gamma(\alpha - \omega + 1)}{\Gamma(\alpha)} \frac{2}{R} \mathrm{sgn}(u) \int_0^{|u|} dx \left( x^2 \frac{n^2}{R^2} + z \right)^{-\alpha+\omega-1} . \tag{2.14}$$

The remaining integral yields a hypergeometric series [15] and we finally obtain

$$J(\alpha, 1) = \pi^{\omega-1}(-1)^\alpha \frac{\Gamma(\alpha - \omega + 1)}{\Gamma(\alpha)} \frac{2u}{R} z^{-\alpha+\omega-1} F\left( \alpha - \omega + 1, \frac{1}{2}, \frac{3}{2}; -\frac{n^2u^2}{R^2 z} \right) . \tag{2.15}$$

The PV result of Konetschny [16] reads

$$\bar{I}(\alpha, 1) = i\pi^\omega (-1)^\alpha \frac{\Gamma(\alpha - \omega + 1)}{\Gamma(\alpha)} \frac{2pn}{n^2} z^{-\alpha+\omega-1} F\left( \alpha - \omega + 1, \frac{1}{2}, \frac{3}{2}; -\frac{(pn)^2}{n^2 z} \right) , \tag{2.16}$$

and this, together with (2.15) and (2.6) yields the basic one-loop integral in the LM prescription.

## 2c Consistency Checks

It is clear from (2.7) that we must have

$$n_\mu \frac{\partial}{\partial p_\mu} J(\alpha, 1) = 0 , \qquad n_\mu \frac{\partial}{\partial n^*_\mu} J(\alpha, 1) = 0 . \tag{2.17}$$

The former is also due to the fact that both the PV and the LM prescription are internally consistent [17]. This latter relation [8], however, is not dictated by internal consistency of the integrals but a specific property of the LM prescription. $J(\alpha, 1)$ is only a function of $u$ $z$ and $R$, all of which vanish under $n_\mu(\partial/\partial p_\mu)$ and $n_\mu(\partial/\partial n^*_\mu)$. Hence, (2.17) is indeed satisfied by (2.15).

## 2d Deriving Integrals Occurring in 1PI Vertices from this Basic Formula

To calculate integrals with double poles $(qn)^{-2}$, one could use

$$\left( \frac{1}{(qn)^2} \right)_{LM} = \mathrm{P} \left( \frac{1}{(qn)^2} \right) + i\pi \mathrm{sgn}(qn^*) \delta'(qn) , \tag{2.18}$$

but assuming invariance under the shift of integration variables – which has also been used to derive $J(\alpha, 1)$ – one can obtain the result directly [17]

$$\int d^{2\omega}q \frac{1}{(q^2+2pq-L)(qn)^2} = -\frac{\partial}{\partial(pn)} \int d^{2\omega}q \frac{1}{(q^2+2pq-L)(qn)} . \tag{2.19}$$

Deriving these basic integrals with respect to $p_\mu$ then yields expressions with $q_\mu$-terms in the numerator.

Integrals occurring in 1PI vertex functions can now be obtained from the expressions just derived via the standard technique of Feynman parameters. For the UV-divergent parts which are polynomials in $m^2$,$p^2$,$pn$,$pn^*$ and $(pn)^{-1}$ this is indeed straightforward. For the finite parts of the integrals, however, this turns out to be a rather cumbersome task. A calculation of a full 1PI vertex for $n^2 \neq 0$ will certainly need a good knowledge how to simplify expressions involving dilogarithms in 1PI vertices.

# 3 UV Divergences in the YM Selfenergy in Axial Gauges

We now study YM theory within a family of gauges comprising the homogeneous axial gauge and the planar gauge. The action of this model reads including the Fadde'ev-Popov (FP) ghosts $c_\pm$

$$S = tr \int d^4x \left[ -\frac{1}{4} F^{\mu\nu} F_{\mu\nu} + B\partial^2(nA) + \frac{\alpha}{2} B\partial^2 B - c_- \partial^2(nD)c_+ \right] \tag{3.1}$$

where $A_\mu = A_\mu^a \tau^a$ etc., $[\tau^a, \tau^b] = if^{abc}\tau^c$, $F_{\mu\nu} = \partial_\mu A_\nu - \partial_\nu A_\mu - ig[A_\mu, A_\nu]$, $D_\mu = \partial_\mu - ig[A_\mu, \cdot]$. The propagator for the gauge field $A_\mu$ is given by

$$\Delta_{\mu\nu}^{ab}(q) = \frac{-i\delta^{ab}}{q^2+i\varepsilon} \left( g_{\mu\nu} - \frac{q_\mu n_\nu + q_\nu n_\mu}{qn} + \frac{(1+\alpha)n^2 q_\mu q_\nu}{(qn)^2} \right) . \tag{3.2}$$

The gauge-dependent parts of (3.2) are proportional to $q_\mu$ or $q_\nu$. This will be important for the cancellation of gauge-dependent terms in the S-matrix (3.1) is Becchi-Rouet-Stora-Tyutin (BRST) invariant, but, as common in axial gauges, ghost loops vanish in dimensional regularization, hence we forget about $c_\pm$ from now on. Canonical quantization [11] starting from the Lagrangian given in (3.1) yields a prescription for the poles $(qn)^{-1}$ and $(qn)^{-2}$ which is just the generalized LM prescription (1.2).

Using the symbolic manipulator program REDUCE to cope with the enormous number of terms we obtain an expression for the UV-divergent part of the gluon selfenergy whose complexity is quite impressive. Introducing the abbreviations

$$R = \sqrt{(nn^*)^2 - n^2(n^*)^2}, \tag{3.3a}$$

$$\hat{n}_\mu = n_\mu - \frac{pn}{p^2} p_\mu, \qquad \hat{n}_\mu^* = n_\mu^* - \frac{pn^*}{p^2} p_\mu, \qquad \check{n}_\mu^* = n_\mu^* - \frac{pn^*}{pn} n_\mu, \tag{3.3b}$$

$$C_F \delta_{ij} = (\tau^a \tau^a)_{ij}, \qquad C_A \delta^{ad} = f^{abc} f^{bcd}, \tag{3.3c}$$

$$\alpha_s = \frac{g^2}{4\pi}, \qquad \varepsilon = 4 - 2\omega, \tag{3.3d}$$

and the following tensors

$$S^1_{\mu\nu}(p) = \left(p^2 g_{\mu\nu} - p_\mu p_\nu\right), \tag{3.4a}$$

$$S^2_{\mu\nu}(p) = \left(\frac{p^2}{pn}\right)^2 \hat{n}_\mu \hat{n}_\nu, \qquad S^3_{\mu\nu}(p) = -\frac{p^2}{R}\left(\hat{n}_\mu \breve{n}^*_\nu + \hat{n}_\nu \breve{n}^*_\mu\right), \tag{3.4b}$$

$$S^4_{\mu\nu}(p) = \left(\frac{n^2}{R}\frac{p^2}{pn}\right)^2 \hat{n}^*_\mu \hat{n}^*_\nu, \qquad S^5_{\mu\nu}(p) = \left(\hat{n}^*_\mu \breve{n}^*_\nu + \hat{n}^*_\nu \breve{n}^*_\mu\right), \tag{3.4c}$$

$$N^1_{\mu\nu}(p) = \frac{p^2}{R}\left(n_\mu \hat{n}^*_\nu + n_\nu \hat{n}^*_\mu\right), \qquad N^2_{\mu\nu}(p) = \frac{p^2}{n^2}\left(n_\mu \hat{n}_\nu + n_\nu \hat{n}_\mu\right), \tag{3.4d}$$

the gluon selfenergy reads [18]

$$\begin{aligned}
\Pi^{ab}_{\mu\nu}(p) = iC_A \frac{\alpha_s}{2\pi\varepsilon}\delta^{ab} \Bigg\{ & \Bigg[ S^1_{\mu\nu}(p)\left(\frac{11}{3} - 2\frac{n^2}{R}\frac{pn^*}{pn}\right) \\
& + S^3_{\mu\nu}(p)\left(2 + \frac{n^2(n^*)^2}{R^2} - \frac{n^2(nn^*)}{R^2}\frac{pn^*}{pn}\right) + S^5_{\mu\nu}(p)\left(\frac{(nn^*)}{R} - \frac{n^2}{R}\frac{pn^*}{pn}\right)\Bigg] \\
& + \alpha \Bigg[ S^1_{\mu\nu}(p)\left(2 - 2\frac{nn^*}{R}\right) \\
& + N^1_{\mu\nu}(p)\left(\frac{n^2(n^*)^2}{R^2} - 2\frac{n^2(nn^*)}{R^2}\frac{pn^*}{pn} + \frac{(n^2)^2}{R^2}\left(\frac{pn^*}{pn}\right)^2\right) \\
& + N^2_{\mu\nu}(p)\Bigg(2\frac{nn^* - R}{R} - \frac{n^2(n^*)^2(nn^*)}{R^3} \\
& \qquad + 2\frac{(n^2)^2(n^*)^2}{R^3}\frac{pn^*}{pn} - \frac{(n^2)^2(nn^*)}{R^3}\left(\frac{pn^*}{pn}\right)^2\Bigg)\Bigg] \\
& + (1+\alpha)^2 \Bigg[ S^1_{\mu\nu}(p)\frac{(n^2)^2}{R}\frac{p^2 pn^*}{(pn)^3} + S^2_{\mu\nu}(p)\frac{(n^2)^2(n^*)^2}{R^3}\frac{pn^*}{pn} \\
& + S^4_{\mu\nu}(p)\frac{n^2}{R}\frac{pn^*}{pn} + \left(S^5_{\mu\nu}(p)\frac{n^2 p^2}{(pn)^2} - 2S^4_{\mu\nu}(p)\right)\frac{nn^*}{R}\Bigg]\Bigg\}.
\end{aligned} \tag{3.5}$$

(Special cases of this formula have been derived by various authors [9],[19],[10].) We note that it satisfies the Ward identity for $p^\mu \Pi^{ab}_{\mu\nu}(p)$ [18]. This is no surprise because the integrals are internally consistent and dimensional regularization is known to preserve the Ward identities. Dealing with such a large number of terms as in (3.5), however, the Ward identity is very useful to check the result.

We observe that (3.5) contains nonlocal terms $(pn)^{-\beta}$. The usual concept of renormalization of 1PI vertices with local counterterms obviously *fails* for these UV-divergences. If the theory is to make sense (and we do believe that it does) these divergences must not affect physical quantities. They must decouple from the physical sector of the theory. We shall see in the next section that this is really the case for the S-matrix element of fermion-fermion scattering.

Before that, however, we want to examine the structure of these nonlocal divergences which, however, *do not define counterterms.* The UV-divergent part of $\Gamma$ to one-loop order, whose two-point contribution is given by (3.5), can be written in the form $\Gamma_{\text{div}} = aF^2 + DF\Omega$ where $\Omega$ contains the gauge-dependent terms, provided the tensors we have just defined originate from (for example) [18]

$$S^2_{\mu\nu}(p) \simeq (DFn)\frac{1}{(nD)^2}(DFn), \qquad S^3_{\mu\nu}(p) \simeq \frac{1}{R}(DFn)\frac{1}{nD}(nFn^*), \tag{3.6a}$$

$$S^4_{\mu\nu}(p) \simeq \frac{(n^2)^2}{R^2}(DFn^*)\frac{1}{(nD)^2}(DFn^*), \qquad S^5_{\mu\nu}(p) \simeq \frac{n^2}{R^2}(DFn^*)\frac{1}{nD}(nFn^*), \tag{3.6b}$$

$$N^1_{\mu\nu}(p) \simeq \frac{1}{R}(DFn^*)(nA), \qquad N^2_{\mu\nu}(p) \simeq \frac{1}{n^2}(DFn)(nA), \tag{3.6c}$$

$$S^1_{\mu\nu}(p) \simeq F^2 \simeq DFA, \qquad \frac{pn^*}{pn}S^1_{\mu\nu}(p) \simeq (DF)^\mu \frac{1}{nD}(Fn^*)_\mu. \tag{3.6h}$$

Actually, however, we do not know if this interpretation of $\Gamma_{\text{div}}$ is the correct one. In the LC gauge $n^2 = 0$ all nonlocal divergent structures come from (3.6b) [20],[21]. For $n^2 \neq 0$, however, n-point 1PI functions $n \geq 3$ are likely to contain additional structures not contributing to the two-point function. Without calculating at least the three-point 1PI vertex for $n^2 \neq 0$ (a task so enormous that nobody has undertaken it as yet) we do not know which $\partial_\mu$'s in (3.6a-h) really correspond to $D_\mu$'s. Moreover, there is yet another reason why the ansatz (3.6a-h) is probably insufficient. It has been shown by Skarke and Gaigg [21] that all nonlocal terms must vanish if we set $pn^*=0$ for all external momenta. (The respective theorem does not rely on $n^2 = 0$) This clearly holds for the selfenergy (3.5) but among others it is no longer true for the three-point contributions of (3.6a) and (3.6c). (Remember that $1/nD$ is actually an infinite series!) Hence, even to one-loop order the underlying structure of the UV-divergent terms of YM theories in axial gauges within the generalized LM prescription is not fully understood as yet. [1]

## 4 Checking the Gauge Independence of the S-Matrix

As we have seen, the structure of UV-divergent terms is quite complicated and no obvious building principle is visible. Nevertheless, we believe that the theory is physically meaningful. Therefore we now study how these terms behave in the S-matrix [22]. Axial gauges in the generalized LM prescription should lead to a "nonstandard" renormalization program where individual 1PI vertices *cannot* be made finite by local counterterms. *The S-matrix, however, can be made finite by a finite number of local counterterms.* We consider the scattering of two fermions of different flavour at one-loop order. Taking the fermions to be very heavy even should eliminate the usual infrared (IR) problems [23]. The gauge-invariant part of the action now reads

$$\Gamma_{\text{inv}} = \int dx \left[ -\frac{1}{4}(F^a_{\mu\nu})^2 + \sum_j \bar{\psi}^j (i\gamma^\mu \mathcal{D}_\mu - m_j)\psi^j \right], \tag{4.1}$$

[1] The three-gluon vertex for $nn^* = 0$ has been presented at this workshop by Bagan and Martin.

where $\mathcal{D}_\mu = \partial_\mu - ig\tau^a A_\mu^a$. The graphs we have to sum up with unit weight are depicted in fig. 2. To this, we must then add the effects of the renormalization of the external lines. Taking the external fermions on shell means that we have

$$k^\mu \bar{u}(p_3)\gamma_\mu u(p_1) = k^\mu \bar{u}(p_4)\gamma_\mu u(p_2) = 0, \tag{4.2}$$

where $k = p_3 - p_1 = p_2 - p_4$ is the momentum transfer. This current conservation annihilates many expressions occuring in the graphs involved. For example, the fermion loop (fig. 2.1) cannot give a gauge-dependent contribution, being transverse to the transferred momentum $k_\mu \Pi_f^{\mu\nu} = 0$ and being gauge-independent itself, hence we can forget about it. The graphs we have to write down explicitely are the gluon loop already considered

$$\Pi_{\mu\nu}(k) = iC_A \frac{\alpha_s}{2\pi\varepsilon} k^2 C_{\mu\nu}, \tag{4.3}$$

where $C_{\mu\nu}$ is given by

$$\begin{aligned} C_{\mu\nu} = \Bigg\{ & \Bigg[ \frac{S^1_{\mu\nu}(k)}{k^2}\left(\frac{11}{3} - 2a_2\right) \\ & - (\hat{n}_\mu \breve{n}^*_\nu + \hat{n}_\nu \breve{n}^*_\mu)\left(\frac{2}{R} + \frac{n^2}{R^3}(n^*\breve{n}^*)\right) + (\hat{n}^*_\mu \breve{n}^*_\nu + \hat{n}^*_\nu \breve{n}^*_\mu)\frac{n^2}{R^3}(n\breve{n}^*) \Bigg] \\ & + \alpha \Bigg[ 2a_1 \frac{S^1_{\mu\nu}(k)}{k^2} + (n_\mu \hat{n}^*_\nu + n_\nu \hat{n}^*_\mu)\frac{n^2}{R^3}(\breve{n}^*)^2 \\ & + (n_\mu \hat{n}_\nu + n_\nu \hat{n}_\mu)\left(-a_1 \frac{2}{n^2} - \frac{(n^*)^2}{R^3}(n\breve{n}^*) + \frac{n^2}{R^3}\frac{kn^*}{kn}(n^*\breve{n}^*)\right) \Bigg] \\ & + (1+\alpha)^2 \frac{k^2 n^2}{(kn)^2} a_2 \Bigg[ \frac{S^1_{\mu\nu}(k)}{k^2} \\ & + \frac{1}{R^2}\left(\hat{n}^*_\mu \hat{n}^*_\nu n^2 - (\hat{n}_\mu \hat{n}^*_\nu + \hat{n}_\nu \hat{n}^*_\mu) n n^* + \hat{n}_\mu \hat{n}_\nu (n^*)^2\right) \Bigg] \Bigg\}, \end{aligned} \tag{4.4}$$

the quark selfenergy

$$\Sigma(p) = C_F \frac{\alpha_s}{2\pi\varepsilon}\left[\not{p} + 2m + \frac{2}{R}\left(\not{n} p n^* - \not{n}^* p n\right) + 2a_1(1+\alpha)(\not{p} - m)\right], \tag{4.5}$$

the quark-quark gluon vertex

$$\Lambda_\mu(p', p) = \Lambda_\mu(p' - p) := \Lambda_\mu(k) = -\frac{\alpha_s}{2\pi\varepsilon}\left(C_F V_\mu(k) + C_A U_\mu(k)\right), \tag{4.6}$$

where

$$V_\mu(k) = \left[\gamma_\mu + \frac{2}{R}\left(\not{n} n^*_\mu - \not{n}^* n_\mu\right) + 2a_1(1+\alpha)\gamma_\mu\right], \tag{4.7}$$

$$U_\mu(k) = \Bigg\{ \left[ -\not{n}\breve{n}^*_\mu \left( \frac{2}{R} + \frac{n^2}{R^3}(n^*\breve{n}^*) \right) + \not{n}^*\breve{n}^*_\mu \frac{n^2}{R^3}(n\breve{n}^*) - a_2\gamma_\nu \frac{S^1_{\mu\nu}(k)}{k^2} \right]$$
$$+\alpha \Bigg[ a_1\gamma_\mu + a_2 \frac{\not{k} n_\mu}{kn} + \not{n}^* n_\mu \frac{n^2}{R^3}(\breve{n}^*)^2$$
$$+ \not{n} n_\mu \left( -a_1 \frac{2}{n^2} - \frac{(n^*)^2}{R^3}(n\breve{n}^*) + \frac{n^2}{R^3}\frac{kn^*}{kn}(n^*\breve{n}^*) \right) \Bigg] \tag{4.8}$$
$$+(1+\alpha)^2 \frac{k^2 n^2}{(kn)^2} a_2 \Bigg[ \gamma_\nu \frac{S^1_{\mu\nu}(k)}{k^2}$$
$$+ \frac{1}{R^2} \left( \not{n}^* \hat{n}^*_\mu n^2 - (\not{n}^* \hat{n}_\mu + \not{n}\hat{n}^*_\mu) nn^* + \not{n}\hat{n}_\mu (n^*)^2 \right) \Bigg] \Bigg\},$$

and the four-point 1PI vertex

$$\Gamma^{(4)}(p_1,p_2,p_3,p_4) = -\frac{C_A}{2}\bar{u}(p_3)\tau^a\gamma^\mu u(p_1)\bar{u}(p_4)\tau^a\gamma^\nu u(p_2)\Xi_{\mu\nu}\,, \tag{4.9}$$

where

$$\Xi_{\mu\nu} = -i\frac{4\alpha_s^2}{\varepsilon}\frac{(n^2)^2\, kn^*}{R^3(kn)^3}(1+\alpha)^2$$
$$\times \left[ R^2 g_{\mu\nu} - nn^* \left( n_\mu n^*_\nu + n_\nu n^*_\mu \right) + n^2 n^*_\mu n^*_\nu + (n^*)^2 n_\mu n_\nu \right]. \tag{4.10}$$

In addition to the abbreviations already defined we have used

$$a_1 = \left( 1 - \frac{nn^*}{R} \right), \qquad a_2 = \frac{n^2}{R}\frac{kn^*}{kn}. \tag{4.11}$$

For the sake of completeness we note that (4.6) satisfies the respective Ward identity [22]. Special cases of (4.5) - (4.8) can be found in the papers of [24].

First we investigate the contributions proportional to $C_A$. If we consider for the moment the special case $\alpha = 0$, we have, according to the Ward-identities, $k^\mu U_\mu(k) = k^\mu \Pi_{\mu\nu}(k) = k^\nu \Pi_{\mu\nu}(k) = 0$. Therefore the $n_\mu$-dependent parts of the gluon propagator (3.2) vanish either at the 1PI-vertices or by current conservation (4.2), leaving only the "Feynman part" $-i\delta^{ab}g_{\mu\nu}/k^2$. We can thus compare directly eqs. (4.4) and (4.8), and set to zero all terms proportional to $k^\mu$,$k^\nu$, and $\not{k}$. In this way we observe that the gauge-dependent part from the gluon selfenergy cancels with corresponding expressions of the quark-quark-gluon vertices, up to a part proportional to $(kn)^{-3}$ which exactly matches the contribution of the four-quark 1PI-vertex (4.10). Hence, for $\alpha = 0$, the S-matrix element to one-loop order contains no gauge-dependent UV-divergences proportional to $C_A$. If we now turn to the case $\alpha \neq 0$, we need only consider the two blocks in eqs. (4.4) and (4.8) having $\alpha$ in front. For these pieces, the effective part of the propagator (after taking into account current conservation) is $(-i\delta^{ab}/k^2)(g_{\mu\nu} - k_\mu n_\nu/kn)$. Being proportional to $n_\mu$ or $n_\nu$, most of these pieces vanish immediately. The remaining contributions from (4.4) and (4.8) cancel one another, thus gauge independence holds also for $\alpha \neq 0$.

Comparing (4.4), (4.8) and (4.10) we observe a one-to-one correspondence of terms $\not{n}$,$\not{n}^*$ to terms $\hat{n}_\mu$,$\hat{n}^*_\mu$ which holds in an analogous way for $k^2\gamma_\mu \leftrightarrow S^1_{\mu\nu}(k)$ and $\not{k} \leftrightarrow 0$. This means that the gauge-dependent divergences proportional to $C_A$ have the form

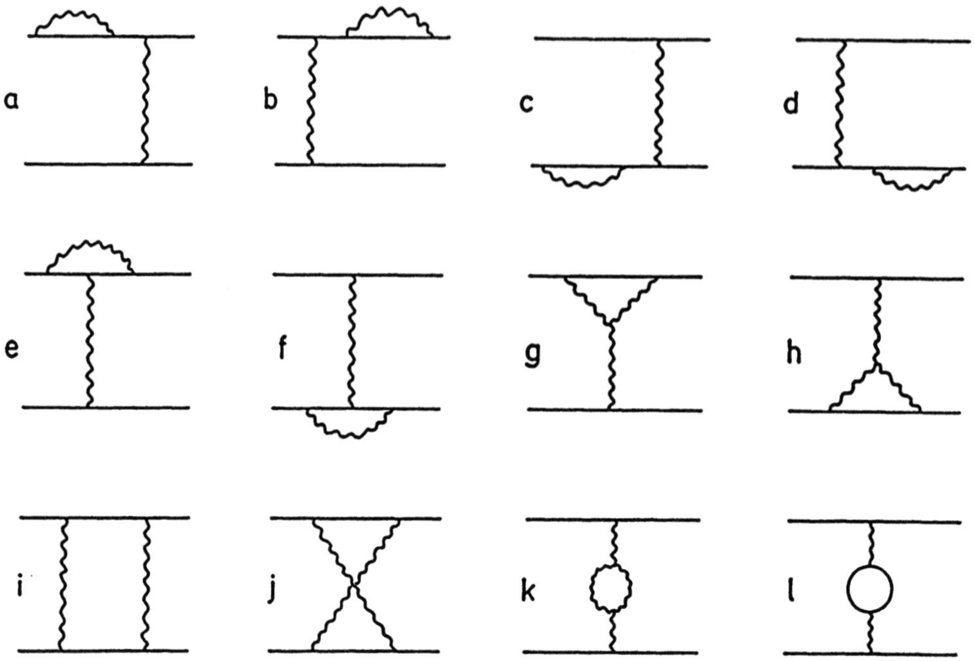


Fig. 2. Graphs contributing to the S-matrix element of quark-quark scattering

$$(D^{ab}_{\mu} F^{b}_{\mu\nu} - g\bar{\psi}\tau^{a}\gamma^{\nu}\psi)\Omega^{a}_{\nu}, \tag{4.12}$$

where $\Omega^a_\nu$ contains nonlocal terms. We recognize the equation of motion in (4.12). As we have just seen by explicit calculation, the cancellation of gauge-dependent terms proportional to $C_A$ is due to the fact that all these divergences combine into (4.12).

All nonlocal divergences have cancelled in the S-matrix, they really need not be renormalized by counterterms. This is not the case for the divergences proportional to $C_F$. In accordance with "nonstandard" renormalization, however, these divergences proportional to $C_F$ are *local.* They only vanish in the *renormalized S-matrix*, i. e. if the renormalization of the external fermion lines is taken into account. The selfenergy and the contribution of the abelian vertex can be written in the form

$$\Sigma(p) = \delta m + \bar{\mathcal{M}}(\not p - m) + (\not p - m)\mathcal{M}, \tag{4.13}$$

$$-C_F V_\mu(k) = -\bar{\mathcal{M}}\gamma_\mu - \gamma_\mu\mathcal{M}, \tag{4.14}$$

where

$$\delta m = m C_F \frac{3\alpha_s}{2\pi\varepsilon}, \tag{4.15}$$

$$\mathcal{M} = C_F \frac{\alpha_s}{2\pi\varepsilon}\left[\left(\frac{3}{2}+\alpha\right) + \frac{1}{R}\left(\not n^* \not n - (2+\alpha)nn^*\right)\right], \tag{4.16}$$

and $\bar{\mathcal{M}} = \gamma_0 \mathcal{M}^\dagger \gamma_0$. Hence the mass renormalization and the field renormalization are given by $m_0 = m - \delta m$ and $\psi^{(0)} = (1 - \mathcal{M})\psi$. *Note, that a gauge dependent matrix* $(1 - \mathcal{M})$ *replaces the familiar renormalization constant* $Z_2^{-\frac{1}{2}}$. Therefore we must take some care in performing the renormalization of the external lines (asymptotic fields) in the Lehmann-Szymanzik-Zimmermann (LSZ) reduction technique [25]. The fermion field $\psi(x)$ reads

$$\psi(x) = \int d^3p \sum_{\pm s} \left[ b(p,s) U_{ps}(x) + d^\dagger(p,s) V_{ps}(x) \right], \tag{4.17}$$

where

$$U_{ps}(x) = \frac{1}{(2\pi)^{3/2}} \sqrt{\frac{m}{E_p}}\, u(p,s) e^{-ipx}, \tag{4.18}$$

thus the creation and annihilation operators are given by

$$b(p,s) = \int d^3x U^\dagger_{ps}(x)\psi(x), \qquad b^\dagger(p,s) = \int d^3x \psi^\dagger(x) U_{ps}(x). \tag{4.19}$$

In a typical step of the LSZ reduction, we have

$$\begin{aligned} &\langle \beta, \mathrm{out} | b^\dagger_{\mathrm{in}}(p) | \alpha, \mathrm{in} \rangle = \\ &\quad \langle \beta, \mathrm{out} | b^\dagger_{\mathrm{out}}(p) | \alpha, \mathrm{in} \rangle + \int d^3x \langle \beta, \mathrm{out} | \psi^\dagger_{\mathrm{in}}(x) - \psi^\dagger_{\mathrm{out}}(x) | \alpha, \mathrm{in} \rangle U_{ps}(x). \end{aligned} \tag{4.20}$$

Hence, we must put $(1 - \bar{\mathcal{M}})u(p)$ and $\bar{u}(p)(1 - \mathcal{M})$ at the external lines. Summing up all contributions to order $g^4$, including the effects of mass and wave function renormalization, we obtain

$$\begin{aligned} \bar{u}(p_3) \Bigg[ &(\delta m + \bar{\mathcal{M}}(\not{p}_3 - m) + (\not{p}_3 - m)\mathcal{M}) \frac{1}{\not{p}_3 - m} \gamma_\mu \\ &+ \gamma_\mu \frac{1}{\not{p}_1 - m} (\delta m + \bar{\mathcal{M}}(\not{p}_1 - m) + (\not{p}_1 - m)\mathcal{M}) \\ &\qquad - \bar{\mathcal{M}}\gamma_\mu - \gamma_\mu \mathcal{M} \\ &- \delta m \frac{1}{\not{p}_3 - m} \gamma_\mu - \gamma_\mu \frac{1}{\not{p}_1 - m} \delta m - \mathcal{M}\gamma_\mu - \gamma_\mu \bar{\mathcal{M}} \Bigg] u(p_1) \Delta_{\mu\nu} \bar{u}(p_4) \gamma^\nu u(p_2) \end{aligned} \tag{4.21}$$

and corresponding expressions from the graphs turned upside down. With the help of

$$\frac{1}{\not{p} - m} \bar{\mathcal{M}}(\not{p} - m) u(p) = \left( \frac{1}{\not{p} + m} \bar{\mathcal{M}}(\not{p} - m) + \bar{\mathcal{M}} \frac{1}{1 + \frac{\not{p} - m}{2m}} \right) u(p) = \bar{\mathcal{M}} u(p) \tag{4.22}$$

the contributions from the upper (lower) fermion line are seen to vanish identically. The renormalized S-matrix element contains no gauge-dependent divergences proportional to $C_F$ to one-loop order. Only the graph depicted in fig. 2.1 contributes the usual gauge-independent divergence to the S-matrix, which is then made finite by the renormalization of the gauge field $A_\mu$ ("charge renormalization") [25].

## 5 Conclusions and Outlook

We have studied QCD in a class of axial gauges, using the generalized LM prescription to regularize the $1/kn$-terms. This prescription (at variance with the inconsistent PV prescription) generates a very complicated structure of UV-divergences whose building principle is yet unknown. Nonetheless, gauge independence of the S-matrix appears to hold for QCD in the class of gauges considered, at least at the one-loop level.

There are several open problems:
*What is the full structure of the UV-divergences to one-loop order?*
*What is the correct renormalization of external boson lines?*
*What happens to the gauge-dependent IR-divergences in the S-matrix?* [26]
These are the easier ones. In the end, however, we will have to tackle two questions, which are likely to involve an even bigger amount of work:
*Does this "nonstandard" renormalization also work at multiloop order and how? (How (4.11) generalizes to multiloop order?)*
*Does the enormous quantity of $n_\mu$- and $n^*_\mu$-dependent terms in the* **finite** *part of the S-matrix cancel and how?*
If we believe that the generalized LM prescription (which naturally emerges from canonical quantization) is the correct way to treat axial gauges, the answer to these questions must of course be positive.

**Acknowledgement:**
This work has been supported by the Austrian "Fonds zur Förderung der wissenschaftlichen Forschung" under contract no. P6827P.

## References

[1] W. Kummer, Acta Phys. Austr. **14** 149 (1961);
W. Kummer, Acta Phys. Austr. Suppl. **XV** 423 (1976);
R. L. Arnowitt and S. J. Fickler, Phys. Rev. **127** 1821 (1962);
J. Schwinger, Phys. Rev. **130** 402 (1963);
Y. P. Yao, J. Math. Phys.**5** 1319 (1964);
G. Leibbrandt, Rev. Mod. Phys. **59** 1067 (1987).

[2] W. Kummer, Acta Phys. Austr. **41** 315 (1975);
W. Konetschny and W. Kummer, Nucl. Phys. **B100** 106 (1975); **B108** 397 (1976);
J. Frenkel, Phys. Rev. **D13** 2325 (1976).

[3] D. M. Capper, J. J. Dulwich and M. J. Litvak, Nucl. Phys. **B241** 463 (1984);
H. C. Lee and M. S. Milgram, J. Math. Phys. **26** 1793 (1985).

[4] G. Leibbrandt, Phys. Rev. **D29** 1699

[5] R. Soldati, these proceedings;
S. Caracciolo, G. Curci and P. Menotti, Phys. Lett. **113B** 311 (1982);
V. F. Müller and W. Rühl, Ann. Phys. (N.Y.) **133** 240 (1981).

[6] S. Mandelstam, Nucl. Phys. **B213** 149 (1983).

[7] P. Gaigg, M. Kreuzer, M. Schweda and O. Piguet, J. Math. Phys. **28** 2781 (1987); M. Kreuzer, G. Pollak, H. Skarke and M. Schweda, Phys. Lett. **196B** 557 (1987).

[8] P. Gaigg and M. Kreuzer, Phys. Lett. **205B** 530 (1988).

[9] G. Leibbrandt, Nucl. Phys. **B310** (1988) 405.

[10] E. Bagan and C. P. Martin, Phys. Lett. **223B** 187 (1989).

[11] I. Lazzizzera, these proceedings; A. Bassetto, M. Dalbosco, I. Lazzizzera and R. Soldati, Phys. Rev. **D31** 2012 (1985); I. Lazzizzera, Phys. Lett. **210B** 188 (1988); Nuovo Cim. **A102** 1385 (1989).

[12] A. Burnel, these proceedings; Phys. Rev. **D40** 1221 (1989); P. V. Landshoff, these proceedings; Phys. Lett. **227B** 427 (1989).

[13] P. V. Landshoff, these proceedings; H. Hüffel, P. V. Landshoff and J. C. Taylor, Phys. Lett. **217B** 147 (1989).

[14] H. C. Lee and M. S. Milgram, Nucl. Phys. **B268** 543 (1986).

[15] I. S. Gradshteyn and I. M. Ryzhik, *Tables of Integrals, Series and Products*, Academic, New York, 1965.

[16] W. Konetschny, Phys. Rev. **D28** 354 (1983).

[17] H. Skarke, J. Math. Phys **30** 1270 (1989).

[18] P. Gaigg, M. Kreuzer and G. Pollak, Phys. Rev. **D38** 2559 (1988).

[19] G. Leibbrandt and S. L. Nyeo, Mod. Phys. Lett. **A3** 1085 (1988); G. Nardelli and R. Soldati, Phys. Lett. **206B** 495 (1988).

[20] A. Bassetto, M. Dalbosco and R. Soldati, Phys. Rev. **D36** 3138 (1987).

[21] H. Skarke and P. Gaigg, Phys. Rev. **D38** 3205 (1988).

[22] G. Pollak, Phys. Rev. **D40** 2027 (1989).

[23] W. Kummer, Phys. Lett. **105B** 473 (1981).

[24] G. Leibbrandt and S. L. Nyeo, Phys. Rev. **D39** 1752 (1989); G. Nardelli, Phys. Rev. **D40** 2662 (1989).

[25] J. D. Bjorken and S. D. Drell, *Relativistic Quantum Fields*, Mc Graw-Hill, 1965.

[26] W. Konetschny, Phys. Lett. **74B** 365 (1978).

# Tests of Gauge Invariance in Quantum Local Gauge Field Theories

Roberto Soldati

Dipartimento di Fisica "A. Righi" Universitá di Bologna, and
Istituto Nazionale di Fisica Nucleare Sezione di Bologna,
Via Irnerio 46 - 40126 BOLOGNA - Italia

**Abstract:** The consistency of the Cauchy principal value prescription in regulating the spurious poles of the axial and planar gauge propagators is analysed, by studying the behaviour of the euclidean Wilson loop up to the fourth order in the coupling constant. A suitable mathematical procedure is proposed to set up the perturbation theory in the presence of the Cauchy principal value distributions, in order to obtain meaningful nonambiguous results. In this way the gauge invariant behaviour of the Wilson loop is recovered in the planar case, where unphysical degrees of freedom are present, whereas it breaks down in the axial case, due to the conflict between reflection positivity and the regularization of the double pole distribution in the propagator.

Gauge field models are theories with an underlying local (i.e. space-time dependent) symmetry group. In order to quantize such kind of theories one has to choose a representative element from each class of equivalent gauge potentials; this is customarily done by imposing so called "subsidiary conditions" which breaks the local symmetry. The requirement that the observables of the theory have to be independent of the choice of a subsidiary condition I will called *Gauge Relativity Principle*. It turns out that some important information may be inferred from an explicit check of the above invariance statement. As a matter of fact, the necessity for the introduction of the unphysical Faddeev-Popov particles in manifestly covariant gauges was originally discovered [1] by analysing the unitarity properties of gauge invariant S-matrix elements. However, in general, a test of the Gauge Relativity Principle using S-matrix elements is usually difficult beyond the tree level, especially in the massless case where the S-matrix is only defined formally, due to the occurrence of severe infrared divergences.

It is the aim of the present contribution to illustrate some properties of invariant phase factors, which turn out to be very useful in probing the consistency of a quantization scheme for Yang-Mills theories in the presence of axial and planar subsidiary conditions. The classical invariant phase factor is a geometrical quantity defined as

$$C(A_\mu;\Gamma) = TrP\exp[-ig\oint_\Gamma A_\mu dx^\mu] \ , \tag{1}$$

where P denotes path-ordering, and $\Gamma$ is a closed contour; it is well known that $C(A_\mu;\Gamma)$ represents a gauge invariant object. At the quantum level the corresponding quantity reads

$$Q(\mathbf{A}_\mu;\Gamma) = \langle O|C(\mathbf{A}_\mu;\Gamma)|O\rangle \ , \tag{2}$$

$\mathbf{A}_\mu$ being the gauge potential quantum operator; now $Q(\mathbf{A}_\mu;\Gamma)$ still is formally gauge invariant, but it is plagued by geometrical as well as dynamical divergences [2]. In perturbation theory with covariant gauge conditions it has been shown [3] that, using dimensional regularization and smooth paths, $Q(\mathbf{A}_\mu;\Gamma)$ can be renormalized by coupling constant renormalization. As gauge invariance has to be satisfied to any order in the coupling constant , we can use the quantum invariant phase factors to obtain a nice perturbative test of the Gauge Relativity Principle for regularized quantum objects.

If we call V(L) the potential energy between two infinitely heavy static (quark-antiquark) fermions separated by a distance 2L, it is well known that the euclidean Wilson loop $\mathcal{W}_\mathcal{E}(\Gamma_0)$, i.e. the phase factor corresponding to the rectangular path $\Gamma_0$ of Fig.(1), is expected to behave as $\exp[-TV(L)]$ when the euclidean time coordinate T becomes very large [4], V(L) being the lowest eigenvalue of the Hamiltonian of the quark-antiquark system.

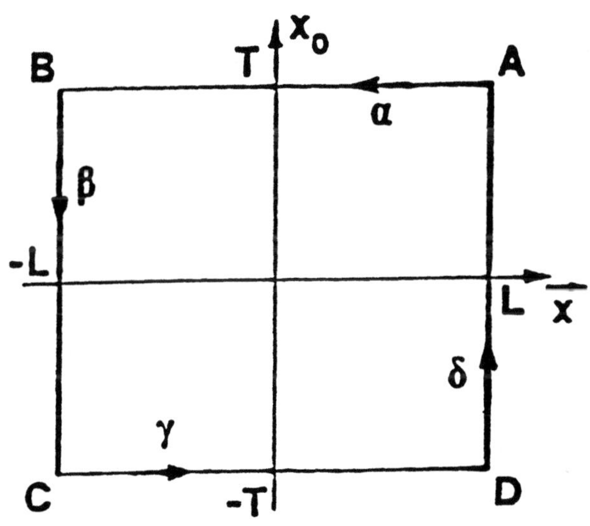


Fig. 1.

The euclidean Wilson loop $\mathcal{W}_\mathcal{E}(\Gamma_0)$ can be expressed as a functional average

$$\mathcal{W}_\mathcal{E}(\Gamma_0) = \mathcal{N}\int [\mathcal{D}A_\mu]\exp[-\mathcal{S}_\mathcal{E}]C(A_\mu;\Gamma_0) \ , \tag{3}$$

$[\mathcal{D}A_\mu]$ being the invariant Faddeev-Popov measure [5] and $\mathcal{S}_\mathcal{E} = \int d^4x\mathcal{L}_\mathcal{E}(x_1,x_2,x_3,x_4)$ the euclidean action. Now the test of gauge invariance works out as follows: one has to verify the formula

$$\mathcal{W}_{\mathcal{E}}(\Gamma_0) \stackrel{largeT}{\approx} \exp[-TV(L)] = 1 - TV(L) + \frac{1}{2}T^2(V(L))^2 + \dots , \tag{4}$$

taking into account that, in perturbation theory, we have

$$V(L) = g^2 V^{(0)}(L) + g^4 V^{(1)}(L) + \dots . \tag{5}$$

In order to obtain a behaviour of $\mathcal{W}_{\mathcal{E}}$ as in eq.(4), one should check if the following <u>necessary</u> conditions are fulfilled:

i) the $O(g^4)$ coefficient of $T^2$ in eq.(4) must be one half the square of the $O(g^2)$ coefficient of T; this requirement entails the *cancellation of the nonabelian contributions* behaving like $T^2$ (or worse) when T becomes large;

ii) a complete cancellation of the nonleading terms (such as $T \log T$) has to take place.

It is worthwhile to note that the conditions i) and ii) have to be satisfied, when $\mathcal{W}_{\mathcal{E}}(\Gamma_0)$ is regularized by means of a gauge invariant cutoff, e.g. the dimensional regularization scheme. To the best of my knowledge, the requirement ii) has never been checked in the literature for any kind of gauge choice. The first explicit test of the necessary condition i) has been performed in [6] within the context of the euclidean general covariant gauge $\partial_\mu A_\mu = \alpha B$, where B is the usual auxiliary field.

In the present contribution the test of gauge invariance represented by the fulfilment of the necessary condition i) is discussed within the context of *planar* and *axial* gauge choices in the euclidean formulation. In order to define the perturbative expansion, the local gauge invariance is broken by means of a *single gauge vector* $n_\mu$; that implies (see the contribution by A. Bassetto to this workshop) the Cauchy principal value (CPV) prescription to define the "spurious" gauge singularities in the Green (or euclidean Schwinger) functions, as originally proposed in ref. [7]. The algebraic noncovariant gauge fixing with a single vector $n_\mu$ entails some important advantageous features, among which:

a) the order-by-order subtractions of the ultraviolet divergences can be accomplished by means of <u>local counterterms</u> [7],[8].

b) In a Minkowski space-time, the quantization can be implemented by means of the canonical Dirac formulation if $n_\mu$ is *space-like*; the only serious problem one has to deal with in that matter relies on the *positivity* of the polarization sum in the axial case [9],[10], as discussed in the sequel.

c) In the presence of only one gauge vector, one is always free to choose a reference frame such that $n_\mu = (0, \boldsymbol{n})$; in so doing the transition from the minkowskian to the euclidean formulations is straightforward, as no interference between covariant and spurious denominators arises in performing Wick's rotation in the Feynman integrals.

d) It follows from the canonical Dirac approach, $n_\mu$ being space-like, that the Faddeev-Popov ghosts are decoupled.

Nevertheless, besides the above listed appealing advantages, one has to take into account some remarkable troubles and complications in working with one gauge vector and CPV prescription; I will try to illustrate those nasty aspects in what follows.

As already mentioned, in the minkowskian space-like axial gauge there is a conflict between a meaningful mathematical definition of the double pole in the vector propagator

$$D_{\mu\nu}(k) = \frac{1}{k^2 + i\epsilon}(-g_{\mu\nu} + \frac{n_\mu k_\nu + n_\nu k_\mu}{[nk]} + n^2 k_\nu k_\mu \frac{d}{d(nk)} \frac{1}{[nk]}) \ , \tag{6}$$

where the poles in square brackets are understood with CPV prescription, and the positive definition of the polarization sum

$$d_{\mu\nu}(k) = \sum_{\alpha=1,2} (e^\alpha_\mu(k))^* e^\alpha_\nu(k) \ , \tag{7}$$

namely the requirement of the absence of unphysical degrees of freedom in the space of states, as it follows from the canonical Dirac quantization. In other words, the propagator of eq.(6) turns out to be different, rigorously speaking, from the Fourier transform of the T-product of two gauge potential operators satisfying $n^\mu A_\mu = 0$; this last expression should exhibit the term $k_\mu k_\nu (\frac{1}{[nk]})^2$ which, unfortunately, is mathematically meaningless.

In the general case, the perturbative expansion of an arbitrary quantity, such as Green functions, Wilson loop, operator products, etc., involves the appearence of mathematical pathologies in the form of ill defined powers of spurious poles, products of spurious poles and $\delta$-distributions and so on. A typical example of that matter is provided by the diagram of Fig. 2: if one tries to write the corresponding integral using naively the Feynman rules, one immediately gets things like $(\frac{1}{[nk]})^r$, r=2,3,4, or $\frac{1}{[np]}\delta(np)$ and similar ones, which are meaningless. Other examples of such kind of troubles will be provided in the study of the Wilson loop.

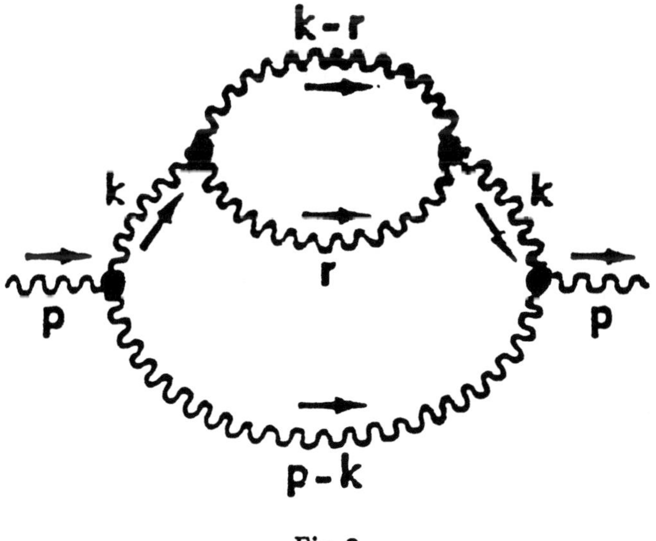


Fig. 2.

Now the question arises of how to deal with those troubles in general; in refs.[10]-[12] a systematic method has been proposed, which can be summarized as follows.

To cope with powers of spurious singularities we follow the general recipe suggested by N.N. Bogoliubov and D.V. Shirkov [13], i.e. in their words: "In developing perturbation theory, we shall have to make use of the operations of *switching on* and *switching off* the interaction. In order to describe this operation mathematically, we shall introduce the function $\xi(x)$ with values in the range (0,1) which represents the extent of switching on the interaction". The smearing function $\xi(x)$ is asigned to any vertex of a diagram: in so doing the $\delta$-distribution ensuring the total momentum conservation is replaced by

$$\delta(\sum_i p_i) = \frac{1}{(2\pi)^4} \lim_{\xi\to 1} \int d^4x \xi(x) \exp[ix \sum_k p_k] \; ; \tag{8}$$

in this way, as long as $\xi \neq 1$, translation invariance does not hold and, consequently, powers of spurious singularities never appear. Translation invariance will be recovered with the transition to the limit $\xi \to 1$ at the end of the calculations (eventually also gauge invariance will be recovered at the end in the same limit).

In order to obtain a well defined limit $\xi \to 1$ it is necessary to consider, order by order in the loop expansion, the Green (or Schwinger) functions as distributions acting on a **Besov space** $\mathcal{T}$ [14] of testfunctions $\phi(nk_1, ..., nk_r)$ *suitably vanishing* when $nk_j = 0, j = 1, 2, ..., r$, namely we set

$$\langle G(k_{1\perp}, ..., k_{r\perp})|\phi\rangle = \lim_{\xi\to 1} \int d(nk_1)... \int d(nk_r) G(k_1, ..., k_r; [\xi])\phi(nk_1, ..., nk_r) \; , \tag{9}$$

for any testfunction $\phi$ belonging to $\mathcal{T}(R^r)$, the $k_{j\perp}, j = 1, ..., r$ being the components of the momenta orthogonal to the the gauge vector $n_\mu$. The above described procedure to regularize the Green functions is iterative in the following sense.

i) First one considers the r-point Green (or Schwinger) functions $G^{(N)}(k_1, ..., k_r; [\xi])$, up to the N-th order in the loop expansion, as $\mathcal{T}$-distributions.

ii) In the limit $\xi \to 1$ one defines the powers of spurious singularities as <u>derivatives</u>; as a matter of fact, the difference between powers and derivatives of CPV-singularities are ill-defined matemathical expressions "concentrated" in $nk_j = 0, j = 1, ..., r$: this basically is the reason for the introduction of the space $\mathcal{T}$, which precisely does away with these pathologies.

iii) The $\mathcal{T}'$-distributions $G^{(N)}(k_1, ..., k_r)$, once calculated in this way, can be continued to $L_2$, as $\mathcal{T}$ is dense in $L_2$ with respect to the $L_2$-topology, subtracted with polynomial counterterms and inserted as subdiagrams in the $G^{(N+1)}(k_1, ..., k_r; [\xi])$ regularized Green (or Schwinger) functions.

It is important to notice that a remarkable consequence of the above outlined procedure i)–iii) is the validity of the *naive splitting formula* for the separation of products of spurious poles. Let us consider in fact the following identity,

$$\frac{1}{[nk]} \cdot \frac{1}{[n(p-k)]} = \frac{1}{np}(\frac{1}{[nk]} + \frac{1}{[n(p-k)]}) - \pi^2 \delta(nk)\delta(np) \; ; \tag{10}$$

Obviously products of this kind always appear in perturbative calculations (in the limit $\xi \to 1$); the introduction of the space $\mathcal{T}$ allows to ignore, or, better, to "regularize to zero", the last Poincare-Bertrand (PB) term on the r.h.s. of eq. (10). I would like to stress that, as it will be more evident in the sequel, it is compulsory to avoid the presence of such

PB-terms, since they would generate ambiguities and/or mathematical nonsense in perturbative calculations.

Although it could appear somewhat cumbersone or involved, the procedure I have described above to handle perturbative calculations in the presence of spurious poles with CPV prescription is nothing but a coherent and rigorous mathematical framework implementing a set of pragmatic and practical rules tacitly assumed in the existing literature on the subject, namely:

(a) to understand the *powers* of spurious singularities as *derivatives*;

(b) to use the *naive splitting formula* for the products of spurious poles;

(c) to give *precise definitions* to otherwise ambiguous quantities.

I would like to add some more comments concerning this set of rules. The proposed regularization procedure allows a formulation of the perturbation theory for algebraic noncovariant gauges which turns out to be free of mathematical pathologies; whether the results obtained in this way are physically meaningful, i.e. whether the subsidiary condition under consideration actually reproduces the physical requirements (such as the correct structure of the physical Hilbert space, the unitarity conditions, etc.), has to be *independently* checked by means, for instance, of gauge invariance tests. To the best of my knowledge, there are no other fully developed methods to cope with the singularities of the perturbation theory for algebraic noncovariant gauges in the presence of only one gauge vector $n_\mu$. I only have to mention a proposal suggested in ref. [7] which, however, although similar to the philosophy I have discussed, still needs a rigorous basis; moreover, it has never been implemented in practical calculations.

The study of the Wilson loop represents a nice pedagogical example to show the relevance of the general rules in perturbation theory I have previously outlined.

Let us first perform the gauge invariance test for the planar gauge [11]. The euclidean Lagrangean is given by

$$\mathcal{L}_\mathcal{E} = \frac{1}{4} G^a_{\mu\nu} G^a_{\mu\nu} - \frac{1}{2n^2} \partial_\mu (n_\nu A^a_\nu) \partial_\mu (n_\rho A^a_\rho) \ , \tag{11}$$

where the euclidean field strengths are defined as

$$G^a_{\mu\nu} = \partial_\mu A^a_\nu - \partial_\nu A^a_\mu - g f^{abc} A^b_\mu A^c_\nu \ , \tag{12}$$

whence the propagator turns out to be

$$D_{\mu\nu}(k) = \frac{1}{k^2} (\delta_{\mu\nu} - \frac{n_\mu k_\nu + n_\nu k_\mu}{[nk]}) \ . \tag{13}$$

In the study of the Wilson loop in the large T limit it is clear that the most singular behaviour is associated to the choice $n_\mu = (\boldsymbol{n}, n_4) = (0, 1)$ for the gauge vector. Let us analyse the most delicate terms.

(a) *The "bubble" diagram.* This diagram corresponds to the insertion of the full one-loop propagator in all the possible ways into the rectangular path, and leads to, up to the fourth order in g and for the SU(N) gauge group,

$$\mathcal{W}_{\text{bubble}} = -\frac{1}{4}g^2(N^2-1)\oint G^{(2)}_{\mu\nu}(x-y)dx_\mu dy_\nu \ , \tag{14}$$

$G^{(2)}_{\mu\nu}(x-y)$ being the full one-loop propagator. In order to evaluate this contribution, one needs the expression for the one-loop selfenergy tensor, or, better, its effective part, namely

$$\Pi_{\mu\nu}(p) = g^2\mu^{4-2\omega}f^{acd}f^{bcd}[\mathcal{A}(\frac{p^2}{\mu^2};h)(p_\mu p_\nu - p^2\delta_{\mu\nu}) + \mathcal{B}(h)p_\mu p_\nu] + \text{irrelevant} \ , \tag{15}$$

where $\mu$ is the dimensional regularization mass parameter, $h^2 = \frac{(np)^2}{n^2p^2}$, and the "irrelevant" terms are those ones not contributing due to a trivial geometrical decoupling from the rectangular contour.

The behaviour of the coefficients $\mathcal{A}$ and $\mathcal{B}$ in the limit $h \to 0$, which is the interesting region, can be explicitly computed. From this calculation it follows that the "bubble" diagram <u>does not contribute</u>, due to delicate cancellations of $T^2$ terms (or worse), to the Wilson loop up to the $g^4$-approximation and in the large T limit.

I would like to remark that, to get the correct result, it is <u>crucial</u> to employ the Besov $\mathcal{T}$ space. As a matter of fact, otherwise, the naive full one-loop propagator *does not exist* in the limit $\xi \to 1$. Let us consider in fact the usual product

$$G^{(2)}_{\mu\nu}(k) = D_{\mu\rho}(k)\Pi_{\rho\sigma}(k)D_{\sigma\nu}(k) \ ; \tag{16}$$

if one does not consider the full propagator as a $\mathcal{T}'$-distribution, then the selfenergy contains the quantity

$$\Pi_{\rho\sigma}(k)|_{PB} = \frac{\pi}{64}\delta(h)(k_\rho k_\sigma - \delta_{\rho\sigma}k^2) \ , \tag{17}$$

which stems from the PB-term in the splitting formula for the products of spurious poles (see eq. (10)). These terms, if retained, would lead to the *ill-defined* formal expression

$$G^{(2)}_{ij}(k)|_{PB} = \mathcal{A}(k)\frac{k^2}{(k^2)^2}\delta(k_4)\cdot(\delta_{ij} - \frac{k^2}{k^2}\frac{k_ik_j}{[k_4]^2}) \ i,j = 1,2,3. \tag{18}$$

Now the product of the distributions in eq. (18) is a pathological formal quantity, which is safely and wisely "regularized to zero" if $G^{(2)}_{\mu\nu}(k)$ is understood as a distribution acting on the space $\mathcal{T}$.

I would like to add one more quite remarkable comment: the Poincare-Bertrand part of the selfenergy $\Pi_{\rho\sigma}|_{PB}$ at one-loop order would generate a <u>finite</u> $T^2$ contribution to the "bubble" diagram, up to the 4-th order in g, namely

$$\lim_{T\to\infty}\frac{1}{T^2}\mathcal{W}_{\text{bubble}}|_{PB} = -\frac{1}{2}\mathcal{N}(\frac{\pi}{2L})^2 \ , \tag{19}$$

with $\mathcal{N} = g^4\frac{(N^2-1)}{4(2\pi)^4}$; however, *coherently* with our meaningful definition of $G_{\mu\nu}(k)$, it <u>**has to be dropped**</u> within the context of the general procedure we are considering.

(b) *The "crossed propagators" diagrams.* This contribution to the Wilson loop arises from the insertion of two free propagators into the rectangular path in all the possible ways. There are thirteen graphs that have to be computed; their sum gives rise to the total contribution

$$\lim_{T\to\infty}\frac{1}{T^2}\mathcal{W}_{\rm cross} = -\mathcal{N}(\frac{\pi}{2L})^2\Phi_\omega(\mu L)\ , \tag{19}$$

where

$$\Phi_\omega(\mu L) = \frac{64}{(2\pi)^4}(2\pi\mu L)^{8-4\omega}\int d^{2\omega-1}\tilde{p}\int d^{2\omega-1}\tilde{q}\frac{p_3\sin(p_3+q_3)\sin(p_3-q_3)}{\tilde{p}^2\tilde{q}^2(p_3-q_3)} \tag{20}$$

is a divergent (when $\omega\to 2$) quantity representing a geometrical singularity of the kind discussed in Ref.[2].

(c) *The "spider" diagrams.* This last $O(g^4)$ contribution corresponds to the insertion of lowest order three-point Schwinger functions into the rectangular path functions in all the possible ways. The sum of all the twelve graphs obtained in this guise leads to the result

$$\lim_{T\to\infty}\frac{1}{T^2}\mathcal{W}_{\rm spider} = \mathcal{N}(\frac{\pi}{2L})^2\Phi_\omega(\mu L)\ . \tag{21}$$

An interesting observation is now in order concerning the behaviour of four of the graphs, the ones drawn in Fig. 3, contributing to the term $\mathcal{W}_{\rm spider}$.

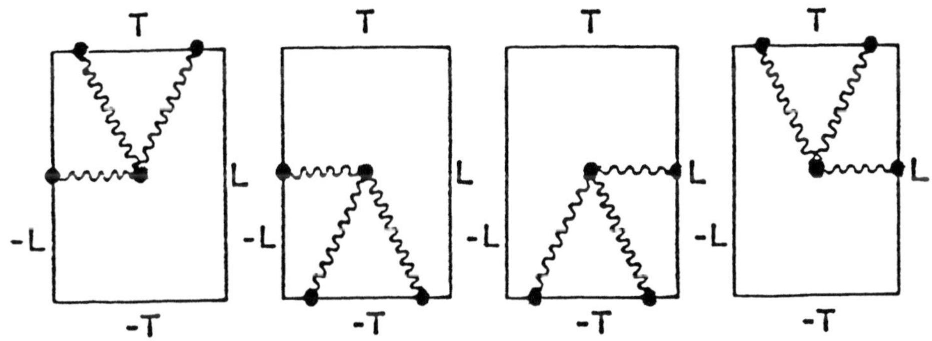


**Fig. 3.**

This very set of graphs is naively proportional, in the large T limit, to the integral

$$I = \int_{-\infty}^{+\infty}dp\int_{-\infty}^{+\infty}dq\int_{-\infty}^{+\infty}dr\frac{\delta(p+q+r)}{[p][q][r]}\frac{\sin r}{r}\sin(p+q)\ . \tag{22}$$

The *integrand* of I turns out to be ambiguous, as it involves an ill-defined product of distributions! As a matter of fact, by formally eliminating the p-variable we get, on the one hand,

$$\int_{-\infty}^{+\infty} dq \int_{-\infty}^{+\infty} dr (\frac{\sin r}{r})^2 \frac{1}{[q][q+r]} = \pi^2 . \tag{23}$$

On the other hand, if we understand the momentum conservation in eq. (22) in terms of the previously described general setting, namely

$$\delta(p+q+r) = \lim_{\xi \to 1} \frac{1}{2\pi} \int_{-\infty}^{+\infty} d\alpha \xi(\alpha) \exp[i\alpha(p+q+r)] , \tag{24}$$

then $I = 0$! Eventually, summing up all the terms, we obtain

$$\lim_{T\to\infty} \frac{1}{T^2} (\mathcal{W}_{\text{cross}} + \mathcal{W}_{\text{spider}}) = 0 , \tag{25}$$

i.e. the ***necessary*** condition for the exponential behaviour of the Wilson loop appears to be fulfilled in the planar gauge. I hope that anyone has appreciated the crucial relevance of the general mathematical procedure needed in order to obtain a precise result which, otherwise, would have been **intrinsically** ambiguous and unreliable.

We can now go through the test for the axial gauge. In the euclidean axial gauge the calculations can be developed in close analogy, mutatis mutandis, with the planar case; however, as we shall see, the final result is opposite, i.e. gauge invariance is lost [12]. The euclidean Lagrangean for the axial gauge is given by

$$\mathcal{L}_E = \frac{1}{4} G^a_{\mu\nu} G^a_{\mu\nu} - n_\mu A^a_\mu \Lambda^a , \tag{26}$$

where the Lagrange multiplier $\Lambda^a$ leads to the constraint

$$n_\mu A^a_\mu = 0 ; \tag{27}$$

the corresponding propagator is given by

$$D_{\mu\nu}(k) = \frac{1}{k^2} (\delta_{\mu\nu} - \frac{n_\mu k_\nu + n_\nu k_\mu}{[nk]} - k_\mu k_\nu n^2 \frac{d}{d(nk)} \frac{1}{[nk]}) , \tag{28}$$

which satisfies, in the sense of distributions, $n_\mu D_{\mu\nu}(k) = 0$. The constraint eq. (27) reduces the number of graphs relevant for extracting the large T behaviour, but the more singular structure of the propagator makes the evaluation of such graphs trickier.

(a) *The "bubble" diagram.* In order to obtain this contribution, one still needs the complete expression for the one loop selfenergy tensor satisfying, in the present case, the simple Lee-Takahashi-Ward identity

$$k_\mu \Pi_{\mu\nu}(k) = 0 . \tag{29}$$

Again, as in the planar case, one can show (see ref. [12] for the detailed calculation) that the "bubble" does not contribute $T^2$-terms (or worse) to the Wilson loop in the large T-limit. Once again, it is essential to implement the $T$-space formalism to get a definite value for the "bubble" term under consideration. Otherwise the selfenergy is affected by pathological quantities of the kind

$$\Delta_{ij} = \delta(p_4) \Phi^{(1)}_{ij}(p) + (\frac{d^2}{dp_4^2} \delta(p_4)) \Phi^{(2)}_{ij}(p) , \tag{30}$$

which, when multiplied by *regular* functions, turn into

$$\Delta_{ij}|_{p_4=0} = \frac{\pi}{64}(p_i p_j - |p|^2\delta_{ij})(\delta(p_4) - \frac{1}{4}\frac{d^2}{dp_4^2}\delta(p_4)) , \tag{31}$$

but, when multiplied by propagators involving the $\frac{1}{[p_4]}$ distribution and its derivative, lead to regularization dependent ambiguous expressions (e.g. depending on the choice of $\xi$) for the Schwinger function $G^{(2)}_{\mu\nu}(p)$. As a consequence, we conclude that, in the absence of a well defined procedure, the result of the present calculation is **arbitrary** or, in other words, the large T behaviour of the Wilson loop fails to be a test of gauge invariance in perturbation theory. A similar conclusion has been obtained in ref. [15], but following a completely different approach, concerning the minkowskian space time temporal gauge.

The occurrence of such a kind of subtlety is evident in the early work of ref. [16], where the authors found that the exponential behaviour can be restored by breaking time translation invariance, as any choice of $\xi$ indeed does. As a matter of fact, in our language, one can always choose a suitable function $\xi$ and a regularization for the product of $\Delta_{i,j}$ and propagators in such a way, for instance, that the exponential behaviour is recovered; this is nothing but to recognize that, without Besov spaces, the result is *arbitrary*, i.e. the test is eluded!

It should be mentioned, for completeness, that the analogous calculations in ref. [17] concerning the CPV prescription unfortunately contain, besides several inaccuracies, an apparent logical inconsistency, i.e. the use of a limiting procedure that would involve, if one was likely to investigate it seriously, the computation of a number of graphs at least ten times bigger than the one actually considered by the author, so that his result is hardly reliable.

(b) *The "crossed propagators" and "spider" diagrams.* In the sum of these four relevant, in the large T-limit, graphs the geometrical divergences cancel out, as in the planar case, but a relic survives equal to $\frac{2}{3}\mathcal{N}(\frac{\pi T}{2L})^2$: therefore the necessary condition for the gauge invariance is **not** satisfied in the axial case!

We are obviously wondering about the reasons for this failure: the "rationale" can be found as follows. In a minkowskian space-time canonical formulation for the space-like axial gauge, leading to CPV, **only physical** degrees of freedom are present, at variance with the planar case, where ghost field quanta are involved. The corresponding euclidean formulation should therefore satisfy the Osterwalder-Schrader *reflection positivity* condition. This means that ,for instance, already to lowest order, the functional

$$\mathcal{F} = (\frac{1}{2\pi})^2 \int \frac{d^4k}{k^2}(\delta_{ij} + \frac{k_i k_j}{[k_4^2]})\phi_i^\dagger(k)\phi_j(k)|\phi(k_4)|^2 \tag{32}$$

should be a Sobolev definite norm; but, again, the only meaningful definition of $\frac{1}{[k_4^2]}$ is by means of the derivative which, unfortunately, spoils the positivity requirement. To put it differently, a ghost degree of freedom is hidden in the double pole, which indeed does not decouple in higher loop calculations [9],[10].

The lesson we can learn from the above analysis is that, in the presence of only one gauge vector to quantize algebraic noncovariant gauges, **locality** of the counterterms is kept but a great care is necessary to handle perturbation theory. Moreover, it is apparent that a quantization scheme for gauge theories, in which only physical degrees of freedom are present, does not seem to be available: some kind of ghost is unavoidable.

The quantization with two "conjugate" gauge vectors has recently become popular; however, many issues are completely open to this concern but for the light-cone case, which is now pretty well understood. As a consequence, the gauge invariance test provided by the euclidean Wilson loop could represent a nice challenge in checking the consistency of the above proposal, at least in the case where Wick's rotation is straightforward.

## References


[1] R.P. Feynman, Acta Phys. Polonica **24** (1963), 697.

[2] A.M. Polyakov, Nucl. Phys. **B164** (1975), 171.

[3] V.S. Dotsenko and S.N. Vergeles, Nucl. Phys. **B169** (1980), 527.

[4] K.G. Wilson, Phys. Rev. **D14** (1974) 2455.

[5] L.D. Faddeev and V.N. Popov, Phys.Lett. **25B** (1967), 29.

[6] W. Fischler, Nucl. Phys. **B129** (1977), 157.

[7] W. Kummer, Acta Phys. Austriaca **41** (1975), 315.

[8] W. Konetschny and W. Kummer, Nucl. Phys. **B100** (1975), 106;
J. Frenkel, Phys. Rev. **D13** (1976), 2325;
D.M. Capper and G. Leibbrandt, Phys. Rev. **D25** (1982), 1002;
A.I. Mil'shtein and V.S. Fadin, Sov. Jour. Nucl. Phys. **35** (1981), 779;
A. Andrasi and J.C. Taylor, Nucl. Phys. **B227** (1983), 494;
P. Gaigg, O. Piguet, A. Rebhan and M. Schweda, Phys. Lett. **175B** (1986), 53;
A. Bassetto, G. Nardelli and R. Soldati, Mod. Phys. Lett. **A3** (1988), 1663.

[9] N. Nakanishi, Prog. Theor. Phys. **67** (1982), 985; Phys. Lett. **131B** (1983), 381.

[10] A. Bassetto, I. Lazzizzera and R. Soldati, Nucl. Phys. **B236** (1984), 319.

[11] A.Bassetto and R.Soldati, Nucl.Phys. **B276** (1986), 517.

[12] G.Nardelli and R.Soldati, preprint DFUB 89/7, Università di Bologna, 1989.

[13] N.N. Bogoliubov and D.V. Shirkov, "Introduction to the theory of quantized fields", J. Wiley, New York, 1980.

[14] H. Triebel,"Theory of function spaces", Basel, Birkhäuser, 1983;
J. Bergh and J. Löfström, "Interpolation spaces", Springer-Verlag, Berlin, 1976.

[15] J.P. Leroy, J. Micheli and G.C. Rossi, Nucl. Phys. **B232** (1984), 511; Nuovo Cim. **A84** (1984), 270.

[16] S. Caracciolo, G. Curci and P. Menotti, Phys. Lett. **113B** (1982), 311.

[17] P.V. Landshoff, Phys. Lett. **169B** (1986), 69.

# The Analytic, Principal-Value and Landshoff Prescriptions for Axial Gauges

H. C. Lee

Atomic Energy of Canada, Chalk River Nuclear Labs.

Chalk River, Ontario K0J 1J0, Canada

The word prescription is itself suggestive: it is a device used to deal with something that, under ideal circumstances, should not be there. In the case of axial gauges [1], the intruders are spurious singularities induced in Feynman integrals by a noncovariant gauge choice. To see how this comes about, consider the pure Yang-Mills Lagrangian with a gauge fixing term $1/(2\alpha)\ (n{\bullet}A)^2$, where n is an arbitrary 4-vector with $n^2 \neq 0$,

$$\mathscr{L} = \int d^4x \left(-\frac{1}{4} F_{\mu\nu}F^{\mu\nu} - \frac{1}{2\alpha} (n_\mu A^\mu)^2\right) , \tag{1}$$

for which the propagator, defined as the reciprocal of the coefficient of the term quadratic in the gauge field A, is in the $\alpha \to 0$ limit

$$D^{(0)}_{\mu\nu}(p) = \frac{1}{p^2} \left[ -g_{\mu\nu} + \frac{p_\mu n_\nu + p_\nu n_\mu}{p{\bullet}n} - \frac{n^2 p_\mu p_\nu}{(p{\bullet}n)^2} \right] , \tag{2}$$

where indices of the gauge group have been suppressed. (We are only interested in the $\alpha \to 0$ limit. In our view, the axial gauge loses much of its merit - such as the decoupling of the Faddeev-Popov ghost - when $\alpha \neq 0$.) This propagator has two types of singularities: (i) at $p^2 = 0$, and (ii) at $p{\bullet}n = 0$. The first type is physical and is the origin of infrared divergence in the amplitudes of certain processes. The second type is induced by the noncovariant gauge fixing and is the source of spurious singularities. We say spurious because they are singularities that should not manifest itself in a physical amplitude. A physical amplitude is supposed to be gauge independent, and had we chosen a covariant gauge to work in, the Feynman integrals for such an amplitude would not have such singularities. The reason that we do encounter such singularities when calculating physical amplitudes in axial gauges

† *Work supported in part by a grant from the Natural Science & Engineering Council of Canada and a NATO Collaborative Research grant.*

has something to do in the way we do calculations, at least in perturbation theory: An amplitude that is not singular is decomposed as a sum of Feynman integrals, some of which may be singular. The singularities ought to cancel among themselves when the Feynman integrals are combined to give the amplitude. A well-known example occurs in the computation of the $\gamma_5$-anomaly (or any other anomaly) in perturbation theory. The anomaly itself is finite, but typically it is decomposed by partial fraction into two easier-to-evaluate but singular integrals. (Taylor [2] has shown us an elegant way to extract the $\gamma_5$-anomaly without hard computation.) The function of a prescription is to isolate the spurious gauge-fixing singularity from the physical infrared singularity. This task is made considerably more complicated by the entanglement of the $p^2=0$ and $p\cdot n=0$ poles at the point $p=0$.

The complication arises because the $p^2=0$ singularity also needs to be regularized by a prescription. One commonly used is the Feynman prescription in which $p^2$ is replaced by $p^2+i\varepsilon'$. This would ensure that a particle propagates forward in time. Furthermore, the ultraviolet divergence at $p\to\infty$ also has to be regularized. All told there are three categories of singularities that need to be separately regularized:

(i) The gauge induced spurious singularities must all cancel at the end of a calculation for any physical amplitude.

(ii) The $p^2=0$ singularities are physical; their noncancellation at the end of the calculation for an amplitude is an indication that that amplitude is infrared divergent. (Here we shall not differentiate between infrared divergence and a zero-mass singularity.)

(iii) The ultraviolet divergences must be collected as data for the renormalization program.

This discussion makes the point clear that not only would a prescription be incorrect if the effect of it were just to throw the infrared and ultraviolet singularities out from each individual integral. It would be equally incorrect if it simply eliminated the spurious singularities in individual integrals.

In this paper we illustrate the point by examining in some detail the analytic properties of three prescriptions for axial gauge integrals -- the principal-value prescription [3], the Landshoff [4] prescription and an analytic prescription. Landshoff's prescription is shown to be the only one that keeps separate accounts of all three categories of singularities mentioned above. The prescription is sometimes said to be ad hoc. We now argue that rather it is a natural generalization of Feynman's prescription for the $p^2=0$ pole.

First we ask how does one get Feynman's prescription if one does not want to simply replace by hand the $p^2$ denominator in (2) by $p^2+i\varepsilon'$. It can be done as follows. One adds a small term $i\varepsilon' A^2$ to the Lagrangian. In Euclidean space, this has the effect of making the functional integral over A well defined for large $A^2$. Of course the added term fails to have a damping effect in Minkowski space. We shall adopt the attitude that it is sufficient for the theory to be well behaved in

Euclidean space -- Minkowski space can be reached by analytic continuation. For a nonAbelian theory there is also the problem of contending with terms cubic in A, which are more divergent than $\mathcal{O}(A^2)$ terms at large A. In perturbation theory this problem is unsolved but evaded; the formalism of the theory only explicitly involves Gaussian functional integrations over A. An important point to know is that the added term has no effect on the spurious singularity of the light-cone gauge [5], which alone among the axial-type gauges can be realized only in Minkowski space, but not in Euclidean space. With the $i\varepsilon'$ term added to the Lagrangian, the propagator, again in the limit $\alpha\to 0$, now becomes

$$D^{(0)}_{\mu\upsilon}(p) = \frac{1}{p^2+i\varepsilon'}\left\{-g_{\mu\upsilon} + \frac{(p_\mu n_\upsilon+p_\upsilon n_\mu)p\cdot n - n^2 p_\mu p_\upsilon + i\varepsilon' n_\mu n_\upsilon}{(p\cdot n)^2 + i\varepsilon' n^2}\right\}. \tag{3}$$

We see that not only is $p^2$ in the denominator replaced by $p^2+i\varepsilon'$, as desired, but also $p\cdot n$ appears in the denominator only in the form of $(p\cdot n)^2 + \eta^2$,

$$\eta^2 \equiv i\varepsilon' n^2 . \tag{4}$$

Before proceeding further, we examine what happens when we send $n^2\to 0$, or when we go to the light-cone gauge. Then $\eta^2$ also goes to zero. Thus the generalized Feynman's prescription does not regularize the spurious light-cone gauge singularity. In what follows, the $\eta^2\to 0$ limit is to be taken after all calculations of the amplitude are done.

The propagator (3) looks very similar to Landshoff's in the $\alpha\to 0$ limit [4]

$$D^{(0)}_{\mu\upsilon}(p)\Big|_{\text{Landshoff}} = \frac{1}{p^2}\left\{-g_{\mu\upsilon} + \frac{(p_\mu n_\upsilon+p_\upsilon n_\mu)p\cdot n - n^2 p_\mu p_\upsilon + \alpha'^2 n^2 n_\mu n_\upsilon}{(p\cdot n)^2 + \alpha'^2 (n^2)^2}\right\}. \tag{5}$$

In fact if we identify $i\varepsilon'$ in (3) with Landshoff's $\alpha'^2 n^2$, then with the exception of the $p^2=0$ pole, (3) and (5) are the same. In what follows we shall also write $\alpha'^2(n^2)$ as $\eta^2$. Of course the $p^2=0$ pole in (5) must still be regularized; in (5) one may separately use Feynman's or another prescription for it. The point is that, as we said before, the $p\cdot n$ and $p^2$ singularities must be kept separate. This can be accommodated when either (3) or (5) is used. For example, in (3) one can choose $\varepsilon'$ and $n^2$ such that $\varepsilon' n^2$ is finite when $\varepsilon'\to 0$.

The principal-value prescription is the replacement, for any nonnegative integer N and s=0 or 1,

$$\left(\frac{1}{p\cdot n}\right)^{2N+s} \longrightarrow \frac{1}{2}\lim_{\eta\to 0}\left[\left(\frac{1}{p\cdot n+i\eta}\right)^{2N+s}+\left(\frac{1}{p\cdot n-i\eta}\right)^{2N+s}\right] = \lim_{\eta\to 0} R(\eta^2)\left[\frac{1}{(p\cdot n)^2+\eta^2}\right]^{N+s}(p\cdot n)^s , \tag{6}$$

where R is a differential operator [6] in $\eta^2$

$$R(\eta^2) \equiv \frac{\Gamma(-N+1/2)}{\Gamma(1/2)}\prod_{L=0}^{N-1}\left(-N + \frac{1}{2} + L - \eta^2\frac{\partial}{\partial\eta^2}\right) . \tag{7}$$

This establishes a relation between PV and Landshoff prescriptions.

We now introduce the analytic prescription [6]. The idea is that all three categories of singularities can be regularized by analytic continuation. Consider the generic one-loop Feynman integral

$$S(p;\ \omega,\kappa,\mu,\upsilon,s;\ \varepsilon',\eta)\ \equiv\ \int d^{2\omega}q[(p-q)^2+i\varepsilon']^{\kappa}\ (q^2+\eta^2)^{\mu}\ [(q\cdot n)^2+\eta^2]^{\upsilon}\ (q\cdot n)^s\ , \tag{8}$$

where $\omega$ and the exponents $\kappa$, $\mu\in\mathbb{R}$, $\upsilon\in\mathbb{Z}$ and $s=0$ or $1$. We are interested in the limit of this integral when $\omega=2$ and $\kappa$, $\mu\in\mathbb{Z}$. All one-loop amplitudes with one nonzero external momentum are linear combinations of (7). The integral has an analytic representation, even when $\varepsilon'$ and $\eta$ are set equal to zero. It is given by [6]

$$S_0\equiv S(\varepsilon'=\eta=0)=\frac{\pi^{\omega}(p^2)^{\alpha_1}(n^2)^{\upsilon}(p\cdot n)^{s}\Gamma(\upsilon+s+1/2)}{\Gamma(-\kappa)\Gamma(-\mu)\Gamma(-\upsilon)\Gamma(-\alpha_0+\beta_1)}\ G^{2,3}_{3,3}\left(y\left|\begin{matrix}1+\alpha_0, 1+\alpha_1,\ 1+\upsilon;\\ 0,\ \beta_1;\ 1/2-s\end{matrix}\right.\right)\ , \tag{9}$$

where

$$\begin{aligned}
y\ &\equiv\ (p\cdot n)^2/p^2n^2,\\
\alpha_0\ &\equiv\ -\omega-\mu-\upsilon-s,\\
\alpha_1\ &\equiv\ \omega+\kappa+\mu+\upsilon,\\
\alpha_2\ &\equiv\ -\omega-\mu-s\ ,\\
\alpha_3\ &\equiv\ -\omega-\kappa,\\
\beta_1\ &\equiv\ \omega+\kappa+\upsilon,\\
\beta_3\ &\equiv\ \upsilon+s+1/2
\end{aligned} \tag{10}$$

and the function G is a Meijer's G-function, in this case expressible in a sum of two ${}_3F_2$ hypergeometric functions. For the general properties of the G-function see [7]. The well-known method of dimensional regularization corresponds to continuing the dimensional parameter $\omega$ alone, but not $\kappa$ and $\mu$. In that case it is not possible to derive (9) unless $\upsilon$ is a nonnegative integer, precisely because of the presence of the p•n pole. To see this, consider the special case $\upsilon=s=0$. Then (9) collapses to

$$S\ (\upsilon=s=0)\ =\ \frac{\pi^{\omega}(p^2)^{\omega+\mu+\kappa}\ \Gamma(\omega+\kappa)\Gamma(\omega+\mu)\Gamma(-\omega-\mu-\kappa)}{\Gamma(-\kappa)\Gamma(-\mu)\Gamma(2\omega+\mu+\kappa)} \tag{11}$$

which is well-defined for any $\kappa$, $\mu\in\mathbb{Z}$. The ultraviolet and infrared divergences of the original integral is retained as poles of the gamma-functions in (10). In this case (11) has poles arising from singularities at to $q=0$, $q=\infty$ and $q=p$ when $\alpha_0=\alpha_2$, $\alpha_1$ and $\alpha_3$ respectively are nonnegative integers. If $\kappa$ and $\mu$ are integers, then the $\alpha$'s are not integers as long as $\omega$ is not an integer. Therefore, for integral $\kappa$ and $\mu$, (11) is holomorphic in the complex $\omega$-plane. Write $\omega=2+\varepsilon$, then the $\alpha_{0,3}$ ($\alpha_1$) poles are of order $\mathcal{O}(-1/\varepsilon)$ ($\mathcal{O}(1/\varepsilon)$). The signs in the denominators are significant; they are direct consequences of a property of analytic regularization that obeys the power law.

We now return to (9), where $\kappa$ and $\mu$ as well as $\omega$ are continuous and, for simplicity, real. The right-hand side of (9) is again a holomorphic function of $\alpha_0$, $\alpha_1$ and $\alpha_3$ with poles of order $\mathcal{O}(1/\varepsilon_0)$, $\mathcal{O}(1/\varepsilon_1)$ and $\mathcal{O}(1/\varepsilon_3)$, respectively, at points in the $\omega$-$\kappa$-$\mu$ hyperplane where $\alpha_0$, $\alpha_1$ or $\alpha_3$ is a nonnegative integer. If one approaches these poles on a line, henceforth called the $\varepsilon$-line, on which $\kappa$ and $\mu$ are integers, then the $\alpha_1$ pole is of $\mathcal{O}(1/\varepsilon)$ and the other poles of $\mathcal{O}(-1/\varepsilon)$. Power counting and a comparison with (11) shows that the $\alpha_1$ pole comes from the ultraviolet divergence and the $\alpha_3$ pole comes from the singularity at $q=p$. Comparison with (11) suggests that the $\alpha_0$ pole comes from the $q=0$ singularity, which can be caused by $q \cdot n=0$ as well as $q^2=0$. The absence of $\alpha_2$ poles implies that the infrared in gauge singularities at $q=0$ are entangled. There are no $q \cdot n=0$, $q \neq 0$ singularities. We therefore conclude: (a) To isolate the ultraviolet divergence it is sufficient to stay on the $\varepsilon$-line. (b) The spurious singularity at $q \neq 0$ is exorcised from every Feynman integral. (c) The spurious singularity at $q=0$ and the $q^2=0$ singularity are entangled and cannot be distinguished anywhere in the $\omega$-$\kappa$-$\mu$ hyperplane.

A generic integral in Landshoff's prescription is still (7) with $\eta^2$ finite. We set $\varepsilon'=0$ and use analytic regularization to treat the $(p-q)^2=0$ and $q^2=0$ singularities. The representation of the integral is known [6] :

$$T \equiv S(\varepsilon'=0,\eta^2 \neq 0) = \pi^{\omega}(p^2)^{\alpha_1}(n^2)^{\upsilon}(p \cdot n)^{s}/[\Gamma(-\kappa)\Gamma(-\mu)\Gamma(-\upsilon)]$$

$$\times \frac{1}{2\pi i}\int_C dt \, \frac{\Gamma(-t)\Gamma(\beta_3-t)(\eta^2/p^2n^2)^t}{\Gamma(-\alpha_0+\beta_1-2t)} \, G^{2,3}_{3,3}\left[y\left|\begin{matrix}1+\alpha_0+t, & 1+\alpha_1-t, & 1+\upsilon-t; \\ 0, & \beta_1-t; & 1/2-s\end{matrix}\right.\right], \tag{12}$$

where the $\alpha$'s and $\beta$'s are the same as in (10). A heavy price is paid in the form of an extra contour integration relative to (9) for retaining the "mass" parameter $\eta^2$. The contour C in (12) goes from $-i\infty$ to $+i\infty$ and closes clockwise. Along the imaginary t axis, it is distorted to enclose all strings of poles in the integrand that extend to $t=\infty$ and exclude all strings of poles that extend to $t=-\infty$. Exterior to the G-function, the integrand has two strings of poles that extend to infinity:

$$t = t_1 = L \,, \tag{13}$$

$$t = t_2 = \beta_3 + L \,, \tag{14}$$

where always L is a nonnegative integer. From the general expansion of a G-function in terms of hypergeometric functions [7]

$$G^{m,n}_{n+r,m+s}\left[z\left|\begin{matrix}a_{\{n\}}; & a_{\{r\}} \\ b_{\{m\}}; & b_{\{s\}}\end{matrix}\right.\right] = \sum_{h \in \{m\}} \frac{\prod_{j \in \{m\}\setminus h} \Gamma(b_j-b_h) \prod_{i \in \{n\}} \Gamma(1+b_h-a_i) z^{b_h}}{\prod_{k \in \{q\}} \Gamma(1+b_h-b_k) \prod_{\ell \in \{r\}} \Gamma(a_\ell-b_h)}$$

$$\times \; {}_{n+r}F_{m+s-1}\left(\begin{matrix}1+b_h - a_{\{n+r\}} \\ 1+b_h - b_{\{m+s-1\}}\end{matrix}\middle| (-)^{r-n} z\right) \tag{15}$$

the G-function in (12) has the form

$$\frac{G(y|\cdots)}{\Gamma(-\mu)\Gamma(\beta_1-\alpha_0-2t)} \sim \frac{\Gamma(\beta_1-t)\Gamma(-\alpha_0-t)\Gamma(t-\alpha_1)\Gamma(t-\upsilon)}{\Gamma(-\mu)\Gamma(\beta_1-\alpha_0-2t)\Gamma(s+1/2)}\; {}_3F_2\left(\begin{matrix}-\alpha_0-t,\ t-\alpha_1,\ t-\upsilon \\ 1-\beta_1+t,\ \ s+1/2\end{matrix}\middle| y\right)$$

$$+\, y^{\beta_1-t}\,\frac{\Gamma(t-\beta_1)\Gamma(-\alpha_3)}{\Gamma(\beta_1-\alpha_3-t)}\; {}_3F_2\left(\begin{matrix}-\alpha_3,\ -\mu,\ \alpha_1-\alpha_0-2t \\ 1+\beta_1-t,\ 1/2+s+\beta_1-t\end{matrix}\middle| y\right), \tag{16}$$

where the hypergeometric ${}_3F_2$ functions have the expansion

$$ {}_mF_n\left(\begin{matrix}a_{\{m\}} \\ b_{\{n\}}\end{matrix}\middle| z\right) = \sum_{\ell=0}^{\infty} \prod_{i\in\{m\}} \prod_{j\in\{n\}} \frac{\Gamma(a_i+\ell)}{\Gamma(a_i)}\,\frac{\Gamma(b_j)}{\Gamma(b_j+\ell)}\,\frac{z^\ell}{\ell!}, \tag{17}$$

In (15) and (17) $a_{\{m\}}$ means the set $a_1,.., a_m$. It is a general property of the G-function that all its poles reside in the factor $\Gamma(1+b_h-a_i)$ on the right-hand side of (15); residues of other apparent poles cancel among themselves exactly. Thus the right-hand side of (16) has only one string of poles extending to $t=\infty$:

$$t = t_3 = -\alpha_0 + L \tag{18}$$

and two strings of poles extending to $t=-\infty$:

$$t = t_4 = -\alpha_1 - L\ , \tag{19}$$

$$t = t_5 = \upsilon - L \tag{20}$$

The contour C therefore looks like

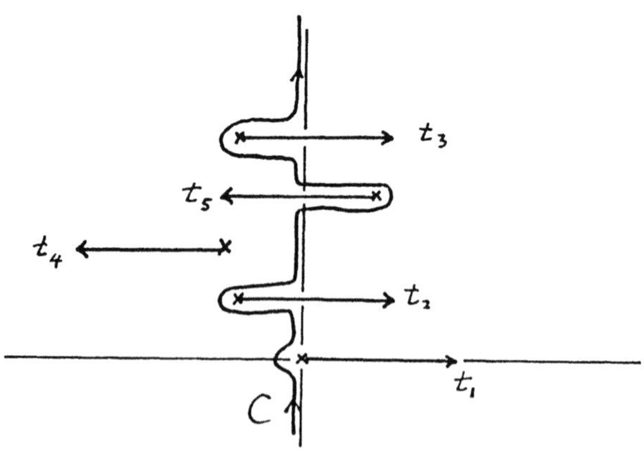


for the first term in (16), and

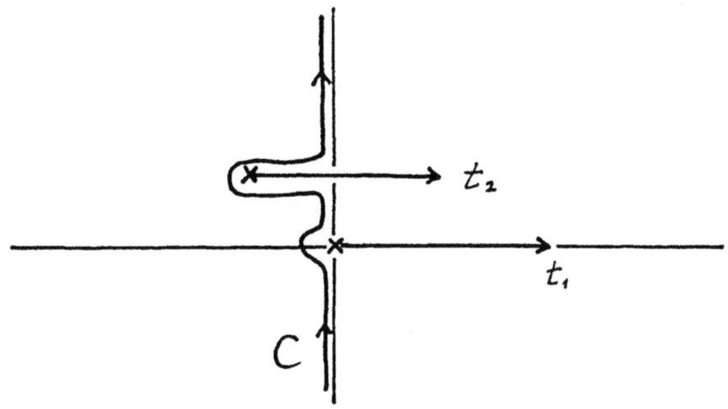


for the second term in (16).

The evaluation of the integral is now reduced to that of evaluating the residues of the integrand at the strings of poles $t=t_i$, $i=1$ to 3. Each of the residues will have a factor $(\eta^2)^{t_i}$. Since we are interested in the limit $\eta^2 \to 0$, only poles at $t_i \leq 0$ need be considered. (In higher-loop calculations involving multiple integrations, terms with positive powers of $\eta$ in a specific integration may also need be considered. Such factors of $\eta$ may be multiplicatively cancelled by factors of negative powers of $\eta$ coming from another integration.) In the following we discuss the contribution from each of the three strings of poles.

(A) The $t_1$ string. The sum of residues at poles in this string is a nonnegative power series in $\eta^2$, with the only nonvanishing (at $\eta=0$) term coming from the pole at $t_1=0$. It is easy to show that this term, which we denote by $T_1$, is identical to the result given by the analytic representation (9), *i.e.*, $T_1=S_0$. Furthermore, only terms from this string can be UV divergent.

(B) The $t_2$ string. Residues from this string may have negative powers of $\eta^2$ when $\beta_3 \leq 0$. Because $\beta_3$ is a half integer, the residue has no other poles from the first term on the right-hand side of (16), but has poles of $\mathcal{O}(1/\varepsilon_3)$ from the $\Gamma(-\alpha_3)$ factor in the second term, due to the $(p-q)^2=0$ singularity in the original integral. The sum of residues from this string, denoted by $T_2$, has the form

$$T_2 = \sum_{L=0} \left\{ \text{functions of } \omega, \kappa, \text{ etc. including possibly poles of } \mathcal{O}(1/\varepsilon_3) \right\} (\eta^2)^{\beta_3+L} . \tag{21}$$

By identifying N+s in (6) with $-\upsilon$ we note [6] that all terms in (21) of negative power in $\eta^2$ are annihilated by the differential operator $R(\eta^2)$ given by (7). On the other hand, $R(\eta^2)$ leaves the $\mathcal{O}((\eta^2)^0)$ term unchanged. Thus

$$\lim_{\eta\to 0} R(\eta)\Big|_{N+s=-\upsilon} (T_1+T_2) \lim_{\eta\to 0} T_1 = S_0 , \tag{22}$$

which says that if the Landshoff integral T were composed of only the residues $T_1$ and $T_2$, then the analytic integral $S_0$ and the PV integral would be equivalent.

In general, however, residues from the strings $t_3$ do not vanish.

(C) The $t_3$ string. There may be terms with negative powers of $\eta$ when $\alpha_0 \leq 0$. The residue is proportional to

$$\sum_{\ell=0} (\eta^2)^{L+\ell-\alpha_0} \frac{\Gamma(\beta_1+\alpha_0-L+\ell)\Gamma(1-\beta-\alpha_1+L_0)\Gamma(-\alpha_0-\alpha_1+L+\ell)\Gamma(-\alpha_0-\upsilon+L+\ell)}{\Gamma(-\upsilon)\Gamma(-\mu)\Gamma(-\kappa)\Gamma(\beta_1+\alpha_0-2L)\Gamma(1-\beta_1-\alpha_0+L)}, \tag{23}$$

whose only poles reside in the factor

$$\Gamma(-\alpha_0-\upsilon+L+\ell)(\eta^2)^{-\alpha_0+L+\ell} \quad . \tag{24}$$

The poles, of $\mathcal{O}(1/\varepsilon_2)$, are due to the singularity of the original integral at $q=0$, excluding the contribution from the factor $[(q\bullet n)^2+\eta^2]^{\upsilon}$. Here we see that the gauge-fixing singularity at $q\bullet n=0$, $q\neq 0$, and the infrared singularity at $q=0$ are completely separated. Herein lies the most important difference between the Landshoff and the other two prescriptions.

We remark in passing that as a result of analytic continuation of $\varepsilon=\omega-2$ and the exponents $\kappa$ and $\mu$, the parameters $\alpha_i$ and $\beta_i$ are not exact integers. This gives meanings to the gamma functions at their respective poles, and yields terms of order $\log(\eta^2)$ when the exponents in (24) are of $\mathcal{O}(0)$. In (9) and (12) (and (26); see below), depending on the values of $\kappa$, $\mu$, $\upsilon$ and s, any integral that is singular can be at most a finite power series in y; only a nonsingular integral may be an infinite series in y. The integral is also an infinite power series in $\eta^2$ with the exponents bounded from below. Thus it is effectively a finite series in $\eta^2$ in the $\eta\to 0$ limit. Tables of integrals (9) are available [8], as are similar tables of integrals for the light-cone gauge [9].

Denote the sums of residues from the strings of poles $t_3$ by $T_3(\eta^2)$, and denote the PV integral by $S_{PV}(\eta^2)$. Then the $S_{PV}(\eta^2)$, the Landshoff integral $T(\eta^2)$ and the $\eta^2$-independent analytic integral $S_0$ are related, from (6) and (22), by

$$\lim_{\eta\to 0} S_{PV}(\eta^2) = \lim_{\eta\to 0} R(\eta^2)\, T(\eta^2) = S_0 + \lim_{\eta\to 0} R(\eta^2)\, T_3(\eta^2) \ . \tag{25}$$

$S_{PV}(\eta^2)$ will in general have terms of negative power in $\eta^2$ except when $T_3(\eta^2)$ has no such terms. In any one-loop calculation, such terms can only cancel linearly. It follows that if an amplitude calculated in the Landshoff prescriptions has a finite $\eta\to 0$ limit at the one-loop order, then the three prescriptions must give the same result for that amplitude at the one-loop order.

We briefly discuss the representation for (8) in the generalized Feynman prescription when both $\varepsilon'$ and $\eta$ are not zero. The result is [10]

$$S(\varepsilon'\neq 0,\ \eta=i\varepsilon' n^2\neq 0) = \pi^{\omega}(p^2)^{\alpha_1}(n^2)^{\upsilon}(p\bullet n)^{s}/[\Gamma(-\kappa)\Gamma(-\mu)\Gamma(-\upsilon)]$$

$$\times \frac{1}{2\pi i}\int_c dt \frac{\Gamma(-t)\Gamma(\beta_3-t)(i\varepsilon'/p^2)^t}{\Gamma(-\alpha_0+\beta_1-2t)} G^{2,3}_{3,3}\left[y\left|\begin{matrix}1+\alpha_0+t, & 1+\alpha_1-t, & 1+\upsilon; \\ 0, & \beta_1-t; & 1/2-s+t\end{matrix}\right.\right], \tag{26}$$

where the notation is the same as in (12). An analysis similar to that given after (12) has been made. Suffice to say that, by converting $n^2$ to $\eta^2/i\varepsilon'$, the integral can be expressed as a power series in $i\varepsilon'$ and $\eta^2$. The integral has singularities coming either from a factor of $\Gamma(-\alpha_1)$ when $\alpha_1$ is close to a nonpositive integer - this is the UV divergence, or from terms of order $(i\varepsilon')^{\alpha}(\eta^2)^{\beta}$ when either $\alpha$ and $\beta$ is close to a nonpositive integer. The infrared singularities are given by terms with $\alpha \le \mathcal{O}(0)$, and the gauge singularities by terms with $\beta \le \mathcal{O}(0)$. The difference between this prescription and the Landshoff prescription is that here the $q^2=0$ and $q=p$ singularities are given by nonpositive powers in $i\varepsilon'$, whereas in the Landshoff prescription (12) they are given by poles in gamma functions. We feel that two prescriptions are equivalent, but that the Landshoff prescription is more elegant. The one-loop order gluon self-energy in a Yang-Mills theory was computed [10] using (26). The result is identical to the PV and analytic result [6]; it has only UV divergences, but not $i\varepsilon'$ and $\eta^2$ singularities. We assume the same would be true if (12) were used. This implies that at least for the one-loop order gluon self-energy, the PV, analytic, Landshoff and generalized Feynman prescription give identical results.

At the two-loop level, finite $\eta\to 0$ terms can be given by a multiplicative cancellation such as

$$(\mathcal{O}(\eta^{-2}) \text{ from inner loop})\ (\mathcal{O}(\eta^2) \text{ from outer loop}) \tag{27}$$

or by a combination of multiplicative and linear cancellations such as

$$\mathcal{O}(\eta^{-4}) \cdot \mathcal{O}(\eta^2) - \mathcal{O}(\eta^{-2}) \,. \tag{28}$$

This is precisely what was demonstrated to happen by Landshoff in his two-loop order calculation of the Wilson loop.[4] Because in (27,28) the $\mathcal{O}(\eta^2)$ term can be UV divergent, it follows that at two-loop order the analytic PV and Landshoff prescriptions will in general give different results not only for the IR divergence, but also for the UV divergence. At higher-loop orders finite $\eta\to 0$ terms can be given by more elaborate cancellations, and it would be unrealistic to assume that such cancellations would not be generic to all higher-loop orders.

We summarize as follows. *(0)* The Landshoff (and generalized Feynman) prescription is the only prescription that cleanly separates the ultraviolet, infrared and gauge singularities. (i) The analytic representation (9) is the $\eta\to 0$ limit of the Landshoff prescription with all the latter's negative power $\eta$ and $\log(\eta)$ terms *truncated.* (ii) The PV prescription is a differentiation operator in $\eta^2$ acting on the Landshoff prescription. For a given Feynman integral, the two prescriptions give distinct power series in $\eta^2$. (iii) At the one-loop order, the analytic, PV and

Landshoff prescriptions all give identical UV divergences which are finite at $\eta \to 0$. (iv) For any amplitude that is finite at $\eta \to 0$, the three prescriptions give identical results at the one-loop order. In this case the analytic prescription is by far the simplest to use. (v) At two-loop order and higher, owing to the difference in the multiplicative cancellation of terms that are divergent at $\eta \to 0$, the three prescriptions will in general give different results that are finite at $\eta \to 0$, including their descriptions of the UV divergences. (vi) Since the Landshoff prescription gives the correct answer for the Wilson loop at the two-loop order, and since the result [4] depends on multiplicative cancellation of $\eta$-dependent terms that are divergent when $\eta \to 0$, it can be surmised that the PV and analytic prescriptions cannot give the correct answer at two-loop order. (vii) From calculations [4,11,12] of the Wilson-loop at two-loop order, it appears that the Landshoff prescription, or something equivalent to it -- such as the Vienna prescription [12] -- is needed for higher-loop order perturbative calculations in the axial gauges. This unfortunately implies that such calculations will be inherently complicated, because the correct result will depend on multiplicative cancellations of terms that are individually divergent when $\eta \to 0$.

I thank Mike Milgram for numerous discussions, Peter Landshoff for a conversation on the importance of separating the $p^2=0$ and $p \cdot n$ singularities at $p=0$, and the Organizers of the Workshop on Physical and Nonstandard Gauges from the Technische Universität Wien for their kind hospitality during the workshop.

## References

[1] W. Kummer, Acta. Phys. Austriaca **15**(1961)149; R. Delbourgo, A. Salam and J. Strathdee, *Nuovo Cimento A23(1974)237;* W. Konetschny and W. Kummer, *Nucl. Phys.* ***B100****(1975)106.*

[2] J.C. Taylor, **Gauge Theories of Weak Interactions**, (Cambridge U. Press, 1976), Chapter 13.

[3] W. Kummer, *Acta Phys. Austriaca* **41***(1975)315.*

[4] P.V. Landshoff, *Phys. Lett.* ***B169****(1986)69.*

[5] S. Mandelstam, *Nucl. Phys.* ***B213****(1983)149;* G. Leibbrandt, *Phys. Rev.* ***D29****(1984) 1699;* H.C. Lee & M.S. Milgram, *Phys. Rev. Lett.* **50***(1985)2122; Nucl. Phys.* **B268** *(1986)5443.*

[6] H.C. Lee & M.S. Milgram, *Ann. Phys.* **157***(1984)408.*

[7] Y.L. Luke, **The Special Functions and Their Approximations**, Vol. 1 *(Academic Press, 1969).*

[8] M.S. Milgram & H.C. Lee, *J. Comp. Phys.* **59***(1985)331.*

[9] M.S. Milgram & H.C. Lee, *J. Comp. Phys.* **71***(1987)316.*

[10] M.S. Milgram, *private communication.*

[11] H. Hüffel, P.V. Landshoff & J.C. Taylor, *Phys. Lett.* **217***(1989)147.*

[12] P. Gaigg & M. Kreuzer, *Phys. Lett.* ***B205****(1988)530;* G. Pollak, TU-Wien preprint (1989), *to appear in Phys. Rev. D.*

# Trouble Even with the Coulomb Gauge

J.C. Taylor

Department of Applied Mathematics and Theoretical Physics,
University of Cambridge, Cambridge CB3 9EW, England

**Abstract**

The Coulomb gauge in non-abelian gauge theories seems to be a very good example of a physical gauge. It is manifestly unitary: although there are ghosts, their propagators have no poles. The Coulomb propagator $(\mathbf{k}^2)^{-1}$ is well defined for all nonzero $\mathbf{k}$. Also, collinear singularities are confined to diagonal transverse gluons.

Nevertheless, there are problems concerned with the convergence of the energy integrals. These are connected with the operator-ordering of the Hamiltonian, as observed by Christ and Lee. These problems would be soluble, were it not for the existence of ordinary ultra-violet divergences along with the above energy divergences. This means that a complete treatment of the Coulomb gauge has not yet been given.

## Introduction

The Coulomb gauge deserves attention at this workshop. The main point in its favour is that none of the problems concerned with the definition of integrals like

$$\int d^4k \frac{1}{(n.k)^2} \cdots \tag{1}$$

appear in the definition of integrals like

$$\int d^4k \frac{1}{\mathbf{k}^2} \cdots \tag{2}$$

in the Coulomb gauge.

Apart from the above obvious point, the advantages of the Coulomb gauge are:- (a) There are no ghost states. (The ghost propagator $(\mathbf{k}^2)^{-1}$ is instantaneous and has no pole.) (b) There is a (positive) Hamiltonian, and unitarity is manifest. (c) Collinear divergences occur only in diagonal diagrams (ladder diagrams), and this fact should facilitate general proofs of factorization etc in perturbative QCD. (d) It is convenient in bound-state problems.

The disadvantages are:- (i) The Hamiltonian ((4) below) is non-local and non-polynomial. (ii) The BRST identities do not reduce to simple QED-like identities. (iii) A complete discussion of renormalization would probably be rather complicated.

These disadvantages seem to me to be complications rather than real obstacles to the use of the gauge. But, to 3-loop order and beyond, there is an obstacle which I do not know how to overcome. The purpose of this talk is to describe this obstacle.

## 2 The coulomb gauge

The gauge is obtained (formally) from the path-integral with $J\delta(\partial.\mathbf{A})$ in it, where $J$ is the Faddeev-Popov determinant, $J = \det(\partial.\mathbf{D})$. One then obtains a transverse propagator

$$(k^2 + i\epsilon)^{-1}\left[\delta_{ij} - \frac{k_i k_j}{\mathbf{k}^2}\right],$$

an instantaneous Coulomb propagator $(\mathbf{k}^2)^{-1}$ and a ghost propagator $(\mathbf{k}^2)^{-1}$, which we shall denote by a wavy line, a continuous line and a dashed line respectively, like this:

In this notation, some of the vertices are

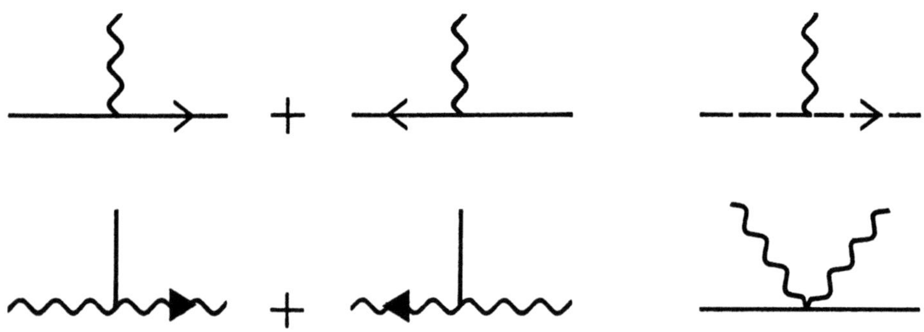

Here an open arrow-head denotes a spatial momentum, and a black arrow-head denotes the time-component of a momentum.
To one-loop order, there are graphs which are linearly divergent in the integral over the time-component of the momentum of integration (i.e. with an integrand which tends to a constant as that component tends to infinity, with the space-components fixed). Examples are

which exactly cancel; and

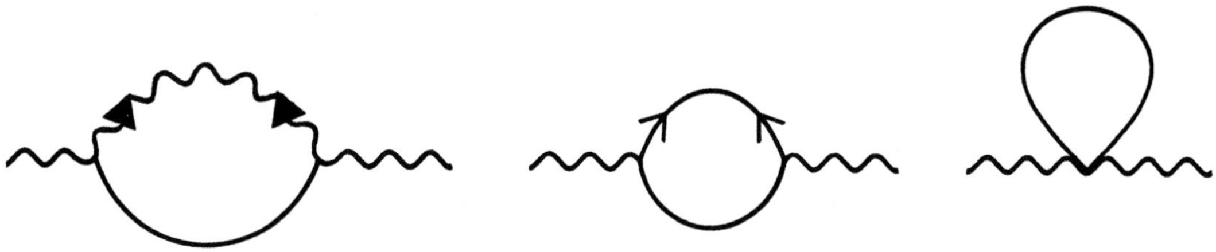

where there is a cancellation resulting in the effective propagator

$$\frac{\mathbf{k}^2}{k^2 + i\epsilon}\left[\delta_{ij} - \frac{k_i k_j}{\mathbf{k}^2}\right] \tag{3}$$

which does make the $k_0$-integral convergent.

One can show diagramatically [1] that these types of cancellation occur generally; but a formal way to implement the cancellations is to pass to the Hamiltonian (phase-space) formalism. Then there is a momentum field $\mathbf{\Pi}$ conjugate to the transverse field $\mathbf{A}$ (colour indices are suppressed). One requires the $\mathbf{A}\,\mathbf{A}$, $\mathbf{A}\,\mathbf{\Pi}$ and $\mathbf{\Pi}\,\mathbf{\Pi}$ propagators. The fourier transform of the last is proportional to (3). There are no ghost lines and no *closed* Coulomb lines. In place of these, the Hamiltonian contains the non-abelian version of the Coulomb interaction, so that the complete Hamiltonian is

$$\frac{1}{2}\int d^3\mathbf{x}[\mathbf{\Pi}^2 + \mathbf{B}^2 - \mathbf{D}.\mathbf{\Pi}(\partial.\mathbf{D})^{-1}\partial^2(\partial.\mathbf{D})^{-1}\mathbf{D}.\mathbf{\Pi}] \tag{4}$$

(where we use a compact and over-simplified notation). Note that $\mathbf{D}.\mathbf{\Pi}$ is the colour charge density ($\mathbf{D}$ is the covariant derivative, and for simplicity we have omitted quarks).

Of course, in perturbation theory the factors $(\partial.\mathbf{D})^{-1}$ in (4) have to be expanded as infinite series in the coupling-constant. This gives rise to terms of which a typical example is represented by the graph

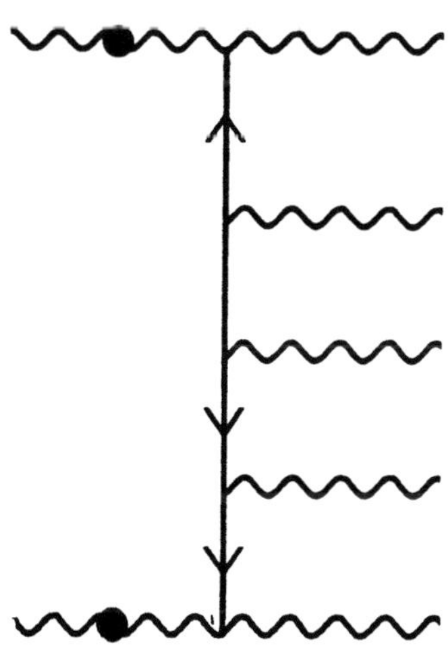

in which the gluon lines with the black blob denote the $\mathbf{\Pi}$ field. Thus the Coulomb term in the Hamiltonian may be considered as giving rise to open Coulomb lines but not closed Coulomb loops.

**2-loop complications**

The Hamiltonian form makes automatic the cancellation of the linear energy-divergences but there are still logarithmic energy-divergences to 2-loop order in graphs like

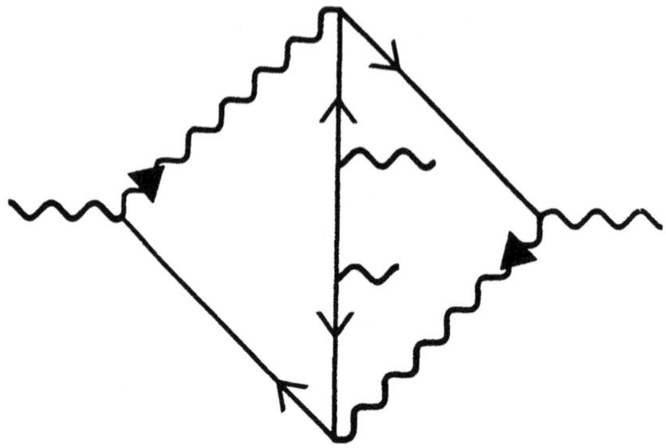

Fig.1

which contains an integral of the form

$$\int d^3\mathbf{p}d^3\mathbf{q}F(\mathbf{p},\mathbf{q})\,W \tag{5}$$

where

$$W = \int dp_0 dq_0\, p_0 q_0\, (p_0^2 - \mathbf{p}^2)^{-1}\,(q_0^2 - \mathbf{q}^2)^{-1}. \tag{6}$$

It can be shown by power counting [2] that this type of divergence occurs just at 2-loop order (or to higher order when there is such a 2-loop graph as a subgraph).

The integral

$$\int dp_0\, p_0\, (p_0^2 - \mathbf{p}^2)^{-1}$$

is obtained from the $\mathbf{A\Pi}$ propagator evaluated at equal times; so it is not too surprising that (6) is connected with an operator-ordering problem in the Hamiltonian (4). In fact it has been shown that (4) is not stricly correct [3,4]. it should be

$$\frac{1}{2}\int d^3\mathbf{x}\, J^{-1}[\mathbf{\Pi}J.\mathbf{\Pi} + \mathbf{B}^2 - \mathbf{D}.\mathbf{\Pi}J(\partial.\mathbf{D})^{-1}\partial^2(\partial.\mathbf{D})^{-1}\mathbf{D}.\mathbf{\Pi}] \tag{7}$$

where $J$ is the Faddeev-Popov determinant (which is a functional of $\mathbf{A}$ and so does not commute with $\mathbf{\Pi}$).

Christ and Lee [3] wrote the Hamiltonian (7) as the sum of a Weyl-ordered Hamiltonian and an extra term $V$. (Weyl-ordering means an averaging over different orders, e.g. $f(q)p^2$ Weyl-ordered is $\frac{1}{4}[f(q)p^2 + 2pf(q)p + p^2 f(q)]$.) They argued that the correct Feynman rules corresponding to the Weyl-ordered Hamiltonian were such as to set the divergent integral (6) to be zero. But the term $V$ was extra. The contributions from $V$ may be denoted graphically, and an example is

$\frac{1}{3}$

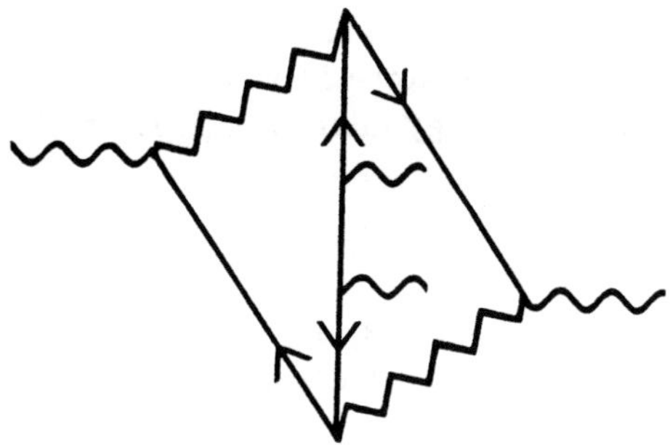

Fig.2

in which we use the zig-zag line to denote

$$\pi\delta(p_0)\left[\delta_{ij}-\frac{k_ik_j}{\mathbf{k}^2}\right].$$

Note that, because of the $\delta(p_0)\delta(q_0)$, Fig.(2) gives only 3-dimensional momentum integrals. Since the Coulomb propagators are instantaneous, the external gluon fields are all evaluated at the same time.

Another approach to the divergence in (6) is to regulate it by going to a "flow-gauge" depending upon a parameter $\theta$ and giving the Coulomb gauge in the limit $\theta \to 0$ [2,4]. One can, for example, arrange that the Coulomb propagator is replaced by

$$(\mathbf{k}^2-\theta^2k_0^2)^{-1}$$

and the transverse propagator by

$$(k^2)^{-1}\left[\delta_{ij}-(1-\theta^2)\frac{k_ik_j}{\mathbf{k}^2-\theta^2k_0^2}\right]$$

(and there are also ghosts for $\theta \neq 0$). This flow-gauge interpolates between the Coulomb gauge and the Feynman gauge, giving the latter at $\theta = 1$.

For $\theta \neq 0$ such a flow-gauge has no advantages over the Feynman gauge; so there is no point in it for practical purposes. However, it may be useful as a tool for proving properties of the Coulomb gauge. Cheng and Tsai [4] used it to show that the Christ-Lee terms like Fig.2 can be obtained from combinations of the $\theta$-regulated integrals like (6). This approach was generalized by Doust [2].

This work can be summarized even without using a flow-gauge [2]. It can be shown that if one combines sets of graphs like Fig.1 , obtained by permuting the attachments of the external gluon lines, the integrals like (6) combine into the convergent integral

$$\int dp_0dq_0dr_0\,\delta(p_0+q_0+r_0)\left[\frac{p_0q_0}{p^2q^2}+\frac{q_0r_0}{q^2r^2}+\frac{r_0p_0}{r^2p^2}\right]=\pi^2. \tag{8}$$

In this way, one gets the Christ-Lee terms like Fig.2. Thus one may use the naive Hamiltonian (4), without attention to operator-ordering, provided that graphs are judiciously combined so that the divergent energy integrals appear in the combination (8).

## 3-loop difficulties

Problems seem to arise when the energy-divergences described in Section 3 are combined with ordinary UV divergences [5]. For definiteness, I shall use dimensional regularization to control the UV divergences.

In order to construct the simplest possible example of these difficulties, I shall now introduce fermions into the theory, and consider UV divergences in fermion loop sub-graphs. Consider the following graph

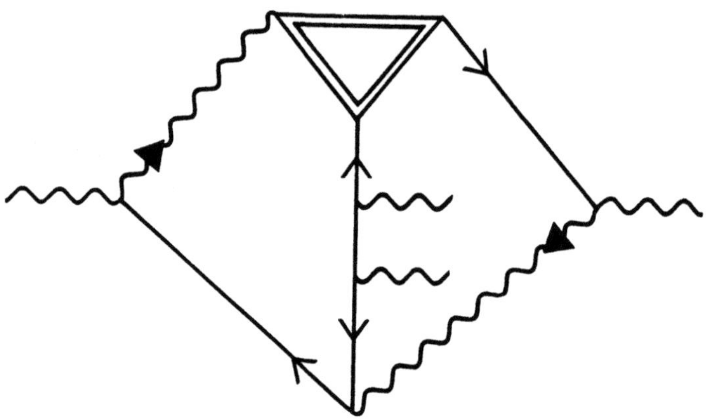

Fig.3

(where the doubled line denotes a fermion). This graph has an UV divergent, fermion-loop vertex part inserted into the graph of Fig.1. We would expect the UV divergence to be cancelled by the renormalization of Fig.1. However, this latter graph contains energy-divergences and is not properly defined on its own. It leads to the Christ-Lee graph Fig.2 ,which is not an ordinary Feynman graph .

Fig.3 on the other hand does not contain energy-divergences. This is certainly true if the $dp_0 dq_0 dr_0$-integrals are done before any of the 3-momentum integrals, because the extra fermion propagator denominators provide convergence factors. Even if the fermion loop integrals are all done first, the result is an integral like (4), with $W$ replaced by

$$W' = \int dp_0 dq_0 \, p_0 q_0 \, f(p_0, q_0, \mathbf{p}, \mathbf{q}) \, (p_0^2 - \mathbf{p}^2)^{-1} (q_0^2 - \mathbf{q}^2)^{-1} \tag{9}$$

where $f$ is the contribution from the fermion loop. In dimensional regularization, with space-time dimension $n < 4$,

$$f \to 0 \quad \text{as} \quad p_0, q_0 \to 0,$$

thus providing a convergence factor in (9).

Hence we encounter the problem of how to absorb the UV divergences of the ordinary Feynman graph in Fig.3 by the renormalization of the Christ-Lee term represented by Fig.2. To do this, one must show that the coefficient of the $(4-n)^{-1}$ term in $f$ in $W'$ in (9) has the Christ-Lee form.

This could be done if one could apply Doust's symmetrization trick (8) to (9) in the same way as we did to $W$ in (6). This would work if the vertex function $f$ in (9) were symmetric in $p, q, r$. But unfortunately this is not the case. For example, one term in $f$ is proportional to (written as an integral over Feynman parameters)

$$\frac{1}{4-n}\int_0^1 dxdydz\delta(1-x-y-z)\left(yzp^2+zxq^2+xyr^2\right)^{\frac{n-4}{2}}x. \tag{10}$$

The pole term is just $\frac{1}{6}\frac{1}{4-n}$ and is indeed symmetrical. However, the rest of (10) is not symmetric; and if Doust's symmetrization procedure is applied to (9) with (10) for $f$, we obtain, instead of (8),

$$\int dp_0dq_0dr_0\,\delta(p_0+q_0+r_0)\frac{1}{4-n}\int_0^1 dxdydz\,\delta(1-x-y-z)$$
$$\left(yzp^2+zxq^2+xyr^2\right)^{\frac{n-4}{2}}\left[\frac{zp_0q_0}{p^2q^2}+\frac{xq_0r_0}{q^2r^2}+\frac{yr_0p_0}{r^2p^2}\right]. \tag{11}$$

The interesting part of (11) is

$$\int dp_0dq_0dr_0\,\delta(p_0+q_0+r_0)\left(p^2q^2r^2\right)^{-1}\frac{1}{4-n}\int_0^1 dxdydz\,\delta(1-x-y-z)$$
$$\left(yzp^2+zxq^2+xyr^2\right)^{\frac{n-4}{2}}\left(xp_0+yq_0+zr_0\right). \tag{12}$$

Here the $dxdydz$ integral has a zero at $n=4$ since $p_0+q_0+r_0=0$. In spite of this, however, I think that in the complete (12) the pole at $n=4$ does survive. To understand this, imagine doing the $dp_0dq_0dr_0$ integrals in (12) first. The result will be of the form

$$\frac{1}{4-n}\int_0^1 dxdydz\,\delta(1-x-y-z)\,g(x,y,z). \tag{13}$$

There is no reason to expect $g(x,y,z)$ to have a zero at $n=4$; so that the pole at $n=4$ in (12) will remain. This pole is not the UV divergence in the fermion loop, but is generated because (12) would contain an energy-divergence if $n=4$.

I have not been able to compute the residue of the pole at $n=4$ in (12), so I will give a simple model to illustrate the phenomena which I think occurs here. Consider the integral

$$\int dp_0dq_0dr_0\,\delta(p_0+q_0+r_0)\,\frac{p_0q_0}{p^2q^2}\frac{1}{4-n}\left[\left(r_0^2\right)^{\frac{n-4}{2}}-\left(p_0^2q_0^2\right)^{\frac{n-4}{2}}\right]. \tag{14}$$

The integrand has no pole at $n=4$. However, when the $dp_0dq_0dr_0$ integrals are done, the contribution from the second term in the square-bracket is zero by symmetry, whereas the contribution from the first term is non-zero and gives the same pole as

$$\frac{1}{4-n}\lim_{\Lambda\to\infty}\int dp_0 dq_0 d\frac{p_0 q_0}{p^2 q^2}\,\theta(\Lambda - |p_0+q_0|)$$

$$= \frac{1}{4-n}\lim\int_{-\Lambda}^{\Lambda}\frac{dp_0}{p_0}\ln\left(\frac{\Lambda+p_0}{\Lambda-p_0}\right) = \frac{1}{4-n}\frac{\pi^2}{3}.$$

So the integral does have a pole.

I conclude that it is far from obvious how to disentangle the UV divergences from the potential energy-divergences in Fig.3, and how to cancel the poles at $n = 4$.

Of course, if *all* the Coulomb-gauge graphs which contribute to a particular S-matrix element are added, it should be possible to combine them so as to get the same as the Feynman gauge, which can be renormalized. But, if one has to do this, there is no point in using the Coulomb gauge in the first place. Also, even granted this about S-matrix elements, it is not clear whether the off-shell Greens functions in the Coulomb gauge are renormalizable or not.

*References*

[1] R.N. Mohapatra, *Phys.Rev.* *D*4 (1971) 378, 1007.
[2] P. Doust, *Annals of Physics* **177** (1987) 169.
[3] N.H. Christ and T.D. Lee, *Phys.Rev.* *D*22 (1980) 939.
[4] H. Cheng and E.C. Tsai, *Phys.Lett* *B*176 (1986) 130; *Phys.Rev.Lett* **57** (1986) 511; "The anomalous Coulomb interaction" (MIT preprint).
[5] P. Doust and J.C. Taylor, *Phys.Lett* **197** (1987) 232.

# Temporal Gauge Propagator from the Coulomb Gauge[1]

A. C. Kalloniatis and R. J. Crewther

Department of Physics and Mathematical Physics,
University of Adelaide, GPO Box 498,
South Australia, 5001, Australia

**Abstract:** We derive the propagator for the abelian gauge field in the temporal gauge using parallel transport from the Coulomb gauge, the quantisation having already been done in the Coulomb gauge. The resulting prescription for the traditionally problematic spurious pole in the longitudinal propagator falls into the class of prescriptions in the literature which break time-translational invariance.

## 1 Introduction

The problem of prescripting the spurious $1/k_0$ pole in the gauge field propagator in the temporal gauge, as with all its axial gauge brethren, has been a vexed question for some years. The Cauchy Principal Value (CPV) prescription was most popular amongst authors until 1982, when Caracciolo, Curci and Menotti (CCM) [1] found that the CPV failed to give the correct exponentiation to fourth order in the coupling constant of the rectangular Wilson loop. Splitting the propagator into its transverse and longitudinal parts, they found the time part of the longitudinal propagator which satisfied the Wilson loop criterion was of the form

$$D(x_0, y_0) = -1/2|x_0 - y_0| \pm 1/2(x_0 + y_0) + \gamma \,, \tag{1}$$

where $\gamma$ may take any value.

Since then a number of authors have derived expressions in agreement with or similar to the CCM propagator[2], which lacks invariance under time translations. In particular, Leroy, Micheli and Rossi (LMR)[3], employ the Feynman kernel formalism derived by Rossi and Testa [4] and use a supplementary gauge-fixing condition $\partial_i A_i(\mathbf{x}, t_0) = 0$ for

[1] Talk presented by A. C. Kalloniatis

some fixed time, $t_0$,to break the residual gauge freedom under time-independent gauge transformations after the temporal gauge choice, $A_0 = 0$. They derive the time-part of the longitudinal propagator to be

$$D(x_0, y_0) = -1/2\big(|x_0 - y_0| - |x_0 - t_0| - |y_0 - t_0|\big) . \tag{2}$$

This expression has also been derived by stochastic methods by Lavelle and Schaden [5]. Note that for $t_0 = \gamma \sim \pm\infty$ (2) agrees with the CCM propagator. LMR claim, contrary to Girotti and Rothe (GR)[6], that to take the time to this limit is a dangerous procedure for calculations greater than one-loop order which, however, has no bearing on the Wilson loop to order $g^4$.

In this paper we rederive the LMR form of the propagator in the abelian theory in this case by treating the problem in terms of a parallel transport (gauge-transformation) of the operators from the Coulomb gauge to the temporal gauge, *where the operators have already been quantised in the Coulomb gauge*. Using the form of the propagators in the former gauge we obtain the temporal gauge boson propagator for the abelian case. In the discussion, we briefly comment on the difference between our approach and that of other authors.

## 2 Derivation of Propagator

Working in abelian theory, we use a metric of signature $(+---)$ and denote the Coulomb gauge potentials by $C_\mu(x)$, satisfying $div\mathbf{C} = 0$, which we know to exist uniquely and whose propagator contains no poles except the standard Feynman pole; we thus have a gauge and Green's function which we claim to know how to handle. The assumption is then that we have in fact performed the quantisation in the Coulomb gauge in the standard way. This is in divergence with standard practice where one shows that a particular gauge choice one is interested in (the temporal gauge) exists in the classical theory, and then proceeds to the quantisation. This approach is, to our knowledge, rather novel, but leads to a simple derivation for the propagator.

The parallel transport along a time-like curve (line) is, in general, given by

$$A_\mu(x) = U^{-1}C_\mu(x)U + igU^{-1}\partial_\mu U , \tag{3}$$

where $U$ may be written

$$U(x_0, \mathbf{x}) = Pexp(-1/ig \int_\gamma^{x_0} d\tau C_0(\tau, \mathbf{x})) . \tag{4}$$

Such a choice automatically ensures $A_\mu(x)$ satisfies the temporal gauge condition, $A_0 = 0$, as $U$ satisfies the differential equation

$$\partial U/\partial x_0 = -(1/ig)C_0 U . \tag{5}$$

In the abelian case, the non-zero components of (1) reduce to

$$A_i(x) = C_i(x) - \partial_i^x \int_\gamma^{x_0} d\tau C_0(\tau, \mathbf{x}) \ . \tag{6}$$

We now construct the temporal gauge propagator in a manner consistent with the Dyson-Wick expansion, as a time-ordered product of the vector fields

$$D_{ij}^{temp}(x,y) \equiv \theta(x_0 - y_0)\langle A_i(x)A_j(y)\rangle + \theta(y_0 - x_0)\langle A_j(y)A_i(x)\rangle \ , \tag{7}$$

where the unordered (Wightman) functions are easily evaluated

$$\begin{aligned}\langle A_i(x)A_j(y)\rangle &= \langle [C_i(x) - \partial_i^x \int_\gamma^{x_0} d\tau C_0(\tau,\mathbf{x})][C_j(y) - \partial_j^y \int_\gamma^{y_0} d\tau' C_0(\tau',\mathbf{y})]\rangle \\ &= \langle C_i(x)C_j(y)\rangle + \partial_i^x\partial_j^y \int_\gamma^{x_0} d\tau \int_\gamma^{y_0} d\tau' \langle C_0(\tau,\mathbf{x})C_0(\tau',\mathbf{y})\rangle \ ,\end{aligned} \tag{8}$$

the "mixed" terms vanishing as the one point functions $\langle C_i(x)\rangle = 0$. In the standard approach, $C_0$ is not expanded in *dynamical* fields. Inserting (8) and its reversed function into the ordered function (7), the various terms conspire, along with the fact that the instantaneous propagator is itself *unordered*, to return the original Coulomb gauge propagator components. Thus

$$D_{ij}^{temp}(x,y) = D_{ij}^{Coul}(x-y) + \partial_i^x\partial_j^y \int_\gamma^{x_0} d\tau \int_\gamma^{y_0} d\tau' D_{00}^{Coul}(x-y) \ . \tag{9}$$

Each term in this expression involves a well-defined Green's function; we can be confident then that the procedure has given an unambiguous propagator. Using well-known momentum space representations of the individual terms,

$$\begin{aligned}D_{ij}^{temp}(x,y) = &-i\int \frac{d^4k}{(2\pi)^4} e^{-ik(x-y)} \frac{1}{k^2 + i\epsilon}\left(g_{ij} - \frac{k_ik_j}{|\mathbf{k}|^2}\right) \\ &- i\partial_i^x\partial_j^y \int_\gamma^{x_0} d\tau \int_\gamma^{y_0} d\tau' \int \frac{d^4k}{(2\pi)^4} e^{-ik(x-y)}\left(\frac{1}{|\mathbf{k}|^2}\right) \ .\end{aligned} \tag{10}$$

The first term here is clearly the propagator for the transverse radiation fields, the second involving only the nondynamical terms. The second piece shall be the focus of our attention, it being the longitudinal propagator if the temporal gauge potentials are decomposed orthogonally. Performing the derivatives and integrations gives

$$D_{ij}^L(x,y) = -i\int \frac{d^4k}{(2\pi)^4}\frac{k_ik_j}{|\mathbf{k}|^2}\frac{1}{k_0^2} e^{i\mathbf{k}.(\mathbf{x}-\mathbf{y})}(e^{-ik_0x_0} - e^{-ik_0\gamma})(e^{ik_0y_0} - e^{ik_0\gamma}) \ , \tag{11}$$

where $\mathbf{a}.\mathbf{b}$ denotes a three-vector dot product. This factorises in a space and time part, the latter, which is of greatest interest, has the form

$$D(x_0, y_0) = \int \frac{dk_0}{(2\pi)} \frac{(e^{-ik_0x_0} - e^{-ik_0\gamma})(e^{ik_0y_0} - e^{ik_0\gamma})}{k_0^2} \ . \tag{12}$$

Note that the integrand in this expression is well-defined at $k_0 = 0$. Performing the energy-integration gives the configuration-space time-part of the longitudinal propagator

$$D(x_0, y_0) = -1/2\big(|x_0 - y_0| - |x_0 - \gamma| - |y_0 - \gamma|\big) \ , \tag{13}$$

which is precisely the expression derived by LMR and Lavelle and Schaden.

We note that only in the regions $x_0, y_0 \gtrless \gamma$ does the propagator agree with the CCM result. On the other hand it disagrees outside these regions: for $x_0 > \gamma > y_0$ and $y_0 > \gamma > x_0$ the longitudinal propagator *vanishes*.

## 3 Discussion

We now comment briefly on the relationship between this approach and that of others. Essentially what all authors who obtain propagators of this form do is break the residual gauge freedom in order to obtain weel-defined Green's functions. Whereas other authors impose a Coulomb-*like* subsidiary condition on the temporal gauge potentials at a particular time, we have considered potentials satisfying the Coulomb gauge condition *for all times* and then parallel transported to a set of potentials satisfying the temporal gauge condition, again, for all times. Moreover we have not quantised the remaining nontransverse degress of freedom, and in that respect we differ again from previous authors who, in doing so, assume commutation relations between "creation" and "annihilation" operators of longitudinal gauge particles consistent with the equations of motion (see also Dahmen, Scholz and Steiner [2]). In the Abelian case at least, our method of obtaining the propagator constitutes the simplest and most direct approach involving minimal assumptions. It also leaves the physical Hilbert space of photon states intact; the only quantum states in this approach to the temporal gauge are precisely the physical transverse states for which Gauss' Law is satisfied as an operator condition. We make no comment at this stage concerning the legitimacy in taking the initial time $\gamma$ (or $t_0$) to $\pm\infty$.

We wish to acknowledge fruitful discussions with our colleagues in Adelaide. One of us (ACK) thanks the organisers of the Vienna Workshop for assistance with travel expenses, and the Australian Commonwealth Government for the support of a CPRA.

## References


[1] S. Caracciolo, G. Curci, P. Menotti: Phys. Lett. **113B** (1982) 311.

[2] H.D. Dahmen, B. Scholz, F. Steiner: Phys.Lett. **117B** (1982) 339;
S.C. Lim: Phys.Lett. **149B** (1984) 201;
A.A. Slavnov, S.A. Frolov: Teor. Mat.Fiz. **68** (1986) 360;
K. Haller: Phys.Rev. D **36** (1987) 1830; 1839.

[3] J.-P. Leroy, J. Micheli, G.-C. Rossi: Z.Phys.C - Particles and Fields **36** (1987) 305.

[4] G.-C. Rossi, M. Testa: Nucl.Phys.B **163** (1980) 109; Nucl.Phys. B **176** (1980) 477; Nucl. Phys. B **237** (1984) 442.

[5] M.J. Lavelle, M.Schaden: Phys.Lett. **203B** (1988) 441.

[6] H.O. Girotti, H.J. Rothe: Z.Phys.C - Particles and Fields **27** (1985) 559; Phys.Rev. D **33** (1986) 3132.

# Regularization and Feynman Rules in Noncovariant Gauges

Su-Long Nyeo

II. Institut für Theoretische Physik der Universität Hamburg,
Luruper Chaussee 149, 2000 Hamburg 50, Fed. Rep. Germany

**Abstract:** A regularization scheme for defining Feynman rules in the path-integral formalism is proposed. It is applied to pure abelian gauge theory and to obtaining regularized propagators in noncovariant gauges. The propagators are more intricate than those commonly adopted, and the known prescriptions for the noncovariant singularities appear not to be immediate consequences of the scheme. Further studies are required to investigate the renormalization and the practicality of the noncovariant-gauge formalisms.

## 1 Introduction

Noncovariant gauges have long been recognized for their usefulness in studying certain aspects of gauge field theories. The most commonly-used noncovariant gauges are the Coulomb gauge and algebraic gauges, such as the temporal, the spatial-axial, and the light-cone gauge. Although considerable work has been devoted to providing practical prescriptions for the spurious singularities in the Feynman rules in these gauges for perturbative calculations, there remain great difficulties in proving the correctness of proposed prescriptions. We can only justify a prescription partially by calculating with it, normally only tractable to one-loop level, physical quantities that have been calculated in a covariant gauge. (For a recent review, see ref. [1])

While the Feynman rules in a covariant gauge can be derived either by the canonical formalism or by the path-integral formalism, those in the noncovariant gauges obtained by the path-integral formalism are shown to be inconsistent [2,3]. Thus, here we propose a regularization scheme for defining Feynman rules in the path-integral formalism, with the aim of determining how practical propagators for gauge field in the noncovariant gauges can be derived in the scheme. We should recall that regularization is a procedure consisting mainly in replacing mathematically ill-defined objects with well-defined ones. We then study the latter and prove that some of the properties of interest, if valid for the regularized version, would carry over to their limit. In this context, we should mention two related regularization methods, which were recently considered to provide well-defined Feynman rules in the noncovariant gauges. One method [4] considers gauges that unify covariant gauge with noncovariant gauges by a tunable parameter, and treats the singularities in the noncovariant gauges as regularized by the parameter. The other [5] involves regularizing noncovariant singularities by adding to the required gauges terms that correspond to some boundary conditions. Finally, it should be mentioned that many studies have been made to establish equivalent formulations of Feynman rules in different gauges [3,6,7].

In the next section, we provide a short introduction to the theory of distributions, and define the role of regularizing functions in the theory. For pedagogical reasons, we shall consider in section 3 some examples of the application of a regularization scheme. In section 4, the scheme is applied to an abelian gauge theory in the linear covariant gauge and in several noncovariant gauges. Section 5 contains our conclusions.

## 2 Regularization and the Theory of Distributions

Since the theory of distributions plays a related role in developing a regularization scheme for defining Feynman rules, we give here a short introduction to the theory [8].

We shall refer to $\phi(x) = \phi(x_1, x_2, ..., x_n)$ as a test function if it is infinitely differentiable. For our purpose, we shall consider the Schwartz space $\mathcal{S}$, which is a space of test functions of rapid descent. If $\phi(x) \in \mathcal{S}$, then $\phi(x)$ and all its partial derivatives decrease to zero faster than every negative power of $|x| = \sqrt{x_1^2 + x_2^2 + \cdots + x_n^2}$.

We define $\widetilde{\phi}(x)$ as a generalized function or distribution and represent it by a sequence $\{\phi_\epsilon; \epsilon = 1/j, j = 1, 2, ..\}$. Then $\phi_\epsilon(x) \in \mathcal{S}$, and

$$\int_\Omega \widetilde{\phi}(x) g(x) dx \equiv \lim_{\epsilon \to 0} \int_\Omega \phi_\epsilon(x) g(x) dx \tag{1}$$

exists for $g(x) \in \mathcal{S}$, $x \in \Omega$. For example, the one-dimensional Dirac $\delta$-function $\delta(x)$ is a generalized function and can be defined as

$$\begin{aligned} \delta(x) &\equiv \lim_{\epsilon \to 0} \delta_\epsilon(x) \\ &= \lim_{\epsilon \to 0} \frac{1}{2\pi} \int_{-\infty}^{\infty} e^{ikx} \varphi_\epsilon(k)\, dk, \end{aligned} \tag{2}$$

where $\varphi_\epsilon(k)$ is commonly referred to as a regularizing function. It should be integrable for $\epsilon \neq 0$ and satisfy $\varphi_\epsilon(k, \epsilon{=}0) = 1$. If we take $\varphi_\epsilon(k) = e^{-\epsilon^2 k^2}$, then

$$\delta_\epsilon(x) \equiv \frac{1}{\sqrt{4\pi\epsilon^2}} e^{-\frac{x^2}{4\epsilon^2}} \tag{3}$$

and $\delta_\epsilon(x) \in \mathcal{S}$, for $\epsilon \neq 0$.

We may also use the regularizing function $\rho_\alpha(k) = e^{-\alpha|k|}$ and get

$$\delta_\alpha(x) \equiv \frac{1}{\pi} \frac{\alpha}{x^2 + \alpha^2}, \; \alpha > 0. \tag{4}$$

The two representations (3) and (4) are *equivalent*, since for $g(x) \in \mathcal{S}$,

$$\lim_{\epsilon\to 0}\int_{-\infty}^{\infty} \delta_\epsilon(x)g(x)dx = \lim_{\alpha\to 0}\int_{-\infty}^{\infty} \delta_\alpha(x)g(x)dx = g(0). \tag{5}$$

We note that the function $\rho_\alpha(k)$ is not differentiable at $k = 0$, but the represented function in $x$-space $\delta_\alpha(x)$ is. Thus, unlike test functions, regularizing functions are not necessarily everywhere differentiable, but should be integrable, both locally and globally. Nevertheless, it is possible to define a regularizing function $\varphi_\epsilon(k)$ such that $\varphi_\epsilon(k) \in \mathcal{S}(k)$.

## 3 Examples

To appreciate the importance of regularization in defining many physical problems, we first consider the simple integral $I = \int_\Omega \frac{1}{r} \exp(i\boldsymbol{k}\cdot\boldsymbol{r})\, d\boldsymbol{r}$, where $\Omega$ is the three-dimensional Euclidean space and $r = |\boldsymbol{r}|$. This integral is undefined but bounded, having an arbitrary value in [0,2]. To give it a definite value, we can define it in a regularized sense as

$$I \equiv \lim_{\alpha\to 0}\int_\Omega \frac{1}{r}\exp(i\boldsymbol{k}\cdot\boldsymbol{r})\, f_\alpha(r)d\boldsymbol{r}\,,\ \alpha > 0, \tag{6}$$

where we have chosen a spherically symmetric $f_\alpha(r)$ as is the potential $1/r$. If we choose $f_\alpha(r) = f(\alpha r) \in \mathcal{S}$, then $I \to 4\pi/k^2$ as $\alpha/|\boldsymbol{k}| \to 0$. (Note that dimensional regularization can be used here and gives the identical result.)

As a second example we consider the derivation of anomalous Ward identities in gauge theories under local chiral transformation in the path-integral approach, as studied notably by Fujikawa [9]. These identities are consequences of noninvariance of fermion measure with respect to local chiral transformation, and a regularization scheme is required to derive them. Fujikawa has used an exponential regularizing function, and argued that we should choose a regularizing function $f_\alpha(x) = f(y{=}\alpha x)$, such that $f(0) = 1, f(\infty) = 0$, and all its derivatives evaluated at $x = \infty$ vanish. Recently, Marozov [10] gave the following sufficient condition on the $n$-th ($n \geq 1$) derivative of $f(y)$:

$$f^{(n)}(y) \leq (constant)/(1+y)^{n+\eta}, \eta > 0, \tag{7}$$

in the limit $y \to \infty$. (For a detailed study of regularization schemes, see ref. [11].)

Finally, we apply the regularization scheme to the field-theoretic $\varphi^4$ model. Quantization of this model in the path-integral formalism amounts to considering the generating functional

$$\mathcal{Z}[J] = \int \mathcal{D}[\varphi] \exp\left(i\int(\mathcal{L}(x) + J(x)\varphi(x))dx\right), \tag{8}$$

$$\mathcal{L}(x) = \frac{1}{2}\left((\partial_\mu\varphi(x))^2 - m^2\varphi^2(x)\right) - \frac{\lambda}{4!}\varphi^4(x)\ ;\ m\,,\lambda > 0.$$

This functional is ill defined in two instances. Firstly, we encounter divergences in perturbative calculations that need be defined by a regularization scheme, such as dimensional regularization, and a renormalization procedure. Secondly, for perturbative calculations, we need a set of Feynman rules that are yet to be specified by some boundary conditions. To specify them, we can define the functional (8) as

$$\mathcal{Z}[J] = \lim_{\epsilon\to 0} \int \mathcal{D}[\varphi]F_\epsilon[\varphi] \exp\left(i\int (\mathcal{L}(x) + J(x)\varphi(x))dx\right), \epsilon > 0, \tag{9}$$

where $F_\epsilon[\varphi]$ is the nonderivative, integrable functional satisfying $F_\epsilon[\varphi, \epsilon=0] = 1$:

$$F_\epsilon[\varphi] = \exp\left(-\frac{\epsilon}{2}\int \varphi^2(x)dx\right). \tag{10}$$

The functional measure is then the weighted measure $\mathcal{D}[\varphi]F_\epsilon[\varphi]$ with weight $F_\epsilon[\varphi]$. The Feynman rules are now well defined, and the propagator for the scalar field reads

$$G(p,m) = \frac{i}{p^2 - m^2 + i\epsilon}, \epsilon > 0. \tag{11}$$

Alternatively, we can define the Feynman rules in Euclidean space by considering (8) as a Wiener integral.

## 4 Application to Abelian Gauge Theory

Here we shall apply the regularization scheme to the pure abelian theory†

$$\mathcal{L}(x) = -\frac{1}{4}(F_{\mu\nu})^2, \tag{12}$$

where $F_{\mu\nu} = \partial_\mu A_\nu - \partial_\nu A_\mu$ is the field strength with $A_\mu$ being the gauge field. In the path-integral approach, we first consider the naive generating functional

$$\mathcal{Z}[J_\mu] = \lim_{\epsilon\to 0} \int \mathcal{D}[A_\mu]F_\epsilon[A_\mu] \exp\left(i\int (\mathcal{L}(x) + J_\mu(x)A_\mu(x))dx\right), \epsilon > 0, \tag{13}$$

where $F_\epsilon[A_\mu]$ is some properly chosen regularizing functional. But, because of local gauge invariance, we need to fix a gauge. It is crucial that the introduction of a regularization scheme does not correspond to fixing a gauge. To show this, we consider

$$F_\epsilon[A_\mu] = \exp\left[\frac{\epsilon}{2}\int (A_\mu \delta_{\mu\nu} A_\nu)\, dx\right], \epsilon > 0, \tag{14}$$

which behaves as a damping factor for space-like $A_\mu$. This functional is not integrable but can be made so by Wick rotation. With (14), we get the singular propagator

$$G_{\mu\nu}(p) = G(p)\left(-\delta_{\mu\nu} - \frac{p_\mu p_\nu}{i\epsilon}\right), \quad G(p) \equiv \frac{i}{p^2 + i\epsilon}. \tag{15}$$

In a general linear gauge, $\mathcal{F}_\mu A_\mu = \phi(x)$, where $\mathcal{F}_\mu$ is a linear operator and $\phi(x)$ is an arbitrary function, the generating functional can be cast in the form [1]

---

† We employ the metric $\delta_{\mu\nu} = \text{diag}(1,\text{-}1,\text{-}1,\text{-}1)$, and summation convention for repeated indices. Greek letters $\mu, \nu$ are Lorentz indices.

$$\mathcal{Z}[J_\mu] = \lim_{\epsilon\to 0} \int \mathcal{D}[A_\mu] F_\epsilon[A_\mu] \exp\left(i\int (\mathcal{L}(x) - \frac{1}{2\alpha}\left(\mathcal{F}_\mu A_\mu\right)^2 + J_\mu(x)A_\mu(x))dx\right), \alpha \neq 0. \tag{16}$$

Using (16), we can now proceed to discuss the theory in the following gauges.

### 4.1 The General Linear Covariant Gauge

This gauge is specified by $\partial_\mu A_\mu(x) = \phi(x)$. The quantized theory is defined by the functional (16). Using the regularizing functional (14), we get the propagator

$$G_{\mu\nu}(p) = G(p)d_{\mu\nu} \equiv G(p)\left[-\delta_{\mu\nu} + \frac{(1-\alpha)p_\mu p_\nu}{p^2 + i\epsilon\alpha}\right], \tag{17}$$

where $G(p)$ is given in (15). An immediate consequence of our regularization scheme is the factorization of the physical and the unphysical poles. If we demand that Wick rotation be possible also with the unphysical poles defined by $p^2 + i\epsilon\alpha = 0$, imposing the familiar boundary conditions on the propagator, we must have $\alpha > 0$. Therefore, we should consider, even in the Landau gauge ($\alpha = 0$), the limits $\epsilon \to 0$ and $\epsilon' \equiv \epsilon\alpha \to 0$ only after momentum integrations.

A slightly more complicated regularizing functional that depends on additional parameters may be considered. For example,

$$F_\epsilon[A_\mu] = \exp\left[\frac{\epsilon}{2}\int \left(A_\mu\left[\delta_{\mu\nu} + \beta n_\mu n_\nu\right]A_\nu\right)dx\right], \epsilon > 0, \tag{18}$$

with $n_\mu = (1,0,0,0)$ and $\beta \geq 0$ may be used. We note that while (14) has a light-like boundary $A_\mu\delta_{\mu\nu}A_\nu = 0$, (18) has a modified boundary $A_\mu[\delta_{\mu\nu} + \beta n_\mu n_\nu]A_\nu = 0$. Clearly, a space-like vector, which is regularized by (14), is not necessarily regularized by (18).

Returning to the functional (18), we get the polarization sum

$$\begin{aligned} d_{\mu\nu} = -\delta_{\mu\nu} + D\{&(1-\alpha)(p^2 + i\epsilon + i\beta\epsilon)p_\mu p_\nu \\ &- i\beta\epsilon p\cdot n(1-\alpha)\left(p_\mu n_\nu + p_\nu n_\mu\right) + i\beta\epsilon(p^2 + i\alpha\epsilon)n_\mu n_\nu\}. \end{aligned} \tag{19}$$

There are four poles defined by the quartic equation

$$D^{-1} \equiv (p^2 + i\alpha\epsilon)(p^2 + i\epsilon + i\beta\epsilon) - i\beta\epsilon(1-\alpha)(p\cdot n)^2 = 0. \tag{20}$$

Thus, the functional (18) offers no practical advantage. We see that, unlike in the scalar field theory, we have a freedom to define the Feynman rules for a vector theory. In particular, the use of a different $F_\epsilon[A_\mu]$ leads to a different representation of a gauge theory, and could lead to calculational difficulties and complicated renormalization.

### 4.2 The Coulomb Gauge

The inhomogeneous Coulomb gauge is specified by $\nabla\cdot\boldsymbol{A} = (\partial_\mu - n\cdot\partial n_\mu)A_\mu = \phi(x)$, where $n_\mu = (1,0,0,0)$. We consider the generating functional (16) with (14) and obtain $G_{\mu\nu}(p) = G(p)c_{\mu\nu}$, where $G(p)$ is defined in (15), and $c_{\mu\nu}$ is given by

$$c_{\mu\nu} = -\delta_{\mu\nu} + D\left\{\left[(\alpha-1)(p^2+i\epsilon) + (p\cdot n)^2\right]p_\mu p_\nu - p\cdot n\left[(p\cdot n)^2 - p^2 - i\epsilon\right](p_\mu n_\nu + p_\nu n_\mu) - i\epsilon(p\cdot n)^2 n_\mu n_\nu\right\}. \tag{21}$$

The spurious poles are defined by

$$D^{-1} \equiv \left[(p\cdot n)^2 - p^2 - i\epsilon\alpha\right](p^2+i\epsilon) - (p\cdot n)^2\left[(p\cdot n)^2 - p^2\right] = 0 \tag{22}$$

or, explicitly, read

$$p_0^{\pm} = \pm(1+i)\sqrt{\frac{1}{2\epsilon\alpha}(p^2-i\epsilon)(p^2-i\epsilon\alpha)}\,, \epsilon > 0. \tag{23}$$

To allow for Wick rotation, the parameter $\alpha$ must be negative and should not be set to zero prematurely. As always, in the path-integral formalism, one should consider the homogeneous Coulomb gauge $\nabla\cdot\boldsymbol{A} = 0$ only in the limit $\alpha \to 0$.

We may also use a different regularizing functional, such as (18). But, we may then have to work with a complicated propagator.

### 4.3 The Temporal Gauge

This gauge, also called the Hamiltonian gauge, is characterized by $A_0 = 0$ [12], or by $n\cdot A = 0$, where $n_\mu = (1,0,0,0)$. The inhomogeneous temporal gauge is defined by $n\cdot A = \phi(x)$.

We define the theory in this gauge as a regularized path integral given by (16) with $F_\epsilon[A_\mu]$ in (14). (Other forms of $F_\epsilon[A_\mu]$ may also be considered.) The propagator in this case reads $G_{\mu\nu}(p) = G(p)t_{\mu\nu}$, where $G(p)$ is given in (15) and

$$t_{\mu\nu} = -\delta_{\mu\nu} - D\left\{\left[\alpha(p^2+i\epsilon)+1\right]p_\mu p_\nu - p\cdot n\,(p_\mu n_\nu + p_\nu n_\mu) - i\epsilon n_\mu n_\nu\right\}, \tag{24}$$

with $D^{-1} = (p\cdot n)^2 + i\epsilon\alpha(p^2+i\epsilon) + i\epsilon$. For $\alpha \approx 0$, the poles are

$$p_0^{\pm} \approx \pm(1-i)\sqrt{\frac{\epsilon}{2}(1-\alpha p^2)}\,, \tag{25}$$

which are seen to permit Wick rotation.

Setting $\alpha = 0$ (the pure temporal gauge), we get the polarization sum

$$t^0_{\mu\nu} = -\delta_{\mu\nu} - D_0\left[p_\mu p_\nu - p\cdot n\,(p_\mu n_\nu + p_\nu n_\mu) - i\epsilon n_\mu n_\nu\right], \tag{26}$$

where $D_0^{-1} = p_0^2 + i\epsilon$. More simply, the propagator reads

$$G_{00}(p) = G_{0j}(p) = G_{j0}(p) = 0,$$
$$G_{ij}(p) = G(p)\left[\delta_{ij} - \frac{p_i p_j}{p_0^2 + i\epsilon}\right], i, j = 1, 2, 3. \tag{27}$$

This propagator can also be found in several publications [7,13]. It can be derived by considering the three-component lagrangian obtained by setting $A_0 = 0$ in (12). Then we introduce the regularizing functional (14) with $A_0 = 0$.

### 4.4 The Spatial-Axial Gauge

Like the temporal gauge, this gauge can be defined as $n \cdot A = 0$, where $n_\mu = (0, 0, 0, 1)$. The inhomogeneous gauge is specified by $n \cdot A(x) = \phi(x)$.

Using $F_\epsilon[A_\mu]$ in (14) and the inhomogeneous gauge, we get the propagator $G_{\mu\nu}(p) = G(p)a_{\mu\nu}$, with $G(p)$ defined in (15) and

$$a_{\mu\nu} = -\delta_{\mu\nu} - D\left\{\left[\alpha(p^2 + i\epsilon) - 1\right] p_\mu p_\nu - p\cdot n\,(p_\mu n_\nu + p_\nu n_\mu) - i\epsilon n_\mu n_\nu\right\}, \tag{28}$$

where $D^{-1} = (p\cdot n)^2 + i\epsilon\alpha(p^2 + i\epsilon) - i\epsilon$. The propagator is more complicated than those commonly used. In particular, it contains additional tensorial terms and its singularities are not defined by any known prescription, such as the principal-value prescription.

If we set $\alpha = 0$ in the polarization sum (28), we get

$$a^0_{\mu\nu} = -\delta_{\mu\nu} + D_0\left[p_\mu p_\nu + p\cdot n\,(p_\mu n_\nu + p_\nu n_\mu) + i\epsilon n_\mu n_\nu\right], \tag{29}$$

where $D_0^{-1} = p_3^2 - i\epsilon$.

Other regularizing functionals that could give practical prescriptions remain to be found.

### 4.5 The Light-Cone Gauge

This gauge is defined as $A_- \equiv (A_0 - A_3) = 0$ or $n \cdot A = 0$, where $n_\mu = (1, 0, 0, 1)$ is a light-like vector. We define the inhomogeneous gauge by $n \cdot A = \phi(x)$.

Adopting the regularizing functional (14), we get the corresponding propagator $G_{\mu\nu}(p) = G(p)d_{\mu\nu}$, with $G(p)$ given in (15) and

$$d_{\mu\nu} = -\delta_{\mu\nu} - D\left[\alpha(p^2 + i\epsilon)p_\mu p_\nu - p\cdot n\,(p_\mu n_\nu + p_\nu n_\mu) - i\epsilon n_\mu n_\nu\right], \tag{30}$$

where $D^{-1} = (p\cdot n)^2 + i\epsilon\alpha(p^2 + i\epsilon)$. For $\alpha \neq 0$, the naive poles given by $(p\cdot n)^2 = 0$ are regularized, but not by the Leibbrandt-Mandelstam prescription [14].

However, setting naively $\alpha = 0$ gives

$$d^0_{\mu\nu} = -\delta_{\mu\nu} + \frac{1}{(p\cdot n)^2}\left[p\cdot n\,(p_\mu n_\nu + p_\nu n_\mu) + i\epsilon n_\mu n_\nu\right], \tag{31}$$

and the singularities $(p\cdot n)^2 = 0$ are not regularized at all. In fact, the functional

$$F_\epsilon[A_\mu] = \exp\left[\frac{\epsilon}{2}\int \left(A_\mu\left[\delta_{\mu\nu} + \beta(n_\mu n^*_\nu + n^*_\mu n_\nu)\right] A_\nu\right) dx\right], \beta \neq 0, \tag{32}$$

with $n^*_\mu = (1,0,0,-1)$, also leads to (31) if we set $\alpha = 0$. With the functional

$$F_\epsilon[A_\mu] = \exp\left[\frac{\epsilon}{2}\int \left(A_\mu\left[\delta_{\mu\nu} + \beta n^*_\mu n^*_\nu\right] A_\nu\right) dx\right], \beta \neq 0, \tag{33}$$

we get, for $\alpha = 0$, a complicated propagator with spurious poles specified by

$$-4i\beta\epsilon[p{\cdot}np{\cdot}n^* + i\epsilon] + (p{\cdot}n)^2(p^2 + i\epsilon) = 0. \tag{34}$$

## 5 Conclusions

By introducing a simple regularization scheme to define Feynman rules in abelian gauge theory quantized in the noncovariant gauges, we found that the propagators are generally more intricate than those commonly used. The propagators contain additional tensorial terms, because these gauges depend on constant vectors. Our preliminary study based on this scheme indicates that there exists a freedom to define Feynman rules for gauge theories. Specifically, the use of a different $F_\epsilon[A_\mu]$ leads to a different propagator. Further studies are, however, required to determine whether the additional tensorial terms found in this scheme play any significant role in perturbative calculations, and whether the propagators are consistent with physics. Also, since the renormalization of these noncovariant-gauge formalisms remains open, it is necessary to perform some calculations.

Eventually we should like to apply the regularization scheme to nonabelian gauge theory, where we also need to regularize the propagator for the Faddeev-Popov ghosts.

Finally, it should be of interest to see how this scheme can be related to the canonical formulation, and how the previously proposed prescriptions can be incorporated in this scheme.

### Acknowledgements

It gives me great pleasure to thank Professor G. Kramer for his warm hospitality at the II. Institut für Theoretische Physik, where this work was carried out. I am also very grateful for his general encouragement and for his useful comments on the manuscript. I should like to thank Professors W. Kummer and M. Schweda, and Dr. P. Gaigg for providing the stimulating atmosphere at the workshop and the opportunity to present this work. The financial support provided by the Alexander von Humboldt Foundation is gratefully acknowledged. In addition, I should like to acknowledge a grant from the Foundation for attending the workshop.

## References


[1] G. Leibbrandt, Rev. Mod. Phys. **59** (1987) 1067.

[2] I. Ichinose and K. Sera, Phys. Rev. **D22** (1982) 982.

[3] H. Cheng and E.-C. Tsai, Phys. Rev. **D36** (1987) 3196.

[4] J. Frenkel and J.C. Taylor, Nucl. Phys. **B109** (1976) 439; T.P. Beven and R. Delbourgo, Lett. Nuovo Cimento **27** (1980) 565; *ibid* **28** (1980) 400; A. Burnel, Phys. Rev. **D36** (1987) 1846 and 1952; S.-L. Nyeo, *ibid* **D36** (1987) 2512.

[5] H.S. Chan and M.B. Halpern, Phys. Rev. **D33** (1985) 540; F. Steiner, Phys. Lett. **B173** (1986) 321; A.A. Slavnov and S.A. Frolov, Theor. Math. Phys. **73** (1987) 1158.

[6] C.R. Handy, Phys. Rev. **D19** (1979) 585.

[7] R. Mills, Phys. Rev. **D3** (1971) 2969.

[8] G.E. Shilov, *Generalized Functions and Partial Differential Equations*, Gordon and Breach, New York (1968).

[9] K. Fujikawa, Phys. Rev. Lett. **42** (1979) 1195; Phys. Rev. **D21** (1980) 2848.

[10] A. Yu. Morozov, Sov. Phys. Usp. **29** (1986) 993.

[11] M. Umezawa, Phys. Rev. **D39** (1989) 3672.

[12] S. Carocciolo, G. Curci, and P. Menotti, Phys. Lett. **113B** (1982) 311; S.C. Lim, Phys. Rev. **D32** (1985) 1384; K. Haller, *ibid* **D36** (1987) 1830 and 1839; A.A. Slavnov and S.A. Frolov, Theor. Math. Phys. **68** (1986) 880; N.V. Antonov, *ibid* **72** (1987) 951; F.M. Saradzhev and V.Ya. Faĭnberg, in *Solitons and Instantons, Operator Quantization*, V.L. Ginzburg (ed.), Nova Science, New York (1987).

[13] K. Huang, *Quarks, Leptons and Gauge Fields*, World Scientific, Singapore (1982).

[14] S. Mandelstam, Nucl. Phys. **B213** (1983) 149; G. Leibbrandt, Phys. Rev. **D29** (1984) 1699; A. Bassetto, M. Dalbosco, I. Lazzizzera, and R. Soldati, *ibid* **D31** (1985) 2012.

# A Criterion of Consistency for Gauge Theories in the Axial Gauge

M. Poljšak

Institute J. Stefan, E. Kardelj University, Jamova 39,
61111 Ljubljana, Yugoslavia

**Abstract:** An attempt to define precise Feynman rules for theories of gauge fields in axial gauges is made by requiring consistency with theories in covariant gauges. The Wilson line operators are used to construct a modified fermion propagator which is invariant under a restricted class of gauge transformations comprising the ones that relate axial to some covariant gauge. Calculating the modified fermion propagator in the axial gauge with the Cauchy principal value prescription for spurious singularities of the gauge field propagator and with the Landshoff's prescription, it is found to one-loop order that at least one of the prescriptions is not consistent with covariant gauge Feynman rules.

One of the most contentious problems in theories of gauge fields, quantized in the axial gauge, is the establishment of consistent Feynman rules. The traditional view of quantization of gauge fields in the axial gauge holds that no ghost fields are required to calculate S-matrix elements [1], that the Ward identities retain their classical form [2], that there is no need to renormalize gauge transformations [3], and that the prescription for the spurious axial gauge singularities of the gauge field propagator is arbitrary, because all observable S-matrix elements are independent of the gauge-fixing vector [4]. These properties of the theory have profitably been employed to obtain general theorems, an example of which is the proof [5] of factorization of long- and short-distance effects in deep inelastic scattering of electrons on hadrons, as described in QCD.

Over the last few years it has become clear, however, that there are problems in defining axial gauges precisely. It has been a long-standing source of worry that the bare propagator of a gauge field in the axial gauge is plagued with singularities, which apparently cannot be made precise by the usual definition of the propagator as the inverse of the quadratic part of the Lagrangian. Until recently the most commonly adopted prescription how to integrate over these spurious axial gauge singularities was to treat them as Cauchy principal values (CPV) [6]. On the assumption that no ghost fields need be introduced or other changes to the traditional Feynman rules made, it was discovered, however, that this prescription leads to incorrect results [7].

This was one instance to prompt a complete overhaul of theories of gauge fields, quantized in the axial gauge. In this process, several suggestions concerning consistent axial gauge Feynman rules were put forward, ranging from careful statements about the general properties these rules should obey [8] up to proposals of definite rules [7],[9],[10]. Many of these suggestions broke away from the traditional formal axial gauge field theory by allowing, for example, translationally noninvariant gauge field propagators [7], or invoking ghosts [9], or introducing more than one gauge-fixing vector [10].

A reliable guide in the search for consistent axial gauge Feynman rules has been the Wilson loop which, as a gauge invariant object, offers a means of testing compatibility with the Feynman rules in other gauges. Morover, such a test simultaneously measures the degree of practicality of new rules. It was the Wilson loop that revealed difficulties with the Cauchy principal value prescription for spurious poles of the axial gauge field propagator [7]. It did so, however, only at the two-loop order of perturbation theory.

In this contribution I explore the possibility of other workable tests of compatibility of Feynman rules in the axial gauge with the Feynman rules in covariant gauges. The starting point of this endeavour is the observation, valid in classical gauge theories, of a restricted invariance of some Wilson lines under gauge transformations that relate the axial gauge to the Lorentz gauge. I discuss implications of this observation in the quantum case. Subject to the proviso of correctness of some technical assumptions I find a criterion of compatibility of Feynman rules in the axial gauge with those in the Landau gauge. Since the Cauchy principal value prescription as well as the $\alpha$-prescription [11] pass this test in one-loop order, I turn my attention to a modified Wilson line operator, which is constructed out of the original Wilson line by replacing a part of it by the bilinear of Dirac fields that defines the propagator of a fermion field from one end of the removed part of the line to the other end. The modified Wilson line has the same transformation properties as the original Wilson line. Testing the Cauchy principal value prescription and the $\alpha$-prescription, I find to one-loop order that at least one of them is not compatible with the covariant Feynman rules.

In principle, any gauge invariant object, constructed out of fields of the theory, may be used as a necessary condition to verify compatibility of Feynman rules in various gauges. It is difficult to imagine a simpler gauge invariant object than a rectangular Wilson loop: calculation of Wilson loops, defined on more complicated closed contours, or of scattering amplitudes for physical processes, as the next most obvious gauge invariant quantities, proved to be a much more taxing demand.

But unrestricted gauge invariance of an object is more than necessary, if we want to employ this object in questioning compatibility of a proposed set of axial gauge Feynman rules with the well-established covariant gauge Feynman rules. A completely gauge-invariant object allows us to give evidence of or to disprove compatibility of a given set of Feynmman rules in the axial gauge with Feynman rules in all other gauges, not only the covariant ones. For our purpose, therefore, it would be sufficient to have an object which is invariant not under all gauge transformations, but under those ones, and only those ones, that relate any gauge potential, satisfying the axial gauge-fixing condition

$$n.A = 0, \tag{1}$$

to a gauge potential, satisfying the Lorentz gauge-fixing condition

$$\partial.A = 0. \tag{2}$$

In a classical gauge field theory the Lorentz condition (2) only constrains $A_\mu$, but does not fix it uniquely. In the abelian case, for example, $A_\mu$ can change by $\partial_\mu \Lambda(x)$, say, and still satisfy (2) provided

$$\Box\, \Lambda(x) = 0. \tag{3}$$

This implies the following

*Proposition*: Given a gauge potential $A_\mu$ which satisfies the Lorentz condition (2), there always exists a gauge equivalent potential $A'_\mu$ in the same gauge (2), that is related to a gauge potential $A''_\mu$, satisfying the axial gauge fixing condition (1), by a gauge transformation that maps a hyperplane, perpendicular to the gauge-fixing vector $n_\mu$, to the identity of the group.

I prove here this proposition separately for the space-like case $n^2 < 0$ and the time-like case $n^2 > 0$.

In the temporal case, we can without loss of generality choose $n = (1,0,0,0)$ and the hyperplane $(T, \mathbf{x})$, where $T$ is a fixed time. Since it is always possible to find a solution to Eq.(3), such that this solution on the hyperplane satisfies the initial condition

$$\partial_0 \Lambda(T, \mathbf{x}) = -A_0(T, \mathbf{x}), \tag{4}$$

where $A_0$ is given by the starting gauge potential in the Lorentz gauge, it follows that

$$A'_0(T, \mathbf{x}) = A_0(T, \mathbf{x}) + \partial_0 \Lambda(T, \mathbf{x}) = A_0(T, \mathbf{x}) - A_0(T, \mathbf{x}) = 0 \tag{5}$$

and so it is, trivially, also in the axial gauge on the hyperplane.

For the space-like case, we can without loss of generality choose $n = (0,0,0,1)$ and the hyperplane $(x^0, x^1, x^2, R)$ with $R$ fixed. It is always possible to make a gauge transformation from a given gauge potential $A_\mu(x)$ to $A''_\mu(x)$ such that $A''_3(x) = 0$. Namely, given $A_\mu(x)$, let us construct the gauge transformation

$$U(P_x) = exp\big[-ig \int_0^{x^3} dx^3\, A_3(x)\big], \tag{6}$$

where $P_x$ denotes the straight line from $(x^0, x^1, x^2, 0)$ to $x = (x^0, x^1, x^2, x^3)$, so that

$$\partial_3 U(P_x) = U(P_x)(-igA_3(x)). \tag{7}$$

If we transform $A_\mu(x)$ by this $U(P_x)$, we get

$$igA_3(x) \to igA_3''(x) = igU(P_x)A_3(x)U(P_x)^{-1} + [\partial_3 U(P_x)]\, U(P_x)^{-1} = 0$$

by Eq.(7).

Instead of constructing $U(P_x)$ out of the original covariant $A_\mu(x)$, we now construct it out of $A'_\mu(x) = A_\mu(x) + \partial_\mu \Lambda(x)$, where $\Lambda$ satisfies Eq.(3) and the boundary condition

$$\Lambda(x^0, x^1, x^2, R) = \Lambda(x^0, x^1, x^2, 0) - \int_0^R dx^3\, A_3(x^0, x^1, x^2, x^3). \tag{8}$$

Since there always exists such a $\Lambda$, for any point $R_x = (x^0, x^1, x^2, R)$ we have

$$U(P_{R_x}) \equiv exp[-ig \int_0^R dx^3\, A'_3(x)] = exp[-ig \int_0^R dx^3\, (A_3 + \partial_3 \Lambda)] =$$

$$exp[-ig \int_0^R dx^3\, A_3(x) - ig\Lambda(x^0, x^1, x^2, R) + ig\Lambda(x^0, x^1, x^2, 0)] = 1,$$

i.e. the transformation to the axial gauge is trivial on the hyperplane.

In the nonabelian case, at least in perturbation theory, i.e. for small $A_\mu$ and $U(x) = 1 + \Lambda(x) = 1 + \Lambda^a(x)T^a$, Eq.(3) gets replaced by

$$\Box\, \Lambda - [A_\mu, \partial_\mu \Lambda] + \ldots = 0, \tag{9}$$

where the dots, denoting higher orders in $\Lambda$, are negligible in perturbation theory. Since Eq.(9) is soluble when suitable boundary conditions on the hyperplane, analogs of (4) or (5), are supplied, we can get the conclusions from the abelian case across to the nonabelian case.

As a consequence of the *Proposition*, any Wilson line

$$Pexp[-ig \int_{R_x}^{R_y} A_\mu(z)\, dz^\mu], \tag{10}$$

which under a gauge transformation

$$igA_\mu(x) \to igA'_\mu(x) = U(x) igA_\mu(x) U^{-1}(x) - [\partial_\mu U(x)]\, U^{-1}(x) \tag{11}$$

transforms into

$$Pexp[-ig \int_{R_x}^{R_y} A'_\mu(z)\, dz^\mu] = U(R_y) Pexp[-ig \int_{R_x}^{R_y} A_\mu(z)\, dz^\mu] U^{-1}(R_x), \tag{12}$$

if it is defined on a curve, both ends of which, $R_x$ and $R_y$, lie on the hyperplane. This Wilson line remains unchanged under some gauge transformation from the axial to the Lorentz gauge. In other words, if we have a Wilson loop, a part of which lies on some hyperplane, that part will contribute in the axial gauge exactly the same amount as in some covariant gauge. Although this covariant gauge is not specified precisely, it is natural to explore whether its mere existence entails any useful consequences, especially for the quantum case.

It is to be expected, at least for the temporal case, that any quantum Wilson loop possesses the property observed in the classical case, because the difference between the classical and the quantum Wilson loop amounts only to additional ordering of the quantum fields according to time, and this ordering does not interfere with any step in the proof of the behaviour under the gauge transformation from the temporal to a covariant gauge of a classical Wilson loop on a space-like surface.

The quantum counterpart of the classical Lorentz condition (2) is the limit of the gauge-fixing part of the Lagrangian of the form

$$\frac{1}{2\lambda}(\partial.A)^2, \tag{13}$$

as $\lambda$ tends to zero. The freedom in the classical case of residual covariant gauge transformations, specified by gauge parameters $\Lambda$ which satisfy Eq.(3) or (9), is now reflected in an unspecified nature of the double pole in $k^2$ of the gauge field propagator

$$D^{ab}_{\mu\nu} = \frac{i\delta^{ab}}{(2\pi)^4}\left[\frac{-g_{\mu\nu}}{k^2+i\epsilon} + (1-\lambda)\frac{k^\mu k^\nu}{(k^2)^2}\right]. \tag{14}$$

If there is some Wilson line or some aspect of it, which is independent of $\alpha$, then this may be used as a diagnosing tool for diseases in axial gauge Feynman rules.

I now examine the Wilson loop of the form of a rectangle whose two sides of length R, one extending from a point $x$ to a point $R_x$ and the other from apoint y to the point $R_y$, are parallel to the gauge fixing (unit) vector $n$ and the other two sides of length $L$ parallel to a unit vector $\xi$ that is perpendicular to $n$: $n.\xi = 0$. To one loop order this Wilson loop is given by the expression

$$W = 1 - 2g^2 \int d^{2\omega}k\, T^a T^b D^{ab}_{\mu\nu}(k)(1-\cos k.nR)(1-\cos k.\xi L)$$

$$\left[\frac{n^\mu n^\nu}{(k.n)^2} - \frac{n^\mu \xi^\nu + \xi^\mu n^\nu}{(k.n)(k.\xi)} + \frac{\xi^\mu \xi^\nu}{(k.\xi)^2}\right], \tag{15}$$

where $D^{ab}_{\mu\nu}(k)$ is the gauge field propagator, and g the coupling constant.

When we calculate (15) in the covariant gauge with the gauge field propagator (14), or in the axial gauge with the gauge field propagator

$$D^{ab}_{\mu\nu} = \frac{i\delta^{ab}}{(2\pi)^4} D_0(k^2)\left[-g_{\mu\nu} + \frac{k_\mu n_\nu + k_\nu n_\mu}{[k.n]} - n^2\frac{k_\mu k_\nu}{[(k.n)^2]}\right], \tag{16}$$

where $D_0(k^2) = \frac{i}{k^2+i\epsilon}$, we get

$$W = 1 + 2g^2 C_F \int \frac{d^{2\omega}k}{(2\pi)^{2\omega}} D_0(k^2)(1-\cos k.nR)(1-\cos k.\xi L)\left[\frac{n^2}{(k.n)^2} + \frac{\xi^2}{(k.\xi)^2}\right], \tag{17}$$

where $C_F = T^aT^a$. This calculation confirms to this order that W is gauge invariant, irrespective of the prescription for spurious singularities $\frac{1}{[k.n]}$ and $\frac{1}{[(k.n)^2]}$.

Next, let us imagine that the side of the loop with the corners $R_x$ and $R_y$ lies on the hyperplane where the gauge transformation from the axial gauge to some covariant gauge was shown to be trivial. Since the contribution of the Wilson line from $R_x$ to $R_y$ is supposed to be the same in the two gauges, I consider only the rest of the Wilson loop, which is given by the expression

$$W_2 = 1 - g^2 T^a T^b \int d^{2\omega}k D^{ab}_{\mu\nu}(1 - \cos k.\xi L)[(1 - \cos k.nR)$$

$$(2\frac{n^\mu n^\nu}{(k.n)^2} - \frac{n^\mu \xi^\nu + n^\nu \xi^\mu}{(k.n)(k.\xi)}) + \frac{\xi^\mu \xi^\nu}{(k.\xi)^2}] \; . \tag{18}$$

The r.h.s. of Eq.(18), when calculated with the covariant propagator (14), differs from its value obtained with the axial propagator (16) by

$$W_{2,A} - W_{2,C} = \int \frac{d^{2\omega}k}{(2\pi)^{2\omega}} D_0(k^2)(1 - \cos k.nR)$$

$$\left[\frac{1-\lambda}{k^2} + n^2\left(\frac{1}{[(k.n)^2]} - \frac{2(1-\cos k.nR)}{(k.n)[k.n]}\right)\right] \; . \tag{19}$$

If further I assume that in the axial case

$$D^{ab}_{\mu\nu} n^\nu \equiv 0, \tag{20}$$

the difference is

$$W_{2,A} - W_{2,C} = \int \frac{d^{2\omega}k}{(2\pi)^{2\omega}} D_0(k^2)(1 - \cos k.\xi L)\times$$

$$\left(\frac{1-\lambda}{k^2} + n^2\left[\frac{1}{[(k.n)^2]} - 2\frac{(1-\cos k.nR)}{(k.n)^2}\right]\right) \; . \tag{21}$$

Although the differences (18) and (19) depend on $\lambda$, their limits as $L$ tends to $\infty$ do not, because

$$\lim_{L\to\infty} \int \frac{d^{2\omega}k}{(2\pi)^{2\omega}} D_0(k^2)\frac{(1-\cos k.\xi L)}{k^2} = 0 \tag{22}$$

in the sense of dimensional regularization, provided that the poles of $k^{-2}$ in (22) lie in the same quadrants of the $k^0$ complex plane as the poles of $D_0(k^2)$. Therefore, subject to the proviso that the unknown covariant gauge that is related to the axial gauge such that this relation is given by the identity gauge transformation on the hyperplane, the prescriptions for the spurious poles of the axial gauge propagator are constrained to satisfy

$$\lim_{L\to\infty} \int \frac{d^{2\omega}k}{(2\pi)^{2\omega}} D_0(k^2)(1 - \cos k.\xi L)\left(\frac{1}{[(k.n)^2]} - \frac{2(1-\cos k.nR)}{(k.n)[k.n]}\right) = 0\, , \tag{23}$$

or, if we assume that Eq.(20) is fulfilled by the axial gauge field propagator,

$$\lim_{L\to\infty}\int\frac{d^{2\omega}k}{(2\pi)^{2\omega}}D_0(k^2)(1-\cos k.\xi L)\left(\frac{1}{[(k.n)^2]}-\frac{2(1-\cos k.nR)}{(k.n)^2}\right)=0\,. \tag{24}$$

The most obvious way to satisfy Eq.(24) is to require

$$\frac{1}{[(k.n)^2]}=2\frac{1-\cos k.nR}{(k.n)^2}. \tag{25}$$

Now that

$$2\frac{1-\cos k.nR}{(k.n)^2}=\left(\frac{1-\cos k.nR}{k.n}\right)^2+\left(\frac{\sin k.nR}{k.n}\right)^2$$
$$=\left(\frac{1-\cos k.nR}{k.n}+i\frac{\sin k.nR}{k.n}\right)\left(\frac{1-\cos k.nR}{k.n}-i\frac{\sin k.nR}{k.n}\right) \tag{26}$$

and that

$$\lim_{R\to\infty}\frac{1-\cos k.nR}{k.n}=PV\frac{1}{k.n}\quad,\qquad \lim_{R\to\infty}\frac{\sin k.nR}{k.n}=\pi\delta(k.n)\,, \tag{27 -- 28}$$

$$PV\frac{1}{k.n}\pm i\pi\delta(k.n)=\lim_{\alpha\to 0}\frac{1}{k.n\mp i\alpha}, \tag{29}$$

one of the possible prescriptions, namely the one obtained from (25) in the limit when $R\to\infty$, would appear equivalent to the $\alpha$-prescription

$$\frac{1}{[(k.n)^2]}=\frac{1}{(k.n)^2+\alpha^2}\,. \tag{30}$$

However, Eqs.(25) and (30) are not equivalent after being multiplied by k.n, for

$$\lim_{R\to\infty}2\left(\frac{1-\cos k.nR}{k.n}\right)(k.n)=2PV\frac{1}{k.n}\,, \tag{31}$$

whereas

$$\lim_{\alpha\to 0}\frac{1}{(k.n)^2+\alpha^2}(k.n)=1PV\frac{1}{k.n}. \tag{32}$$

Eqs.(23) and (24) can serve as criteria of whether a given axial gauge propagator is incompatible with covariant gauge propagator. They may also serve as criteria of conditional incompatibility: if an evaluation of the l.h.s. of Eq.(23) or (24) with $\frac{1}{[(k.n)^2]_1}$ gives the same result as with $\frac{1}{[(k.n)^2]_2}$, this two prescriptions are either both incompatible or both compatible. For this one needs to calculate only

$$\lim_{L\to\infty}\int\frac{d^{2\omega}k}{(2\pi)^{2\omega}}D_0(k^2)(1-\cos k.\xi L)\left(\frac{1}{[(k.n)^2]_1}-\frac{1}{[(k.n)^2]_2}\right)\,. \tag{33}$$

I verified that the $\alpha$-prescription as $\frac{1}{[(k.n)^2]_1}$ and the PV prescription as $\frac{1}{[(k.n)^2]_2}$ satisfy Eq.(33).

The crucial property of the Wilson line operator, which allows straightforward analysis of its behaviour under gauge transformations, is represented by Eq.(12). Exactly the same behaviour under gauge transformations experiences the following binomial of Dirac fields

$$\psi'(y)\bar{\psi}'(x) = U(y)\left[\psi(y)\bar{\psi}(x)\right]U^{-1}(x). \tag{34}$$

Therefore, for the sake of testing compatibility of axial with covariant Feynman rules, we may also study

$$S = <0|T\left[Pexp(-ig\int_x^{R_x} A_\mu dz^\mu)\psi(x)\bar{\psi}'(y)\bar{P}exp(+ig\int_y^{R_y} A_\mu dz^\mu)\right]|0> . \tag{35}$$

Perturbative evaluation of S in a covariant gauge gives

$$\begin{aligned} S = S_0(x-y) + g^2 C_F \int \frac{d^{2\omega}p}{(2\pi)^{2\omega}} e^{-ip.(x-y)} \int \frac{d^{2\omega}k}{(2\pi)^{2\omega}} D_0(k^2) \Big[ & S_0(p)\gamma^\mu S_0(p-k)\gamma_\mu S_0(p) \\ & - S_o(p-k)\ \not{n} S_0(p)\left(\frac{1-\cos k.nR}{k.n} - i\frac{\sin k.nR}{k.n}\right) \\ & - S_0(p)\ \not{n} S_0(p-k)\left(\frac{1-\cos k.nR}{k.n} + i\frac{\sin k.nR}{k.n}\right) \\ & + n^2[S_0(p-k) - S_0(p)]\left(2\frac{1-\cos k.nR}{(k.n)^2} + \frac{1-\lambda}{k^2}\right)\Big] + O(g^4) , \end{aligned} \tag{36}$$

whereas in the axial gauge, where the gauge field propagator is given by Eq.(16) and has the property (20), we get

$$\begin{aligned} S = S_0(x-y) + g^2 C_F \int \frac{d^{2\omega}p}{(2\pi)^{2\omega}} e^{-ip.(x-y)} \int \frac{d^{2\omega}k}{(2\pi)^{2\omega}} D_0(k^2) \Big[ & S_0(p)\gamma^\mu S_0(p-k)\gamma_\mu S_0(p) \\ & - \left[S_0(p-k)\ \not{n} S_0(p) + S_0(p)\ \not{n} S_0(p-k) + 2S_0(p)\ \not{n} S_0(p)\right]\frac{1}{[k.n]} \\ & + n^2\left[S_0(p-k) - S_0(p) + S_0(p)\ \not{k} S_0(p)\right]\frac{1}{[(k.n)^2]}\Big] + O(g^4) . \end{aligned} \tag{37}$$

The expressions (36) and (37) give rise to a compatibility criterion, the analog of Eq.(24), of the form

$$\lim_{L\to\infty} \int \frac{d^{2\omega}k}{(2\pi)^{2\omega}} (1 - e^{-ik.\xi L}) D_0(k^2)\left[S_0(p-k) - S_0(p)\right]\left(\frac{1}{[(k.n)^2]} - 2\frac{1-\cos k.nR}{(k.n)^2}\right) = 0 \tag{38}$$

and a criterion of conditional compatibility of the form

$$\lim_{L\to\infty} \int \frac{d^{2\omega}k}{(2\pi)^{2\omega}} (1 - e^{-ik.\xi L}) D_0(k^2)\left[S_0(p-k) - S_0(p)\right]\left(\frac{1}{[(k.n)^2]_1} - \frac{1}{[(k.n)^2]_2}\right) = 0 . \tag{39}$$

This last condition is not satisfied by

$$\frac{1}{[(k.n)^2]_1} = \frac{1}{(k.n)^2+\epsilon^2} \quad \textit{and} \quad \frac{1}{[(k.n)^2]_2} = \frac{(k.n)^2-\epsilon^2}{((k.n)^2+\epsilon^2)^2} \; .$$

For both the time-like and the space-like $n$ the l.h.s. of Eq.(39) is $O(\frac{1}{\epsilon})$.

## References


[1] J. Frenkel, Phys. Rev. **D13** (1976) 2325.

[2] R. Delbourgo, A. Salam, and J. Strathdee, Nuovo Cimento **23A** (1974) 54.

[3] J.A. Dixon and J.C. Taylor, Nucl. Phys. **B78** (1974 553.

[4] W. Kummer, Acta Phys. Austriaca, Suppl **XV** (1976) 423.

[5] D. Amati, R. Petronzio, and G. Veneziano, Nucl. Phys. **B140** (1978) 54.

[6] Y.P. Yao, J. Math. Phys. **5** (1964) 1319; W. Kainz, W. Kummer, and M. Schweda, Nucl. Phys. **B79** (1974) 484;

[7] S. Caracciolo, G. Curci, and P. Menotti, Phys. Lett. **113B** (1982) 331.

[8] A. Andraši and J.C. Taylor, Nucl. Phys. **B310** (1988) 222.

[9] H. Cheng and E.C. Tsai, Nucl. Phys. **D36** (1987) 3196.

[10] P. Gaigg, M Kreuzer, M. Schweda, and O. Piguet, J. Math. Phys. **28** (1987) 2781; M. Kreuzer, G. Pollak, H. Skarke, and M. Schweda, Phys. Lett. **196B** (1987) 557.

[11] P. V. Landshoff, Phys. Lett. **169B** (1986) 69.

# A Check of the Lee-Ward Identities in the Planar Gauge with the Leibbrandt-Mandelstam Prescription

Giuseppe Nardelli

Dipartimento di Fisica, Università di Trento, 38050 Povo (Trento), Italy, and
Istituto Nazionale di Fisica Nucleare, Gruppo Collegato di Trento

The set of algebraic noncovariant subsidiary (i.e. gauge) conditions of the kind $n^{\mu} A_{\mu} = \phi$, where $n_{\mu} = (n_0, \boldsymbol{n})$ is a constant four vector, and $\phi$ is a given field in the adjoint representation of a semisimple Lie algebra, has been widely and increasingly adopted in the treatment of many aspects of quantum nonabelian gauge theories [1]. The appealing features of that class of gauge choices are well known to be:

a) the decoupling of the Faddeev-Popov ghosts sector;

b) the freedom of Gribov's ambiguities;

c) the possibility of a simple "parton-like" interpretation of the theory due to the apparent decoupling (or even absence) of the unphysical degrees of freedom.

On the other hand, the use of algebraic noncovariant gauges entails, besides the breakdown of manifest Lorentz covariance, several delicate mathematical problems, due to the presence of the spurious gauge singularities of the kind $[nk]^{-1}$, $[nk]^{-2}$ in the free gauge field propagator. If such singularities are defined by means of the Cauchy principal value (CPV) prescription (or, in general, by means of prescriptions depending on only one single gauge vector) several pathologies arise in perturbation theory, namely:

* the positivity of the sum over the polarization vectors is lost [2] in the (homogeneous) axial gauge $nA = 0$; this stems from the conflict between the positivity of the "square" of the spurious pole and the requirement that its definition must be mathemathically consistent in the sense of the theory of distributions;

* the axial gauge does not reproduce the expected exponential behaviour of the Wilson loop in the large T limit, neither in the temporal [3], nor in the space-like [4] case;

* in the light-cone case ($n^2 = 0$), inconsistencies arise in loop calculations [5], essentially due to pinches between spurious poles and Feynman poles in the evaluation of Feynman integrals.

In order to solve the problem of the light cone gauge, S.Mandelstam and G.Leibbrandt independently proposed a clever prescription [6],[7], such, that the spurious poles and the Feynman poles are always located in the same quadrants of the complex $k_0$ plane; this prescription is given by

$$\frac{1}{[nk]} \equiv \frac{1}{nk + i\epsilon\sigma(n^*k)} = \frac{n^*k}{(nk)(n^*k) + i\epsilon} , \tag{1}$$

where $\sigma(x)$ is the sign distribution and $n^*_\mu = (n_0, -\boldsymbol{n})$ is usally called the dual (or conjugate) vector of $n_\mu$. The first expression is Mandelstam's proposal, the second one is Leibbrandt's one; as they are equivalent in the sense of the theory of distributions, we will refer to eq.(1) as to the Mandelstam-Leibbrandt (ML) prescription. Subsequently this prescription received a canonical justification in Ref.[8].

If the ML prescription is used to handle spurious singularities, the Wick rotation in Feynman integrals can be performed in the usual way (i.e. without extra $\delta(nk)$-terms), so that the powercounting criterion holds for the evaluation of the UV divergences.

As it is apparent from eq.(1), the price to be paid for this nice feature is a further violation of the Lorentz covariance, due to the introduction of the dual gauge vector $n^*_\mu$; this fact has serious consequences on the renormalization of the theory: several nonlocal structures in the external momenta appear in the evaluation of the divergent parts (in the framework of the dimensional regularization) of the proper vertices, so that renormalization of the light-cone gauge theory is only obtained by means of nonlocal counterterms in the effective action. The three main points to be taken into account in order to renormalize the light-cone theory can be summarized as follows [9]:

1) all the nonpolynomial (in the external momenta) structures arising from the computation of a proper vertex are always proportional to at least one gauge vector with a free vector index; as a consequence all those structures disappear in the corresponding Green's function, as the free propagator is orthogonal to the gauge vector $n_\mu$;

2) all the divergent parts of the proper vertices with more than four external legs are always proportional to (at least) one gauge vector with a free vector index; this fact allows to study only the (nonlocal) counterterms of the 2, 3, and 4 point one particle irreducible (1PI) diagrams;

3) it exists only one gauge invariant nonlocal structure $\Delta_\Omega$ which generates all the nonpolynomial (in the external momenta) coefficients of the divergent parts of the proper vertices:

$$\Delta_\Omega = n^\mu D^\nu F_{\mu\nu}\, \Omega , \tag{2}$$

$F_{\mu\nu}$ being the field-strength tensor, $D^\nu$ the covariant derivative, and $\Omega$ being defined as the solution of the equation

$$(nD)\,\Omega = \frac{n^\mu n^{*\nu}}{n^*n} F_{\mu\nu} . \tag{3}$$

Another nice feature of the ML prescription is that it can be extended outside the light cone in a natural (and analytic) way, as was first shown by P. Gaigg and coauthors [10] and G. Leibbrandt [11], respectively. If we consider the prescription in eq.(1) without any constraints on the values of $n^2$ and $n^{*2}$ the limit $n^2 \to 0$ on the regularized Feynman integrals exactly reproduces the light-cone gauge results. We remark that this continuity property does not hold for prescriptions defined by one gauge vector: in this case the limit is very singular (and often does not even exist).

The continuity property of the ML-prescription allows to consider the axial and/or planar gauge with ML-prescription as continuous deformations outside the light-cone of the light-cone gauge: both the axial and the planar gauge tend to the light-cone gauge if the extended ML-prescription is used to handle the spurious singularities.

One may wonder which is the most convenient (simpler) deformation: the axial or the planar one? It is not easy to answer this question because both gauge choices have good and bad features. In the axial case, the simplicity of the Lee-Ward-Takahashi (LWT) identities and the orthogonality condition of the Green's functions with respect to $n_\mu$ simplifies the renormalization program; however, the structure of the free propagator (and the troublesome presence of the double spurious pole) causes some technical difficulties in the evaluation of Feynman diagrams. In the planar case just the opposite occurs: the propagator is simpler (and consequently it is easier to evaluate Feynman diagrams), but the structure of the LWT identities is quite cumbersome: to check them, we have to compute "extra graphs".

Here we demonstrate this feature in a specific example, namely in the divergent parts of the quark selfenergy and quark-quark-gluon (qqg) vertex in the planar gauge with the ML-prescription [12]. The same extension has been performed in the axial direction in Ref.[15] and, in a more general treatment, in Ref.[16]. We shall choose, without loss of generality, $SU(3)$ as gauge group. We shall work in the framework of the dimensional regularization in a $2\omega$ dimensional space-time. The free gluon propagator in the planar gauge is given by

$$D^{ab}_{\mu\nu}(p) = -\frac{i\delta^{ab}}{p^2+i\epsilon}\left(g_{\mu\nu} - \frac{n_\mu p_\nu + n_\nu p_\mu}{[np]}\right) \equiv -\frac{i\delta^{ab}}{p^2+i\epsilon} d_{\mu\nu}(p) \ , \tag{4}$$

where the square bracket denotes ML-prescription. The quark selfenergy turns out to be [12]

$$\Sigma(p) = -\frac{2}{3} i \frac{g^2}{8\pi^2(2-\omega)} \{2m + \not{p} + \frac{2}{D}(\not{n}\, n^*p - \not{n}^*\, np)\} \ , \tag{5}$$

where $D = [(n^*n)^2 - n^2 n^{*2}]^{1/2}$. It is quite remarkable that, in this case, the only difference with the corresponding light-cone gauge result [15] is the replacement $D \to n^*n$.

The divergent part of the qqg vertex consists of two contributions, the abelian ($\Gamma^{(1)a}_\mu$) and the nonabelian ($\Gamma^{(2)a}_\mu$) one. They have been computed in Ref.s [12], [14] and the results read

$$\Gamma^{(1)a}_\mu(p,q) = -\frac{i}{12}\phi^a\{-\gamma_\mu + \frac{2}{D}(n_\mu \not{n}^* - \not{n}\, n^*_\mu)\} \ , \tag{6}$$

$$\Gamma^{(2)a}_\mu(p,q) = \frac{3}{4} i\phi^a \left\{\gamma_\mu \left[1 - \frac{2n^*n}{D} + \frac{2n^2}{D}\frac{n^*k}{nk}\right] + \frac{2}{D^3}\left(n^2\frac{n^*k}{nk} - n^*n\right)[n^2 n^*_\mu \not{n}^* - \right.$$

$$-(n^*n)(n_\mu \not{n}^* + \not{n}\, n^*_\mu)] + \frac{2\not{n}\, n_\mu}{D}\left[\frac{n^*k}{nk}\left(\frac{(n^*n)^2}{D^2} - 3\right) + (n^*)^2\left(\frac{2}{n^*n + D} - \frac{n^*n}{D^2}\right)\right]\Bigg\} \ , \tag{7}$$

where $\phi^a = -g^3 T^a/[8\pi^2(2-\omega)]$, $T^a$ ($a = 1, \ldots, 8$) being the (hermitean) generators of $SU(3)$, and $k = q - p$ is the external momentum carried by the vector line. Eq.(7) obviously tends to the corresponding light-cone gauge expression [15] in the limit $n^2 \to 0$.

Now we face the problem of checking the LWT identities on the results obtained. If $Z(j)$ is the generating functional of the connected Green's functions, the Ward-Takahashi identities read

$$(n\partial)\partial^2 \frac{n_\mu}{n^2}\frac{\delta Z}{\delta J^a_\mu(x)} + g f^{acb}\frac{n_\mu n_\nu}{n^2}\frac{\delta Z}{\delta J^c_\nu(x)}\partial^2 \frac{\delta Z}{\delta J^b_\mu(x)} - ig f^{acb}\frac{n_\mu n_\nu}{n^2}\partial^2_x \frac{\delta^2 Z}{\delta J^c_\nu(y)\delta J^b_\mu(x)}\Bigg|_{x=y} -$$

$$-g f^{acb} J^b_\mu(x)\frac{\delta Z}{\delta J^c_\mu(x)} - \partial_\mu J^{a,\mu}(x) + ig\bar{\eta}(x)T^a \frac{\delta Z}{\delta\bar{\eta}(x)} - ig\frac{\delta Z}{\delta\eta(x)}T^a\eta(x) = 0 \ , \tag{8}$$

where $J_\mu$, $\bar{\eta}$ and $\eta$ are the source terms for (respectively) the gauge potential $A_\mu$ and the fermion fields $\Psi$, $\bar{\Psi}$. We want to point out the presence of the term proportional to the second derivative of $Z(j)$: this means that in the check for the three point Green's functions we shall also have to compute "four-point" diagrams. This contribution does not appear in the axial gauge, and its presence leads to similar complications in the structure of the identities for the proper vertices. As a matter of fact, if we perform the usual Legendre transformation we obtain the following Lee identities for the generating functional $\Gamma$ of the 1PI diagrams, viz.

$$(n\partial)\partial^2 \frac{(n\mathcal{A})^a}{n^2} + g f^{acb}\frac{n_\mu n_\nu}{n^2}\frac{1}{i}\left(\partial^2_x\left[\frac{\delta^2\Gamma}{\delta\mathcal{A}^c_\nu(y)\delta\mathcal{A}^b_\mu(x)}\right]^{-1}\right)_{x=y} + g f^{acb}\mathcal{A}^c_\mu \frac{\delta\Gamma}{\delta\mathcal{A}^b_\mu} +$$

$$+\partial_\mu \frac{\delta\Gamma}{\delta\mathcal{A}^a_\mu} + g f^{acb}\frac{n_\mu n_\nu}{n^2}\mathcal{A}^c_\nu \partial^2 \mathcal{A}^b_\mu - ig\frac{\delta\Gamma}{\delta\psi}T^a\psi + ig\bar{\psi}T^a\frac{\delta\Gamma}{\delta\bar{\psi}} = 0 \ , \tag{9}$$

where $\mathcal{A}^a_\mu$ , $\bar{\psi}$ and $\psi$ are the "classical fields" variables corresponding to $A^a_\mu$, $\bar{\Psi}$ and $\Psi$.

The presence of the the second term in eq.(9) yields serious problems in the diagrammatic interpretation of the 1PI diagram: differentiations with respect to the classical fields variables generate "mixed" identities, in the sense that the Lee identities (for the proper vertices) also involve connected Green's functions. A possible way to get rid of this difficilty has been shown in Ref.[16] and subsequently tested in a specific example in Ref.[14]. Here, however, we prefer to check the LWT identities for the connected Green's functions, i.e. to use eq.(8). If we differentiate this equation twice with respect to the fermion sources $\bar{\eta}$ and $\eta$ and perform the Fourier transformation, we obtain the equation symbolically drawn in Fig.(1); here $m$ is the order of the loop expansion, the black dot corresponds to multiplication with $r^2$ and the dashed line means that we have to "short-circuit" the two lines, namely to set $x = y$ in the Fourier transform.

As it is apparent from Fig.(1), in order to check the LWT identities we have to calculate the extra graphs concerning the four point Green's function. These have been computed in Ref.[10] and give rise to two contributions, an abelian ($M^a(p,q)$) and a

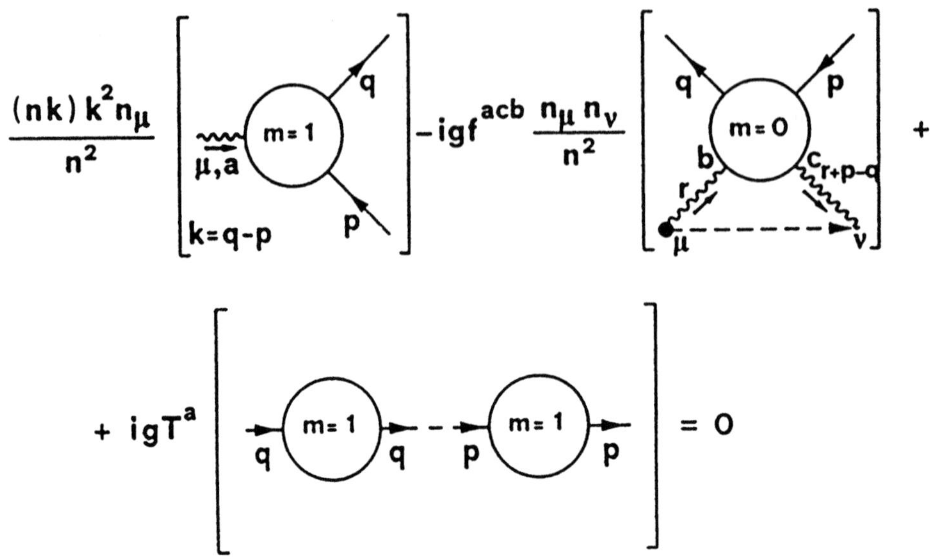


**Fig.(1)** LWT identity for the qqg connected Green's function.

nonabelian one ($L^a(p,q)$). It is convenient to calculate directly the diagrams multiplied by $-igf^{acb}n_\mu n_\nu/n^2$; in this way we obtain ($k=q-p$)

$$M^a(p,q)=\frac{3}{4}\phi^a n^2\frac{\not p+m}{p^2-m^2}\left\{\frac{6\not k}{D}\left(\frac{n^*k}{nk}-\frac{(n^*)^2}{n^*n+D}\right)-\frac{2\not n}{n^2}\left[\frac{n^*n}{D^3}\cdot\right.\right.$$

$$\left.\left(\frac{n^2(n^*k)^2}{nk}+(n^*)^2nk-2(n^*n)\,n^*k\right)++\frac{2}{D}\left(n^*k-\frac{(n^*)^2\,nk}{n^*n+D}\right)\right]+$$

$$\left.+\frac{2\not n^*}{D^3}\left[\frac{n^2(n^*k)^2}{nk}+(n^*)^2\,nk-2(n^*n)\,n^*k\right]\right\}\frac{\not q+m}{q^2-m^2}\,. \tag{10}$$

$$L^a(p,q)=-\frac{8}{3}\phi^a n^2\frac{\not p+m}{p^2-m^2}\left\{\frac{2\not k}{D}\left(\frac{n^*k}{nk}-\frac{(n^*)^2}{n^*n+D}\right)-\frac{\not n}{n^2}\left[\frac{n^*n}{D^3}\cdot\right.\right.$$

$$\left.\cdot\left(\frac{n^2(n^*k)^2}{nk}+(n^*)^2nk-2(n^*n)\,n^*k\right)+\frac{2}{D}\left(n^*k-\frac{(n^*)^2\,nk}{n^*n+D}\right)\right]+$$

$$\left.+\frac{\not n^*}{D^3}\left[\frac{n^2(n^*k)^2}{nk}+(n^*)^2\,nk-2(n^*n)\,n^*k\right]\right\}\frac{\not q+m}{q^2-m^2}\,, \tag{11}$$

The Green's functions corresponding to the 1PI expressions given in eqs. (5), (6), (7) are easily computed attaching propagators to the external lines; concerning the contribution to the Green's function coming from eqs. (6) and (7), we calculate directly the expressions multiplied by $n_\mu$, as required by Fig.(1). In so doing it is easy to obtain, respectively

$$igT^a \mathcal{G}_2(p) = \frac{2}{3}\phi^a \frac{\not p + m}{p^2-m^2}\{2m + \not p + \frac{2}{D}[\not n\, n^*p - \not n^*\, np]\}\frac{\not p + m}{p^2-m^2} \ , \tag{12}$$

$$\mathcal{G}_3^{a(1)}(p,q) = -\frac{1}{12}\phi^a \frac{n^2}{k^2(nk)}\frac{\not p + m}{p^2-m^2}\{-\not k + \frac{2}{D}(nk\, \not n^* - \not n\, n^*k)\}\frac{\not q + m}{q^2-m^2} \ , \tag{13}$$

$$\mathcal{G}_3^{a(2)}(p,q) = \frac{3}{4}\phi^a \frac{n^2}{(nk)k^2}\frac{\not p + m}{p^2-m^2}\left\{\not k\left[1 - \frac{2n^*n}{D} + \frac{2n^2}{D}\frac{n^*k}{nk}\right] + \frac{2}{D^3}\cdot\right.$$
$$\cdot\left(n^2\frac{n^*k}{nk} - n^*n\right)[n^2(n^*k)\,\not n^* - (n^*n)nk\,\not n^* - (n^*n)(n^*k)\,\not n] + \frac{2\,\not n\, nk}{D}\cdot$$
$$\left.\cdot\left[\frac{n^*k}{nk}\left(\frac{(n^*n)^2}{D^2} - 3\right) + (n^*)^2\left(\frac{2}{n^*n + D} - \frac{n^*n}{D^2}\right)\right]\right\}\frac{\not q + m}{q^2-m^2} \ . \tag{14}$$

As a last step we need the contributions to the qqg Green's function produced by the quark selfenergies ($\mathcal{G}_3^{a(3)}(p,q)$, $\mathcal{G}_3^{a(4)}(p,q)$) and by the gluon selfenergy ($\mathcal{G}_3^{a(5)}(p,q)$). They have been computed in Ref.[12] with the following result (the expressions are multiplied by $n_\mu$, as suggested by Fig.(1))

$$\mathcal{G}_3^{a(3)}(p,q) + \mathcal{G}_3^{a(4)}(p,q) = \frac{2}{3}\phi^a \frac{n^2}{k^2(nk)}\frac{\not p + m}{p^2-m^2}\left\{\frac{2}{D}(\not n\, n^*k - \not n^*\, nk) + \right.$$
$$+2\,\not k + \frac{[(3m + \frac{2}{D}(\not n\, n^*p - \not n^*\, np))(\not p + m)(\not q - m)]}{p^2-m^2} -$$
$$\left. - \frac{[(\not p - m)(\not q + m)(3m + \frac{2}{D}(\not n\, n^*q - \not n^*\, nq))]}{q^2-m^2}\right\}\frac{\not q + m}{q^2-m^2} \ . \tag{15}$$

$$\mathcal{G}_3^{a(4)}(p,q) = -\frac{1}{3}\phi^a \frac{n^4}{k^2(nk)}\left\{\frac{2}{D}\left(\frac{n^*k}{nk} - \frac{(n^*)^2}{n^*n + D}\right)\left[\not k - \frac{(nk)\,\not n}{n^2}\right] + \right.$$
$$\left. + \frac{1}{D^3}\left[\frac{n^2(n^*k)^2}{nk} + (n^*)^2 nk - 2(n^*n)\, n^*k\right]\left(\not n^* - \frac{(n^*n)\,\not n}{n^2}\right)\right\} \ . \tag{16}$$

Summing up all this terms as required by Fig.(1) it is easy to check that

$$\frac{(nk)k^2}{n^2}\left(\sum_{i=1}^{4}\mathcal{G}_3^{a(i)}(p,q)\right) + L^a(p,q) + M^a(p,q) - igT^a[\mathcal{G}_2(p) - \mathcal{G}_2(q)] = 0 \ , \tag{17}$$

so that the LWT identities are indeed satisfied.

In conclusion, we want to make a brief comment on the renormalization of "extended light-cone theories". First we notice that the divergent parts of the Green's functions, at variance with the light-cone gauge, are still plagued by nonpolynomial terms. This feature manifests itself also in the axial extension: this means that in the planar and/or axial gauge with the ML- prescription, the locality of the theory will be recovered only at the S-matrix level.

Furthermore, in the light-cone gauge, all the nonlocal contributions are generated by a single nonlocal structure, namely $\Delta_\Omega$. In the extended case, the condition $n^2 \neq 0$ ($n^{*2} \neq 0$) allows the presence of several structures which could generate nonlocal contributions: as a consequence, given a nonlocal divergence, it could be a nontrivial task to find the nonlocal structure which renormalizes it.

Finally, in the light-cone case all the Feynman diagrams with more than four external legs are always proportional to (at least) one gauge vector with a free vector index, so that these contributions vanish in the corresponding Green's functions. In the extended case this does not happen, and the number of the divergences are expected to increase with the number of the external legs. At present, this is the most severe obstacle in the renormalization of the extended light-cone theories: it is necessary to find a criterion to control the divergences of the Green's functions with more than four legs in the multiloop expansion.

## References


[1] For a review, see for example G.Leibbrandt, Rev. Mod. Phys. **59** (1987) 1067.

[2] N.Nakanishi, Phys. Lett. **131B** (1983) 381.

[3] S.Caracciolo, G.Curci and P.Menotti, Phys. Lett. **113B** (1982) 311.

[4] G.Nardelli and R.Soldati, Preprint DFUB 7 (1989) - Bologna.

[5] D.M.Capper, J.J.Dulwich and M.J.Litvak, Nucl. Phys. **B241** (1984) 463.

[6] S.Mandelstam, Nucl. Phys. **B213** (1984) 149.

[7] G.Leibbrandt, Phys. Rev. **D29** (1984) 1699.

[8] A.Bassetto, M.Dalbosco, I.Lazzizzera and R.Soldati, Phys. Rev. **D31** (1985) 2012.

[9] A.Bassetto, M.Dalbosco and R.Soldati, Phys. Rev. **D36** (1987) 3138.

[10] P. Gaigg, M. Kreuzer, O. Piguet and M. Schweda, J. Math. Phys. **28** (1987) 2781; P. Gaigg and M.Kreuzer, Phys. Lett. **205B** (1988) 530.

[11] G. Leibbrandt, Nucl. Phys. **B310** (1989) 405.

[12] G.Nardelli, Phys. Rev. **D40** (1989) 2662.

[13] G.Leibbrandt and S.L.Nyeo, Phys. Lett. **140B** (1984) 417.

[14] G.Pollak, Phys. Rev. **D40** (1989) 2057.

[15] G.Leibbrandt and S.L.Nyeo, Phys. Rev. **D39** (1989) 1752.

[16] W.Kummer, Phys. Rev. **D37** (1988) 454.

# III

# Renormalization and Gauge(in)dependence of Yang-Mills Theories in Noncovariant Gauges

# Introduction

The ultimate goal in quantizing gauge theories is the proof that a finite number of local counterterms to the action suffices for the compensation of UV-divergences in physically observable quantities like the S-matrix. In addition, such observables must not depend on the specific way the gauge has been fixed. For noncovariant gauges of Yang-Mills theories, characterized by a fourvector $n_\mu$ presently consensus exists that the case of light-cone gauges ($n^2 = 0$) alone is amenable to a reduction to the well-known "standard" approach of renormalization. Although nonlocal counterterms appear in this case as well, they can be controlled by a (nonlocal) redefinition of the gauge fields (cf. also the review of Bassetto in part I). In order to attain the rigor of argument achieved for covariant gauges, certain ingredients of the proof of renormalization must be verified to hold also here. For the Leibbrandt gauge the UV-behaviour is investigated (Becchi) whereas the "extended" BRS-technique has been exploited successfully to determine possible counterterms (Skarke) even for $n^2 \neq 0$ and to discuss the relation to the ones in covariant gauges (Piguet). The gauge-dependence of noncovariant counterterms was also checked in explicit one-loop calculations (Andraši, Bagan-Martin). A first attempt towards a proof of the action principle in noncovariant gauges was presented (Balasin).

In the last years the Vienna group has developed a concept of noncovariant supergauges because e.g. the Wess-Zumino gauge in supersymmetric gauge theories can be understood as a special case of a noncovariant gauge fixing in superspace (W. Kummer et al.).

The gauge-independence of "observable" quark condensates in QCD (Elias, Lavelle-Schaden) and other special applications (Wanders) round off this part.

# Nonstandard Gauges: A Personal Point of View

C. Becchi

Dipartimento di Fisica, Università di Genova, and
Istituto Nazionale di Fisica Nucleare, Sezione di Genova, I-16146 Genova, Italy

**Abstract:** A minimal dimensional approach to the pure Yang-Mills theory in the strong light-cone gauge is constructed generalizing a method developed by Speer, Breitenlohner and Maison in the standard, local and covariant, framework. The known results on the ultraviolet renormalizability of the theory are confirmed.

## 1 Introduction

The construction of fully quantized gauge theories with noncovariant gauge fixings is plagued with many relevant difficulties. The basic ones, but not necessarily the hardest to overcome, are the lack of covariance and that of locality. Indeed, as it is well known, covariance and locality are the crucial tools of renormalization theory. Locality, through power counting, and covariance, as a universal symmetry, constrain the counterterms of a renormalizable theory in the framework of Lagrangian vertices. Giving up them, both the nature of counterterms remains widely undetermined thus spoiling completely the power of the Lagrangian approach.

In spite of this threatening prospect the subject of quantization of nonstandard gauge theories, from the Coulomb gauge up to the light-cone, temporal and planar ones, is still alive. The present workshop certainly testifies this.

The problem we are facing now is to push the analysis beyond the traditional formal level reaching the structure of a systematic, to all orders, renormalization procedure . The main hints in this direction are to exploit the residues of the lost locality and covariance. These could consist in some weaker, possibly anisotropic, power counting rules, in the residual Lorentz symmetry, and, last but by no means least,in the gauge symmetry. Willing to make use of gauge invariance to replace partially locality and covariance one has to rely on an invariant regularization procedure thus automatically guaranteeing the persistence of this symmetry beyond the tree approximation.

The use of an invariant regularization scheme turns out to be less compulsory in the standard, local-covariant, situation. Indeed in this case power counting is sufficient to ensure that the counterterms belong to the finite dimensional space of the local vertices with canonical dimension up to four. It is then possible in many instances to show by a purely algebraic, regularization independent, analysis that in this space there is a subspace of counterterms implementing order by order the wanted (gauge) symmetry properties. The subspace, whose coordinates are all the parameters of the Green's functions can be considered as the tangent space to the functional manifold of the Lagrangians of our theory.

In the nonstandard case the algebraic method turns out to be unfeasible since the functional space of the counterterms which are compatible with the residual covariance and power counting is in general infinite dimensional. Therefore the use of an invariant regularization scheme is necessary.

Presently gauge invariant regularization is synonymous for the dimensional one whose more rigorous analysis has been performed by Speer and completed within a minimal subtraction scheme by Breitenlohner and Maison (SBM)[1]. In our approach to nonstandard gauges we shall follow as much as possible this method.

Referring to the standard situation SBM have been able to show that the minimal renormalization scheme is additive, in the sense that it can be traced back, order by order, to a system of local Lagrangian counter-terms. They have also shown that their scheme ensures the validity of the quantum action principle. This essentially means that one can make use of the field equations within the Green functions of the fields and of composite operators and that to derive a Green function with respect to a parameter of the theory is equivalent to the insertion of an integrated vertex corresponding to the derivative of the classical action with respect to the same parameter. In the case of massless theories they also prove that the Green functions are distributions.

Unfortunately the analysis of SBM does not apply directly to the noncovariant case. To give some hints on how this difficulty could be overcome is the aim of the present paper.

In principle it would be particularly nice if we could present a general scheme suitable to treat the whole family of noncovariant gauges. As it is apparent from many interesting reports presented at this workshop, this intention is still too ambitious. We shall therefore limit ourselves to the discussion of the strong light-cone case. In this case Bassetto, Soldati and coworkes (BS)[2] have given an almost complete series of arguments in favour of renormalizability.

Thus our effort appears drastically reduced. It consists in a check of BS suggestions after their translation into SBM framework.

The main arguments discussed by BS are the following:

First of all, even if the vertices of the theory are known to develop nonlocal divergences already at one loop [3], and hence, in order to define them, one has to introduce nonlocal counterterms, the Green's functions of the fields do not seem to need any such term. Taking into account the convenience of dealing with vertices in the renormalization procedure, BS introduce the concept of the effective vertex.

A very simple definition of the effective vertices can be given in the functional framework by replacing the gauge field variable $a$ appearing in the vertex generator $\Gamma[a]$ with its projection $a_t$ orthogonal to the light-cone gauge parameter $n_\mu$. Calling $n^*_\mu$ the parity-reflected image of $n_\mu$ one has:

$$a_{\mu T} = a_\mu - \frac{n_\mu n^*_\nu + n_\nu n^*_\mu}{n.n^*} a_\nu, \tag{1}$$

$$a_{\mu t} = a_{\mu T} + n_\mu \frac{n^*.a}{n.n^*}. \tag{2}$$

Then one defines the effective vertex generator:

$$\Gamma_{eff}[a] = \Gamma[a_t]. \tag{3}$$

This new functional should be free of nonlocal divergences, in other words, it should not require any nonlocal counterterm to be defined.

A second point in BS analysis is the remark that the ultraviolet degree of the light-cone propagator introduced by Leibbrandt and Mandelstam is anisotropic. Writing a Minkowskian vector $p$ as $p = p_T + p_1 n + p_2 n^*$, we have degree $-1$ in the hyperplane $p = p_T + p_1 n$ (BS-transverse), while we have $-2$ in the whole D-dimensional space-time. BS suggest consequently that the Feynman integrals should have different degrees of divergence in the BS-transverse and longitudinal directions. They also notice that the derivations with respect to the BS-transverse components of the external momenta lower both degrees of a vertex. From this they conclude that the divergent part of a primitive vertex (without divergent proper subdiagrams) is a polynomial in the BS-tranverse components of the external momenta.

Finally they introduce an argument excluding the dipendence of the divergent parts of the effective vertices on dimensionless ratios of longitudinal components of external momenta. This argument is based on a comparison of the light-cone with the planar gauge.

## 2 The Dimensional Scheme

Let us now come to the scheme of SBM. In the standard (covariant) framework, this is based on the Feynman-Schwinger parametric representation of the propagator:

$$\frac{g^{\mu\nu}}{p^2 + i\epsilon} = i g^{\mu\nu} \int_0^\infty d\alpha e^{i\alpha(p^2+i\epsilon)}. \tag{4}$$

After Gaussian integration over the internal momenta a regularized amplitude in D space-time dimensions becomes:

$$A(p_1, ..., p_N, D) = \lim_{\epsilon \to 0} A_\epsilon(p_1, ..., p_N, D), \tag{5}$$

and:

$$A_\epsilon(p, D) = \int_0^\infty \prod_L d\alpha_l \left[ I_\epsilon(p, u, \alpha, D)|_{u=0} \right], \tag{6}$$

where:

$$I_\epsilon(p,u,\alpha,D) = d(\alpha)^{-\frac{D}{2}} Z(\frac{-i\partial}{\partial u}) exp[iV(p,u,\alpha) - \epsilon \sum_L \alpha_l], \tag{7}$$

and the quadratic form V is given by:

$$V(p,u,\alpha) = \sum_V \frac{d_{ij}(\alpha)}{d(\alpha)}(p_i + \sum_L e_{il}\frac{u_l}{2\alpha_l})(p_j + \sum_L e_{jl}\frac{u_l}{2\alpha_l}) - \sum_L \frac{u_l^2}{4\alpha_l}. \tag{8}$$

In these formulae $L$ and $V$ are the sets of lines and that of external vertices of the diagram. Therefore the integration variables $\alpha_l$ and the auxiliary D-vectors $u_l$ are associated to the lines, while the momenta $p_i$ are attached to the external vertices, the matrix $e_{il}$ defines the flow of the momentum of the line l into the vertex i, its elements are either null or +/-1. The elements of the matrix $d_{ij}(\alpha)$ and $d(\alpha)$ are the so called Symanzik functions of the diagram, they are homogeneous polynomials of degree h+1 and h respectively, h being the number of loops. The differential operator $Z$ corresponds to the numerator of the integrand in the representation of the amplitude as a momentum integral.

The crucial step in SBM analysis is based on the division of the domain of integration in $\alpha$-space into sectors corresponding to families of subgraphs H of the diagram G. In each sector one introduces scaling variables $t_H$ as well as line variables $\beta_l$ so that:

$$\alpha_l = \beta_l \prod_{H \supset l} t_H^2. \tag{9}$$

In $D = 4 - \nu$ dimensions, the contribution of a given sector to $A(p, D)$, integrated over $t_G$, yields a finite number of terms of the form:

$$\Gamma((\nu h_G - \omega_G + r)/2) \int_0^1 \prod_{H \neq G} dt_H t_H^{\nu h_H - \omega_H - 1} \int_0^1 \prod d\beta_l g_r(p,t,\beta,D)$$
$$. [-iW(p,t,\beta) + \epsilon S(t,\beta)]^{(\nu h_G - \omega_G + r)/2} . \tag{10}$$

In this formula:

$$W(p,t,\beta) = -\frac{V(p,u,\alpha)}{t_G^2}, \tag{11}$$

and:

$$S(t,\beta) = \frac{\sum_L \alpha_l}{t_G^2}. \tag{12}$$

The ultraviolet divergence of the subdiagram $H$ $(G)$ is given by $\omega_H$ $(\omega_G)$. and $g_r(p,t,\beta,D)$ is a homogeneous polynomial in $p$ of degree $r$ holomorphic in D; it is generated by the action of the differential operator Z. The functions $g_r, W$ and $S$ are $C^\infty$ in $t$ and $\beta$ and $S \geq 1$, therefore, for $\epsilon > 0$, Eq.(10) gives a meromorphic function of D.

Always keeping $\epsilon > 0$, it is important to notice that the poles of this functions are related with sets of nonoverlapping, 1-particle irreducible, subdiagrams of $G$, that is with forests of $G$. The recursive subtraction of the pole due to the divergent subdiagrams leads to a result for the whole amplitude which is divergent only if $\omega_G \geq 0$. In this case, the left over singularity is, essentially for dimensional reasons, a homogeneous polynomial in the external momenta of $G$ and $\epsilon^{\frac{1}{2}}$. This pole term represents the counter term corresponding to $G$ in the construction of amplitudes related with diagrams containing $G$ as a subdiagram.

The analysis of SBM also gives an exhaustive knowledge of the infrared limit, $\epsilon \to 0$, of the amplitude, showing in particular that in the pure renormalizable case this limit defines a distribution. In the following, trying to translate the light-cone amplitudes into the SBM formalism, we shall disregard this infrared problem which is however one of the remaining major difficulties with nonstandard gauges.

## 3 The Light-Cone Gauge in the Dimensional Scheme

The dimensionally regularized light-cone amplitudes will be constructed following step by step the SBM formulation described above. The first point is the parametric representation of the propagator. Taking into account the Leibbrandt-Mandelstam [2,4] prescription and defining $p_T$ as in Eq.(1) and

$$p_L = p - p_T, \tag{13}$$

the propagator is written:

$$\begin{aligned}
&\frac{1}{(n.n^*)}\frac{1}{p^2+i\epsilon}\frac{1}{p_L^2+i\eta}(n^*.p)\left[g^{\mu\nu}(n.p)-n^\mu p^\nu-n^\nu p^\mu\right]= \\
&\frac{1}{(n.n^*)}\int_0^\infty d\alpha d\beta\left[(n^*.\partial_u)\left[g^{\mu\nu}(n.\partial_u)-n^\mu\partial_u^\nu-n^\nu\partial_u^\mu\right]e^{i(p.u)}\right]|_{u=0} \\
&\qquad .exp\left[i\alpha(p^2+i\epsilon)+i\beta(p_L^2+i\eta)\right]= \\
&\frac{1}{(n.n^*)}\int_0^\infty d\alpha d\beta\left[(n^*.\partial_u)\left[g^{\mu\nu}(n.\partial_u)-n^\mu\partial_u^\nu-n^\nu\partial_u^\mu\right]e^{i(p.u)}\right]|_{u=0} \\
&\qquad .exp\left[i\alpha(p_T^2+i\epsilon)+i(\alpha+\beta)(p_L^2+i\eta)\right].
\end{aligned} \tag{14}$$

Now one has to insert this propagator into the formal expression for the amplitude in D dimensions and to perform the Gaussian integration with respect to the D-2 dimensional components of $p_T$ and the two dimensional ones, in the plane $(n-n^*)$, of $p_L$. The resulting amplitude is:

$$A_{LC\epsilon}(p,D)=\int_0^\infty\prod_L d\alpha_l\prod_L d\beta_l\left[I_{LC\epsilon}(p,u,\alpha,\beta,D)|_{u=0}\right], \tag{15}$$

where:

$$I_{LC\epsilon}(p,u,\alpha,\beta,D) = d(\alpha)^{-\frac{D-2}{2}} d(\alpha+\beta)^{-1}$$
$$Z(\frac{-i\partial}{\partial u}) exp[iV_{LC}(p,u,\alpha,\beta) - \epsilon \sum_L \alpha_l - \eta \sum_L \beta_l], \tag{16}$$

and the quadratic form $V_{LC}$ is now given by:

$$V_{LC}(p,u,\alpha,\beta) = \sum_V \frac{d_{ij}(\alpha)}{d(\alpha)} (p_{Ti} + \sum_L e_{il} \frac{u_{Tl}}{2\alpha_l})(p_{Tj} + \sum_L e_{jl} \frac{u_{Tl}}{2\alpha_l})$$
$$+ \sum_V \frac{d_{ij}(\alpha+\beta)}{d(\alpha+\beta)} (p_{Li} + \sum_L e_{il} \frac{u_{Ll}}{2(\alpha+\beta)_l})(p_{Lj} + \sum_L e_{jl} \frac{u_{Ll}}{2(\alpha+\beta)_l})$$
$$- \sum_L \frac{u_{Tl}^2}{4\alpha_l} - \sum_L \frac{u_{Ll}^2}{4(\alpha+\beta)_l}. \tag{17}$$

The symbols appearing in Eq.'s(15,16,17) have exactly the same meaning of those in Eq.'s(6,7,8).

Trying to perform the next step, i. e. the division of the domain of integration in $\alpha$- and $\beta$-space into sectors, we understand the nature of the ultraviolet difficulties in the light-cone gauge. Indeed now the number of these sectors is enormously increased and in general we lose the correspondence of these sectors with families of subdiagrams of $G$. Although here, in much the same way as in the standard case, the regularized amplitudes turn out to be meromorphic functions of D and their singularities also come from the integration around the origin of subsectors of the parametric space, ***a priori*** in the light-cone case the interpretation of the source of divergences in terms of subdiagrams might be lost. In particular together with those due to the singular behavior of the integrand when the $\alpha$ and $\beta$ parameters vanish simultaneously there is the possibility of new divergences coming from the $\alpha$-integration alone. In this case the subdiagram responsible of a divergence would not appear any more in the singular part of the amplitude as shrunk to a point as it does in the standard case.

To give a better picture of this phenomenon and of its possible consequences, let us consider the following integral which reproduces in a simplified form the critical properties of $A_{LC\epsilon}$:

$$\int_0^\infty d\alpha d\beta \frac{\alpha^{\mu-1}\beta^{N-1}}{(\alpha+\beta)^\lambda} exp\left[i(\alpha p_T^2 + \beta p_L^2) - \epsilon(\alpha+\beta)\right] \tag{18}$$

The integration over a global scale variable yields:

$$\Gamma(\mu+N-\lambda) \int_0^1 \frac{dx}{x(1+x)^\lambda} \times$$

$$\times\left[\frac{x^N}{[p_T^2+(1+x)p_L^2]^{\mu+N-\lambda}}+\frac{x^\mu}{[xp_T^2+(1+x)p_L^2]^{\mu+N-\lambda}}\right]. \tag{19}$$

In Eq.'s (18) and (19) $N$ is a positive integer which mimics the number of lines. Hence the first term in Eq.(19) leads to a pole only if $\mu + N - \lambda$, the global degree of the integrand in Eq.(18), is a nonpositive integer and the residue of this pole is a polynomial in the momenta $p_L$ and $p_T$. However, when $\mu$ tends to a nonpositive integer, the second term develops an independent singularity and in general the corresponding residue is not a polynomial in the momenta. In the realistic case this should correspond to a nonlocal divergence which, as shown by our example, might stem from the the integration around the origin of some $\alpha$-sector. We know from explicit calculations that such nonlocal divergences already appear in the vertices at one loop; the question is if they appear in the effective vertices.

Confirming the results of BS we are going to show that in the case of the effective vertices the $\alpha$- degrees ($deg_\alpha$) of the corresponding sectors at the origin are always positive. Let us consider, first of all, a dimensionally regularized primitive effective vertex $\Gamma_{\mu_1,\ldots,\mu_k}$. In its momentum representation the vertex appears as the linear combination of terms of the form:

$$\int\prod_{\mathcal{L}} dq_a^D I_{\mu_1,\ldots,\mu_k}(p)\prod_L \frac{1}{(p_l^2+i\epsilon)((n.p)(n^*.p)+i\eta}(n.p_l')(n^*.p). \tag{20}$$

Here $\mathcal{L}$ is the set of loops of the diagram, $q_a$ is a loop momentum, $p_l$ is the momentum carried by the line l and $p_l'$ a linear combination of line momenta. The vertex factors are collected in $I_{\mu_1,\ldots,\mu_k}$ which depends only on the $p_l$'s. The effectiveness of the vertex $\Gamma$ is exhibited by the absence of uncontracted $n$ vectors. Now the smallest $deg_\alpha$ appearing in the parametric version of Eq.(20) is given by the number of lines $|L|$, minus $\frac{D-2}{2}$ times the number of loops $|\mathcal{L}|$ plus the contribution of the numerator. The last term comes only from $I_{\mu_1,\ldots,\mu_k}$, since the rest of the numerator contains only longitudinal components of the line momenta, and it is not smaller than $-\frac{1}{2}$ times the degree of the vertex factor. This degree, in turn, does not exceed the number of vertices $|V|$.

Thus we have:

$$deg_\alpha = |L| - \frac{D-2}{2}|\mathcal{L}| - \frac{1}{2}|V| \geq \frac{1}{2}|V| - \frac{D-2}{2} - \frac{D-4}{2}|L| \tag{21}$$

This is positive for diagrams with more than two vertices and for $D$ not too far from 4. For the few diagrams with one or two vertices, the locality of counterterms can be directly verified [5]. In general the only possible divergences in $\Gamma_{\mu_1,\ldots,\mu_k}$ come from the origin of its parameter space. They are poles whose residues are polynomials in the external momenta and $\epsilon$.

It remains to discuss the case when the effective vertex, not being primitive, contains divergent subdiagrams. The argument used above is easily generalized to this case by a standard iterative procedure, owing to the fact that the divergent subdiagrams are effective vertices and the corresponding counterterms are local.

Acknowledgments: The author is deeply indebted to professors W. Kummer and M. Schweda for having invited him to attend the present workshop thus urging him to enter into the subject of nonstandard gauges.

## References

[1] E. Speer, *Generalized Feynman Amplitudes*, Princeton University Press 1969.
E. Speer, Ann. Inst. Henri Poincare, **A23** (1975), 1.
P. Breitenlohner and D. Maison, Commun. Math. Phys. **11** (1977), 11.
P. Breitenlohner and D. Maison, Commun. Math. Phys. **11** (1977), 38.

[2] G. Leibbrandt, Phys. Rev. **D29** (1984), 1699.

[3] A. Bassetto, M. Dalbosco and R. Soldati, Phys. Rev. **D36** (1987), 3138.
A. Bassetto, G. Nardelli and R. Soldati Mod. Phys. Lett. **A3** (1988), 1663.

[4] S. Mandelstam, Nucl. Phys. **B213** (1983), 149.

[5] M. Dalbosco, Teorie di Yang-Mills nella gauge di cono luce", Ph. D. Thesis, Padova (1986), unpublished.

# Gauge Independence Through Extended BRS Symmetry: Application to the Light-Cone Gauge

Olivier Piguet

Dép. de Physique Théorique, Université de Genève,
CH-1211 Genève 4, Switzerland

**Abstract:** The gauge independence of the observables of a gauge theory can be ensured by a construction based on an extension of the BRS symmetry. Here the set of gauge fixing parameters transforms under BRS into a set of BRS-invariant Grassmann parameters and the corresponding extended Slavnov-Taylor identity is required to hold. After a review of the procedure, we shall apply it to the problem of the nonlocal divergences encountered in the renormalization of gauge theories in the light-cone gauge. A regularization of these singularities is first introduced. Their cancellation in physical quantities then results from the regulator being a gauge parameter. A few words will also be said about the Lorentz symmetry breaking caused by the gauge condition: the covariance of the physical quantities follows from the fact that the breaking is controlled by gauge parameters, too.

## 1 Extended BRS Invariance

Extended BRS invariance is a method for controlling the independence of the physical quantities of a gauge theory of the parameters defining the gauge condition [1,2,3]. Let us describe this method on the example of pure Yang-Mills theory in a covariant gauge. The results presented in this section generalize straightforwardly to theories coupled with matter fields [3], to supersymmetric theories [4,5] and to gauges involving more than one parameter [4].

The invariant action is

$$\Gamma_{\text{inv}} = \text{Tr} \int d^4x \left\{ -\frac{1}{4g^2} F_{\mu\nu} F^{\mu\nu} \right\} , \tag{1}$$

where $A_\mu(x) = A_\mu{}^i(x)\tau_i$, $Tr(\tau_i\tau_j) = \delta_{ij}$, $[\tau_i, \tau_j] = if_{ijk}\tau_k$, and $F_{\mu\nu} = \partial_\mu A_\nu - \partial_\nu A_\mu - i[A_\mu, A_\nu]$ ; the gauge group is chosen to be semi-simple. The gauge fixed classical action reads

$$\Gamma_{\text{class}} = \Gamma_{\text{inv}} + \text{Tr}\int d^4x\, \Big\{ B\partial^\mu A_\mu + \frac{1}{2}\alpha B^2 + \partial^\mu \bar{c} D_\mu c + \gamma^\mu s A_\mu + \sigma s c \Big\} , \tag{2}$$

where $\alpha$ is the *gauge parameter;* $B = B^i\tau_i$ is a Lagrange multiplier field for the gauge condition, $\bar{c} = \bar{c}^i\tau_i$ and $c = c^i\tau_i$ are the Faddeev-Popov ghost fields. We have included terms coupling the BRS transformations of $A^i_\mu$ and $c^i$ — to be defined below – with external fields $\gamma^i_\mu$ and $\sigma^i$, respectively.

The classical action (2) is invariant under the usual BRS transformations

$$\begin{aligned} &sA_\mu = D_\mu c , \quad sc = ic^2 , \quad s\bar{c} = B , \quad sB = 0 , \\ &s^2 = 0 . \end{aligned} \tag{3}$$

We observe that the dependence of the classical action on the gauge parameter $\alpha$ is given by an insertion which is a BRS variation:

$$\partial_\alpha \Gamma_{\text{class}} = \frac{1}{2}\text{Tr}\int d^4x\; B^2 \equiv \Delta = \frac{1}{2}s\Big[\text{Tr}\int d^4x\; \bar{c}B\Big] \equiv s\Delta^- . \tag{4}$$

This implies – in the classical approximation – the gauge independence of the physical quantities, e.g. of the gauge invariant operators. In order to maintain this features to higher orders of perturbation theory, we impose an extension of BRS symmetry which also acts on the gauge parameters (here $\alpha$):

$$s\alpha = \alpha_+ , \quad s\alpha_+ = 0 , \tag{5}$$

in a way which preserves the nilpotency of the BRS transformations and transforms the parameters into Grassmann numbers (here $\alpha_+$).

The effect of extended BRS invariance in the classical approximation is to add to the action (2) the $\alpha_+$-dependent term

$$\text{Tr}\int d^4x\; \frac{\alpha_+}{2}\bar{c}B = \alpha_+\Delta^- , \tag{6}$$

so that (4) may be written as

$$\partial_\alpha \Gamma_{\text{class}} = s\; \partial_{\alpha_+}\Gamma_{\text{class}} . \tag{7}$$

In higher orders, the extended BRS invariance is expressed on the generating functional of Green's functions $Z$ via the extended Slavnov-Taylor identity

$$\begin{aligned} &\mathcal{S}Z(J_A, J_c, J_{\bar{c}}, J_B, \gamma, \sigma) \\ &\equiv \text{Tr}\int d^4x\, \Big\{ J_{A^\mu}\frac{\delta}{\delta\gamma_\mu} + J_c\frac{\delta}{\delta\sigma} + J_{\bar{c}}\frac{\delta}{\delta J_B}\Big\} Z + \alpha_+\frac{\partial}{\partial\alpha}Z \\ &= 0 , \end{aligned} \tag{8}$$

where $J_A$, etc. are the classical sources of the quantum fields $A$, etc. Differentiating this with respect to $\alpha_+$ yields the identity

$$\partial_\alpha Z = \mathcal{S}[\partial_{\alpha_+} Z] \ , \tag{9}$$

which is the quantum generalization of (4) or (7) and implies the gauge independence of the physical quantities, e.g. of the Green's functions of gauge invariant operators [2,3].

The validity of the extended Slavnov-Taylor identity (8) to all orders follows from the observation that the part of the extended BRS cohomology which depends on the gauge parameter, is trivial. The last statement concerns the general solution of

$$b\Delta(A_\mu, c, \bar{c}, B; \alpha, \alpha_+) = 0 \ , \tag{10}$$

in the space of local (i.e. polynomial) functionals, where $b$ is the nilpotent "linearized" Slavnov-Taylor operator

$$\begin{aligned} b &= \mathrm{Tr} \int d^4x \left\{ \frac{\delta W}{\delta A}\frac{\delta}{\delta \gamma} + \frac{\delta W}{\delta \gamma}\frac{\delta}{\delta A} + \frac{\delta W}{\delta c}\frac{\delta}{\delta \sigma} + \frac{\delta W}{\delta \sigma}\frac{\delta}{\delta c} \right\} + \alpha_+ \frac{\partial}{\partial \alpha} \ , \\ b^2 &= 0 \ , \end{aligned} \tag{11}$$

with

$$W = \Gamma_{\mathrm{class}}\Big|_{B=0} \ . \tag{12}$$

($b$ concides with $s$ when applied to the fields $A$, $c$ and to the gauge parameter $\alpha$).

It can be quite generally shown [4] that any "BRS doublet" ($X, Y$: $sX = Y$, $sY = 0$) of fields or gauge parameters contributes to the solution of (10) only in its trivial part, i.e. the part which has the form of a $b$-variation. Hence the general solution of (10) reads

$$\Delta = \Delta_{\mathrm{nontrivial}}(A_\mu, c) + s\hat{\Delta}(A_\mu, c, \bar{c}, B; \alpha, \alpha_+) \ . \tag{13}$$

In the ghost number 1 sector the nontrivial cohomology consists of the usual chiral anomaly and, in the absence of the latter, the extended Slavnov-Taylor identity therefore holds.

Let us finally consider the consequence of the result (13) on the BRS invariant counterterms – which correspond to the "infinite" counterterms in a BRS invariant subtraction procedure – and on the gauge independence of the theory. First, the *gauge invariant* counterterms – here the Yang-Mills action (1) – are the BRS invariants which, beyond having their dimension bounded by powercounting, are cohomologically nontrivial and thus $\alpha$-independent. But this implies that the normalization conditions which define them must be taken at some fixed value $\alpha_0$ of the gauge parameter: then the physical quantities, which are independent of $\alpha$, will depend on $\alpha_0$. This latter dependence however is controlled by a renormalization group equation [2,3] – like the dependence on the momentum space normalization point.

The remaining BRS invariant counterterms are cohomologically trivial. More explicitly [2,3], they are of the form (constrained by powercounting and ghost number conservation):

$$\begin{aligned} b\hat{\Delta} &= b\mathrm{Tr} \int d^4x \ [\gamma_\mu F^\mu(A, n) + \sigma G(c, n)] \ , \\ &= \mathrm{Tr} \int d^4x \left[ \frac{\delta W}{\delta A^\mu} F^\mu + \frac{\delta W}{\delta c} G - \gamma_\mu s F^\mu + \sigma s G \right] \ , \end{aligned} \tag{14}$$

with

$$F^\mu(A,\alpha) = z_A(\alpha)A^\mu \ , \quad G(c,\alpha) = z_c(\alpha)c \ . \tag{15}$$

One can interpret the first two terms in the second line of (14) as a field redefinition:

$$A^\mu \to A^\mu + F^\mu(A,\alpha) \ , \quad c \to c + G(c,\alpha) \ . \tag{16}$$

We note that, since the counterterms (14) are gauge dependent, the normalization conditions needed to fix them must be taken for any value of $\alpha$.

## 2 The Nonlocal Divergences of the Light-Cone Gauge

Four-dimensional gauge theories quantized in the light-cone (LC) gauge

$$n^\mu A_\mu = 0 \ , \quad n^2 = n_0^2 - \boldsymbol{n}^2 = 0 \ , \tag{17}$$

are known to develop ultraviolet divergences which are nonlocal, i.e. nonpolynomial in the momenta [6-15]. The purpose of the remainder of this talk is to present a general proof [16] of the cancellation of these nonlocal divergences in the computation of physical quantities. The proof proceeds in two steps: first a regularization of these divergences is introduced. Then it is shown that the ad hoc regulator is a gauge parameter in the sense of the preceding section, i.e. a nonphysical parameter. The procedure I am explaining here avoids the difficulties of a front attack on the singular light-cone gauge (17). (For recent progress in the latter framework, see [12-15,17,18,39])

The standard LC gauge fixing action

$$\Gamma_{\rm gf} = {\rm Tr} \int d^4x \ (B n^\mu A_\mu) + \text{ghost contr.} \ , \tag{18}$$

added to the gauge invariant Yang-Mills action (1) leads to the gluon propagator

$$\Delta_{\mu\nu}(k) = \frac{1}{k^2}\left[g_{\mu\nu} - \frac{n_\mu k_\nu + n_\nu k_\mu}{nk}\right] \ . \tag{19}$$

Beyond the usual $1/k^2$ it possesses the ghost propagator singularity

$$\Delta_{\rm g}(k) \equiv \frac{1}{nk} = \frac{1}{k_3 + ik_4} = \frac{k_3 - ik_4}{k_3^2 + k_4^2} \ . \tag{20}$$

We have chosen here $n^\mu(\mu = 0,\cdots,3) = (1,0,0,1)$ and have written the expression in Euclidean space, related to Minkowski space by Wick's rotation $k_0 = ik_4$. Equation (20) corresponds in Minkowski space to the Leibbrandt-Mandelstam prescription (LM) [7,9,10,11,19]

$$\left[\frac{1}{nk}\right]_{\rm LM} = \frac{n^* k}{(nk)(n^* k) + i\epsilon} \ , \quad n^* = (n^0, -\boldsymbol{n}) = (1,0,0,-1) \ . \tag{21}$$

The effect of the spurious singularity (20) is to invalidate the usual powercounting arguments because of the resulting "anisotropy" of the Euclidean large momentum behaviour [12]. This is what leads to the presence of nonlocal divergences.

The simplest example for the latter is provided by the (dimensionally regularized) integral [7,9]

$$
\begin{aligned}
J(p) &= \int d^{2\omega}k\, \Delta_g(k)\Delta_g(p+k)\frac{1}{(p+k)^2} \\
&= -i\pi^2\Gamma(2-\omega)\frac{2}{nn^*}\frac{pn^*}{pn} + \text{finite terms} ,
\end{aligned}
\tag{22}
$$

which contributes to the $n_\mu n_\nu$ terms of the gluon self-energy.

## 3 Regularization of the Nonlocal Divergences

A possible way to regularize the nonlocal divergences is to replace $n^\mu A_\mu$ in the gauge fixing action (18) by

$$
n^\mu A_\mu + \frac{1}{\xi}\frac{\partial^\mu}{n^*\partial}A_\mu , \quad \xi \geq 0 . \tag{23}
$$

(The apparent nonlocality of this Ansatz will be taken care of later on with the help of "auxiliary" fields.)

The new gauge condition (23) represents an interpolation between the LC gauge (17) ($\xi \to \infty$) and the covariant Landau gauge $\partial^\mu A_\mu = 0$ ($\xi \to 0$), as it can readily be seen inspecting the new gluon propagator (written in Euclidean space)

$$
\Delta_{\mu\nu}(k,\xi) = \frac{1}{k^2}\left[g_{\mu\nu} - \frac{(k_\mu n_\nu + k_\nu n_\mu)\xi kn^*}{k^2+\xi(kn)(kn^*)} - \frac{k^2 k_\mu k_\nu}{(k^2+\xi(kn)(kn^*))^2}\right] . \tag{24}
$$

The spurious singular term (20) is replaced now by

$$
\Delta_g(k,\xi) = \frac{\xi(kn^*)}{k^2+\xi(kn^*)(kn)} = \frac{\xi(k_3-ik_4)}{k_1^2+k_2^2+(1+\xi)(k_3^2+k_4^2)} . \tag{25}
$$

One verifies from the improved asymptotic behaviour of (24) and (25) that the usual powercounting rules are valid, as long as the interpolating parameter $\xi$ remains finite, real and nonnegative[1]. And indeed in the example (22) the right-hand side becomes finite, diverging only if the regularization is removed, i.e. for $\xi \to \infty$:

$$
J(p,\xi) \sim -i\pi^2\frac{2}{nn^*}\frac{pn^*}{pn}\log\xi . \tag{26}
$$

[1] The gauge fixing condition (23) has some similarity with that proposed in Ref. [20]. In the latter, however, $1/\xi$ is an imaginary infinitesimal parameter participating to the "$\epsilon$ prescription" for the propagator in Minkowski space.

# 4 Gauge Independence and the Cancellation of the Nonlocal Divergences

The validity of the usual powercounting rules in the interpolating gauge for finite $\xi$ ensures the general theorems of renormalization theory, in particular the quantum action principle [21-23]. This allows for a standard treatment of the problem of renormalizability [24,25].

We have to show that the renormalized theory is gauge invariant and gauge independent in the sense that it obeys an extended Slavnov-Taylor identity of the type (8). In particular, this implies the independence of the physical quantities of the regulator $\xi$, which amounts to the cancellation of the nonlocal divergences in the nonregularized theory. More precisely we shall realize in this way the construction of a physical theory which remains free of nonlocal divergences in the LC limit $\xi \to \infty$. This is the most one can achieve without being involved in the difficulties of the pure light-cone gauge mentioned in Sect. 2.

As in Sect. 1 we introduce extended BRS transformations, which are taken in the classical approximation as

$$\begin{aligned} &sA_\mu = D_\mu c \ , \quad s\bar{c} = B \ , \quad sc = ic^2 \ , \\ &sK_- = K \ , \quad sD = D_+ \ , \\ &s\xi = \xi_+ \ , \quad sn = n_+ \ , \quad s\alpha = \alpha_+ \ , \\ &s^2 = 0 \ . \end{aligned} \tag{27}$$

$(K_-, K)$ and $(D, D_+)$ are two BRS doublets of "auxiliary" fields enabeling us to rewrite the gauge fixing action in a local form:

$$\Gamma_{gf} = s\Big\{ Tr \int d^4x \ \Big[ K_- \left( \, \xi(n^*\partial)D - (\partial A) \, \right) + c_- D + c_-(nA) + \frac{\alpha}{2} K_- K \Big] \Big\} \ . \tag{28}$$

The BRS transforms of the gauge parameters $\xi$, $n$ and $\alpha$ are the Grassmann numbers $\xi_+$, $n_+$ and $\alpha_+$ (recall that $n^0$ and the components of $n^*$ are not independent parameters – see [20,26,27] for a discussion in the case of two independent vectors $n$ and $n^*$). One notes the presence of the supplementary gauge parameter $\alpha$, allowed by powercounting ($K$ has dimension 2).[2]

An extended Slavnov-Taylor identity finally holds, as it was explained in Sect. 1:

$$\begin{aligned} &SZ(J_A, J_c, J_{\bar{c}}, J_B, \gamma, \sigma) \\ &\equiv \mathrm{Tr} \int d^4x \ \Big\{ J_{A^\mu} \frac{\delta}{\delta \gamma_\mu} + J_c \frac{\delta}{\delta \sigma} + J_{\bar{c}} \frac{\delta}{\delta J_B} \Big\} Z \\ &\quad + \left( \xi_+ \frac{\partial}{\partial \xi} + n_+ \frac{\partial}{\partial n} + \alpha_+ \frac{\partial}{\partial \alpha} \right) Z = 0 \ . \end{aligned} \tag{29}$$

[2] In Ref.[16] $\alpha$ was fixed to the value 1 in order to be able to implement Zimmermann's $\epsilon$ - prescription [28,29] for the propagators and to be able to perform Wick rotations. This is in fact not necessary, the denominators remaining positive definite in Euclidean space for any positive value of $\alpha$. This is sufficient for the physical, gauge independent theory to be well defined in Minkowski space. I am indebted to Carlo Becchi for a discussion on this point.

This proves the for independence of the physical quantities from the gauge parameters $\xi$, $n$ and $\alpha$.

## 5 Lorentz Invariance

The symmetry-breaking of Lorentz invariance being due to the gauge condition, physical quantities are expected to be Lorentz covariant – in the absence of the gauge anomaly [30]. We shall see that this is indeed the case [31].

The breaking, parametrized by the vectors $n_\mu$ and $n^*_\mu$, is controlled as usually [32] by coupling it to external vector fields, $N_\mu$ and $N^*_\mu$, and by requiring the Ward identity

$$W_{[\mu\nu]}Z \equiv \int d^4x \left\{\sum_J \delta_{[\mu\nu]} J \frac{\delta}{\delta J} + \delta_{[\mu\nu]}(N_\sigma + n_\sigma)\frac{\delta}{\delta N_\sigma} + \delta_{[\mu\nu]}(N^*_\sigma + n^*_\sigma)\frac{\delta}{\delta N^*_\sigma}\right\}Z = 0 \,. \tag{30}$$

The summation over $J$ runs on all the quantum field sources $J_A$, $J_c$, etc. and on the external fields $\gamma^\mu$, $\sigma$ (as defined in (2)). The infinitesimal Lorentz transformations $\delta_{[\mu\nu]}$ are of the "spin + orbit" form

$$\begin{aligned}\left[\delta_{[\mu\nu]}\phi\right]_a &= \Sigma_{\mu\nu a}{}^b\phi_b + (x_\mu\partial_\nu - x_\nu\partial_\mu)\phi_a \,, \\ \phi_a &= J_{A^\mu},\ J_c,\ \cdots,\ \gamma_\mu,\ \cdots,\ N^*_\mu \,.\end{aligned} \tag{31}$$

The inhomogeneous terms ($n_\sigma$ and $n^*_\sigma$ terms) of (30) represent the breaking.

In order to show the physical Lorentz invariance we extend once more the BRS symmetry by allowing the external field $N^\mu$ to transform, too:

$$sN^\mu = N^\mu_+ \,, \quad sN^{*\mu} = N^{*\mu}_+ \,, \quad sN^\mu_+ = 0 \,, \quad sN^{*\mu}_+ = 0 \,, \tag{32}$$

which means adding the new contributions

$$\int d^4x \left\{N^\mu_+\frac{\delta}{\delta N^\mu} + N^{*\mu}_+\frac{\delta}{\delta N^{*\mu}}\right\}Z$$

and (33)

$$\int d^4x \left\{\delta_{[\mu\nu]}(N^\sigma_+ + n^\sigma_+)\frac{\delta}{\delta N^\sigma_+} + \delta_{[\mu\nu]}(N^{*\sigma}_+ + n^{*\sigma}_+)\frac{\delta}{\delta N^{*\sigma}_+}\right\}Z \,,$$

to the Slavnov-Taylor identity (29) and to the Lorentz Ward identity (30), respectively. (We notice that the commutativity of the operators $\mathcal{S}$ and $W_{[\mu\nu]}$ is preserved).

It becomes clear that the breaking terms in the Lorentz Ward identity are now Slavnov-Taylor variations:

$$\frac{\delta Z}{\delta N^\mu} = \mathcal{S}\frac{\delta Z}{\delta N^\mu_+} \,, \quad \frac{\delta Z}{\delta N^{*\mu}} = \mathcal{S}\frac{\delta Z}{\delta N^{*\mu}_+} \,, \tag{34}$$

and thus they do not affect the physical quantities.

As we see, the whole argument rests on the validity of the Lorentz Ward identity to all orders. The proof of the latter is a cohomology problem for the Lorentz algebra (31). It can be solved [31] by reducing it to the cohomology of the "spin" part of the algebra, i.e. to the cohomology of the Lie algebra of SO(1,3), which is known to be trivial [33].

## 6 Invariant Counterterms

The BRS invariant counterterms appearing at each order of the perturbation series can be deduced from the discussion at the end of Sect. 1. The gauge invariant counterterm is still the Yang-Mills action, and the gauge dependent ones are still given by (14) or by the field redefinition (16), but with the functions (15) replaced by

$$\begin{aligned} F^\mu(A,\xi,n,\alpha) &= (\delta^\mu_\nu z_1 + n^\mu n_\nu z_2 + n^\mu n^*_\nu z_3 + n^{*\mu} n_\nu z_4 + n^{*\mu} n^*_\nu z_5) A^\nu \ , \\ G(c,\xi,n,\alpha) &= z_6 c \ , \end{aligned} \tag{35}$$

with

$$z_a = z_a(\xi, n\cdot n^*, \alpha)\ , \quad a = 1,\cdots,6\ . \tag{36}$$

The dependence on $n$ and $n^*$ has been constrained by the broken Lorentz Ward identity (30).

In the cases of the pure light-cone and axial gauges a general structure for the counterterms, analogous to the one above and also based on extended BRS invariance, was given in Refs. [9,10,27,34-42].

### Acknowledgments

I have benefitted greatly from discussions with the collaborators of the "Vienna group", with whom most of the work reported in Sects. 2 to 6 was done: H. Balasin, P. Gaigg, G. Pollak, A. Rebhan, M. Schweda, H. Skarke, M. Stierle. I have also profited of their remarks the writing of these notes.

I would also like to express my gratitude to the Austrian Physical Society and to the City of Vienna for their financial support, and to the organizors of this workshop for their very kind hospitality.

This work was supported in parts by the Swiss National Science Foundation.

## References


[1] H. Kluberg-Stern and J.B. Zuber, Phys. Rev. **D12** (1975) 467, 482, 3159.

[2] O. Piguet and K. Sibold, Nucl. Phys. **B253** (1985) 517.

[3] O. Piguet, *Comments on the gauge (in)dependence of gauge theories*, University of Geneva preprint UGVA-DPT 1988/12-600, to be published in the proceedings of the "XIII Semester in the Stefan Banach International Mathematical Center", 1988.

[4] O. Piguet and K. Sibold, *Renormalized Supersymmetry*, series "Progress in Physics", vol. 12, Birkhäuser Boston Inc. (1986).

[5] W. Kummer, *Superaxial gauges*, these proceedings.

[6] D.M. Capper, J.J. Dulwich and M.J. Litvak, Nucl. Phys. **B241** (1984) 463.

[7] G. Leibbrandt, Phys. Rev. **D29** (1984) 1699.

[8] G. Leibbrandt and S.-L. Nyeo, Z. Phys. **C30** (1986) 501.

[9] G. Leibbrandt, Rev. Mod. Phys. **59** (1987) 1067.

[10] G. Leibbrandt, *Practical look at noncovariant gauges*, "Proc. XXIV Int. Conf. on High Energy Physics", p.782, eds.: R. Kotthaus and J. Kühn, Springer-Verlag Berlin, Heidelberg 1989.

[11] A. Bassetto, M. Dalbosco, I. Lazzizzera and R. Soldati, Phys. Rev. **D31** (1985) 2012.

[12] A. Bassetto, M. Dalbosco and R. Soldati, Phys. Rev. **D36** (1987) 3138.

[13] A. Bassetto and R. Soldati, Phys. Rev. **D37** (1988) 3065.

[14] A. Bassetto, G. Nardelli and R. Soldati, Mod. Phys. Lett. **A3** (1988) 1663.

[15] A. Bassetto, *Recent developments in algebraic noncovariant gauges*, these proceedings.

[16] O. Piguet, G. Pollak and M. Schweda, Nucl. Phys. **B328** (1989) 527.

[17] C. Becchi, *The axial gauges – a personal point of view*, these proceedings.

[18]. M. Ögren, *Cancellation of $(1/pn)$-factors in light-cone gauge Feynman diagrams*, these proceedings.

[19] S. Mandelstam, Nucl. Phys. **B213** (1983) 149.

[20] A. Andraši and J.C. Taylor, Nucl. Phys. **B302** (1988) 123.

[21] Y. M. P. Lam, Phys. Rev. **D6** (1972) 2145, 2161.

[22] T. E. Clark and J. H. Lowenstein, Nucl. Phys. **B113** (1976) 109.

[23]. H. Balasin, *Comments on the proof of the renormalized action principle*, these proceedings.

[24]. C. Becchi, A. Rouet and R. Stora, Ann. Phys. **98** (1976) 287.

[25]. O. Piguet and A. Rouet, Phys. Reports **76** (1981) 1.

[26]. P. Gaigg and M. Kreuzer, Phys. Lett. **205B** (1988) 530.

[27] G. Pollak, *Checking the S-matrix within the generalized LM prescription*, these proceedings.

[28] W. Zimmermann, Commun. Math. Phys. **11** (1968) 1.

[29] J. H. Lowenstein, Commun. Math. Phys. **47** (1976) 53.

[30] D.M. Capper, D.R.T. Jones and N. Linden, Phys. Lett. **181B** (1986) 106.

[31] H. Balasin, O. Piguet, M. Schweda and M. Stierle, Phys. Lett. **215B** (1988) 328.

[32]. C. Becchi, A. Rouet and R. Stora, *Renormalizable theories with symmetry breaking*, in: Field theory, quantization and statistical physics", ed. E. Tirapegui (Reidel, Dordrecht, 1981).

[33]. N. Jacobson, *Lie Algebras*, Dover Books on Advanced Mathematics (1962).

[34]. A. Andraši and J. C. Taylor, DAMTP 86-11 (unpublished).

[35] A. Andraši, *The Piguet-Sibold identities and the light-cone gauge*, these proceedings.

[36] J.C. Taylor and H.C. Lee, Phys. Lett. **185B** (1987) 363.

[37] P. Gaigg, O. Piguet, A. Rebhan and M. Schweda, Phys.Lett. **175B** (1986) 53.

[38] O. Piguet, M. Schweda and H. Skarke, Phys. Lett. **210B** (1988) 159.

[39] H. Skarke and P. Gaigg, Phys. Rev. **D38** (1988) 3205.

[40] H. Skarke, *Extended BRS symmetry and renormalization of Yang-Mills theories in the axial gauge*, these proceedings.

[41] E. Bagan and C.P. Martin, Phys. Lett. **223B** (1989) 187 and DAMTP 89-12.

[42] G. Leibbrandt, *Two-loop integrals in a unified gauge formalism*, these proceedings.

# Extended BRS-Symmetry and Renormalization in the Axial Gauge

Harald Skarke

Institut für Theoretische Physik, Technische Universität Wien,
Wiedner Hauptstraße 8-10, 1040 Wien, Austria

**Abstract:** The concept of extended BRS-symmetry is presented. With the help of this technique the structure of the counterterms of Yang-Mills theory in the axial gauge with the principal value prescription is determined to all orders in perturbation theory.

## 1 Introduction

For a long time the only prescription used for axial-type gauges [1] has been the principal value (PV) prescription [2]. The invention of the so-called Leibbrandt-Mandelstam-(LM-) prescription [3] for the light-cone gauge and technical problems in certain PV-calculations have led to generalizations of the LM-prescription [4] which are applicable in the whole range of axial-type gauges. Still, however, the PV-prescription possesses a number of advantages. It is the only prescription with consistent one-loop integrals depending only on a single gauge vector $n_\mu$ [5], and due to the validity of powercounting [6] it is free of the nonlocal divergences which plague integrals regularized by the LM-prescription or one of its generalizations. In the following it will be shown that the locality of possible counterterms and the so-called extended BRS-symmetry [7] determine uniquely the structure of possible counterterms to all orders in perturbation theory.

## 2 Extended BRS-Symmetry

We shall now define the so-called extended BRS-Symmetry [7] and deduce the corresponding Slavnov-Taylor-identity. We work in pure Yang-Mills theory [8] with

$$S_{\rm YM} = \frac{-1}{4} tr \int d^4x F^{\mu\nu} F_{\mu\nu} \tag{1}$$

and a gauge of the type

$$S_{gb} = tr \int d^4x (-\frac{\lambda}{2} BfB + BF^\mu A_\mu - c_- F^\mu D_\mu c_+) \, . \tag{2}$$

$A_\mu = A^a_\mu \tau^a$, $tr(\tau^a \tau^b) = \delta^{ab}$, $D_\mu = \partial_\mu - ig[A_\mu, \cdot]$, and $F_{\mu\nu} = \partial_\mu A_\nu - \partial_\nu A_\mu - ig[A_\mu, A_\nu]$ denotes the field-strength tensor. Different types of axial gauges are characterized by $f = -1$, $F^\mu = n^\mu$ or $f = -1$, $F^\mu = n^\mu \frac{n\partial}{n^2}$ or $f = \partial^2$, $F^\mu = n^\mu \partial^2$. $S = S_{\mathrm{YM}} + S_{gb}$ is invariant under the infinitesimal transformation

$$\delta A_\mu = D_\mu c_+ \delta\psi,\ \delta c_+ = igc_+c_+\delta\psi,\ \delta c_- = B\delta\psi,\ \delta B = 0. \tag{3}$$

The ordinary BRS-operator $s$ [9] is the right derivative with repect to $\delta\psi$, i.e. $sA_\mu = D_\mu c_+$ etc. In the gauge (2) with the Lagrange-multiplier field $B$ it fulfils $s^2 = 0$. The fact that the derivative of $S$ with respect to $n$ can be written as

$$\frac{\partial S}{\partial n_\mu} = s\, tr \int d^4x\ c_- \frac{\partial F_\nu}{\partial n_\mu} A^\nu \tag{4}$$

suggests an extension of $s$ by $s_{\mathrm{ext}}\, n_\mu = \chi_\mu$ and $s_{\mathrm{ext}}\, \chi_\mu = 0$ [10], where $\chi_\mu$ is anticommuting. Under this new transformation our action $S$, extended by

$$S_{\mathrm{ext}} = \chi_\mu\, tr \int d^4x\ c_- \frac{\partial F_\nu}{\partial n_\mu} A^\nu, \tag{5}$$

is invariant. We introduce sources $\rho_\mu$, $\rho_+$ with $s\rho_\mu = s\rho_+ = 0$ for $sA_\mu$ and $sc_+$ and define

$$S_{\mathrm{S}} = tr \int d^4x\ [\rho^\mu D_\mu c_+ + \rho_+ igc_+c_+ + J^\rho A_\rho + J_B B + \xi_+ c_+ + c_- \xi_-] \tag{6}$$

and $S = S_{\mathrm{YM}} + S_{gb} + S_{\mathrm{ext}} + S_{\mathrm{S}}$. I the path integral-formalism the generating functional of Green's functions $Z$ s

$$Z(J, J_B, \xi_-, \xi_+, n, \chi, \rho, \rho_+) = tr \int \mathcal{D}(A, B, c_-, c_+) e^{iS}. \tag{7}$$

An ordinary (that is, not extended) BRS-transformation leaves $\mathcal{D}$, $S_{\mathrm{YM}}$ and $S_{\phi\Pi}$ invariant and therefore leads to

$$\int \mathcal{D}(A, B, c_-, c_+) s(S_{\mathrm{ext}} + S_{\mathrm{S}}) e^{iS} = 0, \tag{8}$$

which implies

$$\int \mathcal{D}(A, B, c_-, c_+) \Big( tr \int d^4x (J^\mu D_\mu c_+ + \xi_+ igc_+c_+ + B\xi_-) + i\chi^\mu \frac{\partial}{\partial n^\mu} \Big) e^{iS} = 0. \tag{9}$$

Following the usual steps in deriving Slavnov-Taylor-identities for the generating functional $\Gamma$ of one-particle-irreducible vertex functions (see, e.g., [11]) we get

$$\mathcal{B}(\Gamma) = tr \int d^4x \left( \frac{\delta\Gamma}{\delta A_\mu} \frac{\delta\Gamma}{\delta\rho^\mu} + \frac{\delta\Gamma}{\delta c_+} \frac{\delta\Gamma}{\delta\rho_+} + B \frac{\delta\Gamma}{\delta c_-} \right) + \chi^\mu \frac{\partial\Gamma}{\partial n^\mu} = 0. \tag{10}$$

This is the extended Slavnov-Taylor-identity.

We shall also use the concepts of ghost numbers $N_{\mathrm{g}}$:

$$
\begin{aligned}
&N_g(A_\mu) = N_g(B) = N_g(n_\mu) = 0,\\
&N_g(c_+) = N_g(\chi) = 1,\\
&N_g(c_-) = N_g(\rho_\mu) = -1,\\
&N_g(\rho_+) = -2,\\
&N_g(\Gamma) = 0
\end{aligned}
\tag{11}
$$

and mass dimensions:

$$
\begin{aligned}
&\dim(A_\mu) = 1,\\
&\dim(\rho_\mu) = 3,\\
&\dim(c_+) = 0,\\
&\dim(\rho_+) = 4,\\
&\dim(c_-) = \dim(B),\\
&\dim(n_\mu) = \dim(\chi_\mu),\\
&\dim(\Gamma) = 4.
\end{aligned}
\tag{12}
$$

# 3 Renormalization in the Axial Gauge

If we try to calculate $\Gamma = \Gamma(S)$ we find that the corresponding Feynman integrals can be divergent. By introducing a regularization parameter it is possible to render these integrals finite. In dimensional regularization [12], which leaves all the symmetries of the theory invariant, this parameter is $\varepsilon = 2 - \omega$, where $2\omega$ is the dimension of space-time. The poles in $\varepsilon$ are then called "infinite parts". It is the aim of renormalization to make the functional $\Gamma$ finite by formally redefining $S$ to $S_{\text{ren}} = S + \hbar C_1 + \hbar^2 C_2 + \cdots$ with infinite counterterms $C_i$ such that $\Gamma(S_{\text{ren}}) = S + \hbar\Gamma_1^{\text{ren}} + \hbar^2\Gamma_2^{\text{ren}} + \cdots$ is finite. With the notations $S_{\text{ren}}^{(n)} = S + \hbar C_1 + \cdots + \hbar^n C_n$ and $\Gamma_{\text{ren}}^{(n)} = S + \hbar\Gamma_1^{\text{ren}} + \cdots + \hbar^n\Gamma_n^{\text{ren}}$ the counterterms are determined by $\Gamma(S_{\text{ren}}^{(n)}) = \Gamma_{\text{ren}}^{(n+1)} - \hbar^{n+1}C_{n+1} + O(\hbar^{n+2})$. In [13] it has been shown that $\mathcal{B}(\Gamma(S)) = 0$ always implies $\mathcal{B}(S_{\text{ren}}^{(n)}) = O(\hbar^{n+1})$, i. e. $\mathcal{B}(S_{\text{ren}}) = 0$.

With the help of this relation we now try to determine the structure of the counter terms in the axial gauge; i. e. we look order by order in perturbation theory for solutions of (10), where the zeroth order solution is the classical action $S$. As long as we work in the principal value prescription all divergences are polynomial. We could write down all possible terms with the correct mass dimensions and ghost numbers and derive relations for their coefficients by inserting this ansatz in (10). Since this would be rather tedious we shall try to find the solution step by step.

We first note that the insertion of expressions containing $B$ or $c_-$ into (10) leads again to expressions of the same type. Therefore we shall henceforth neglect these terms which are irrelevant for the renormalization of the gauge field. An alternative way is to eliminate them by explicitly solving the equation of motion for $B$ and redefining the source $\rho_\mu$ [10]. Next we note that $n$- and $\chi$-dependent terms in $S$ will lead to the same sort of terms when inserted into (10). Therefore as a first step we shall look for solutions of

$$
tr \int dx \left[\frac{\delta S}{\delta\rho_\mu}\frac{\delta S}{\delta A^\mu} + \frac{\delta S}{\delta\rho_+}\frac{\delta S}{\delta c_+}\right] = 0\,, \tag{13}
$$

which are local functionals independent of $B, c_-, n$ and $\chi$. The answer to this problem is well known: It corresponds to counterterms with the same structure as the terms in $S$, coming from a multiplicative renormalization of $\rho, A, \rho_+$ and $c_+$ [11].

Let us now examine terms which, when they are inserted into (10), yield expressions which contain only $A, n$ and $\chi$ and are linear in $\chi$. Because of ghost number and mass dimension only terms of the form $\mathcal{G}(A,n)$ (depending only on $A$ and $n$) or $tr \int d^4x \rho^\mu \mathcal{M}_\mu$, where $\mathcal{M}_\mu(A,n,\chi)$ is linear in $\chi$, contribute. Then (10) implies

$$tr \int d^4x \frac{\delta \mathcal{G}}{\delta A_\mu} \mathcal{M}_\mu + \chi_\mu \frac{\partial \mathcal{G}}{\partial n_\mu} = 0. \tag{14}$$

Since we work in a perturbative expansion we have

$$\mathcal{G} = \mathcal{G}_0 + \hbar \mathcal{G}_1 + \cdots \text{ with } \mathcal{G}_0 = -\frac{1}{4}\mathcal{L}_1, \tag{15}$$

$$\mathcal{L}_1 = tr \int d^4x F^{\mu\nu} F_{\mu\nu} \tag{16}$$

$$\text{and} \quad \mathcal{M}^\mu = \mathcal{M}_0^\mu + \hbar \mathcal{M}_1^\mu + \cdots \text{ with } \mathcal{M}_0^\mu = 0. \tag{17}$$

With

$$\frac{\delta \mathcal{L}_1}{\delta A^\mu} = -4 D^\nu F_{\nu\mu} \tag{18}$$

we get at first order

$$tr \int d^4x D^\mu F_{\mu\nu} \mathcal{M}_1^\nu + \chi^\mu \frac{\partial \mathcal{G}_1}{\partial n^\mu} = 0, \tag{19}$$

which implies

$$\mathcal{G}_1 = -\frac{a_1}{4}\mathcal{L}_1 + b_1 tr \int d^4x D^\mu F_{\mu\nu} \mathcal{Y}^\nu \tag{20}$$

and

$$\mathcal{M}_1^\mu = -b_1 \chi_\nu \frac{\partial \mathcal{Y}^\mu}{\partial n_\nu}. \tag{21}$$

The only expression with the correct mass dimension and ghost number is

$$\mathcal{Y}_\mu = \frac{1}{n^2} n_\mu n A. \tag{22}$$

With

$$\mathcal{L}_2 := \frac{1}{n^2} tr \int d^4x \, DFn\, nA \tag{23}$$

and

$$\mathcal{K}_\mu := \chi_\nu \frac{\partial \mathcal{Y}_\mu}{\partial n_\nu} = \chi_\mu \frac{nA}{n^2} + n_\mu \frac{\chi A}{n^2} - n_\mu \frac{2n\chi}{n^4} nA\,, \tag{24}$$

we can write

$$\mathcal{G}_1 = -\frac{a_1}{4}\mathcal{L}_1 + b_1 \mathcal{L}_2 \tag{25}$$

and

$$\mathcal{M}_1^\mu = -b_1 \mathcal{K}^\mu. \tag{26}$$

This is the structure of the counterterms at first order. We note that the terms without the source can be written as the zeroth and first order contributions of $\mathcal{L}_1(\hat{A}_\mu, \hat{g})$ with

$$\begin{aligned}\hat{A}_\mu =&(1+\hbar a_1)^{\frac{1}{2}}A_\mu + b_1 n_\mu \frac{nA}{n^2},\\ \hat{g} =&(1+\hbar a_1)^{-\frac{1}{2}}g.\end{aligned} \tag{27}$$

The transformation

$$\begin{aligned}\mathcal{F}: A_\mu \to& \alpha A_\mu + \beta n_\mu \frac{nA}{n^2},\\ g \to& \alpha^{-1} g\end{aligned} \tag{28}$$

possesses the following properties: The composition of two such functions with $(\alpha_1, \beta_1)$ and $(\alpha_2, \beta_2)$ leads to the same sort of transformation, namely

$$\begin{aligned}A_\mu \to& \alpha_1\alpha_2 A_\mu + (\alpha_1\beta_2 + \alpha_2\beta_1 + \beta_1\beta_2) n_\mu \frac{nA}{n^2},\\ g \to& \alpha_1^{-1}\alpha_2^{-1} g.\end{aligned} \tag{29}$$

There is an inverse, given by

$$\begin{aligned}\mathcal{F}^{-1}: A_\mu \to& \alpha^{-1}(A_\mu - \frac{\beta}{\alpha+\beta} n_\mu \frac{nA}{n^2}),\\ g \to& \alpha g.\end{aligned} \tag{30}$$

We see that such a redefinition removes the first order infinities in the gauge field sector. In the framework of ordinary BRS-analysis one would redefine $A_\mu$ with an appropriate function $\mathcal{F}$ and the source $\rho$ with its inverse $\mathcal{F}^{-1}$ (and both with an additional factor), which would leave the ordinary Slavnov-Taylor-identity invariant. Here this is not possible: Firstly this redefinition would not remove the term $\rho^\mu \mathcal{M}_\mu$, and secondly the extended Slavnov-Taylor-identity would not remain invariant because of

$$\chi_\mu \frac{\partial S}{\partial n_\mu} \to \chi_\mu \left( \frac{\partial S}{\partial n_\mu} + \int dx \left[ \frac{\delta S}{\delta \hat{A}_\nu} \frac{\delta \hat{A}_\nu}{\delta n_\mu} + \frac{\delta S}{\delta \hat{\rho}_\nu} \frac{\delta \hat{\rho}_\nu}{\delta n_\mu} \right] \right). \tag{31}$$

At second order (14) implies

$$tr \int d^4x \left( \frac{\delta \mathcal{G}_0}{\delta A^\mu} \mathcal{M}_2^\mu + \frac{\delta \mathcal{G}_1}{\delta A^\mu} \mathcal{M}_1^\mu \right) + \chi_\mu \frac{\partial \mathcal{G}_2}{\partial n_\mu} = 0, \tag{32}$$

which leads, after some algebra, to

$$\mathcal{G}_2 = -\frac{a_2}{4}\mathcal{L}_1 + b_2\mathcal{L}_2 + c_2\mathcal{L}_3 \text{ with } c_2 = b_1^2, \tag{33}$$

$$\mathcal{L}_3 := \frac{1}{2n^4} tr \int d^4x \, nA \left(n^2 D^2 - (nD)^2\right) nA \tag{34}$$

$$\text{and } \mathcal{M}_2^\mu = b_1^2 n^\mu \frac{n\mathcal{K}}{n^2} + (a_1 b_1 - b_1^2)\mathcal{K}^\mu. \tag{35}$$

Similar, but even more tedious calculations can be done at third order. One notices that the terms we have found, when inserted into (14), always give rise to the same terms again. Therefore we can make the general ansatz

$$\mathcal{G} = -\frac{a}{4}\mathcal{L}_1 + b\mathcal{L}_2 + c\mathcal{L}_3, \tag{36}$$

$$\mathcal{M}^\mu = d\mathcal{K}^\mu + en^\mu \frac{n\mathcal{K}}{n^2}, \tag{37}$$

which leads to

$$tr \int d^4x \left( \left( aD^\mu F_{\mu\nu} + b\frac{\delta \mathcal{L}_2}{\delta A^\nu} + c\frac{\delta \mathcal{L}_3}{\delta A^\nu} \right) \left( d\mathcal{K}^\nu + en^\nu \frac{n\mathcal{K}}{n^2} \right) + bDF\mathcal{K} \right) + c\chi^\mu \frac{\partial \mathcal{L}_3}{\partial n^\mu} = 0 \tag{38}$$

$$\text{and } c = \frac{b^2}{a},\ d = -\frac{b}{a},\ e = \frac{b^2}{a(a+b)}. \tag{39}$$

Writing $a$ and $b$ as $1 + \hbar a_1 + \cdots$ and $\hbar b_1 + \cdots$ and developing $c, d$ and $e$ after $\hbar$ yields the values for the $c_i$ and $\mathcal{M}_i^\mu$ that we have found above.

In a similar way the complete solution of (10) (without $B$- or $c_-$-dependent terms) is found to be

$$\mathcal{G} + tr \int d^4x \left( \rho^\mu \mathcal{M}_\mu + \alpha \Big( \rho^\mu (D_\mu - \frac{b}{a+b} \frac{n_\mu}{n^2}(n\partial))c_+ + ig\rho_+ c_+ c_+ \Big) \right). \tag{40}$$

All terms except $tr \int d^4x \rho^\mu \mathcal{M}_\mu$ are created by the following nonmultiplicative redefinition:

$$\begin{aligned} A_\mu &\to a^{\frac{1}{2}}(A_\mu + \frac{b}{an^2} n_\mu nA), \\ g &\to a^{-\frac{1}{2}} g, \\ \rho_\mu &\to \gamma(\rho_\mu - \frac{b}{(a+b)n^2} n_\mu n\rho), \\ c_+ &\to \alpha\gamma^{-1} c_+, \\ \rho_+ &\to a^{\frac{1}{2}} \alpha^{-1} \gamma^2 \rho_+. \end{aligned} \tag{41}$$

## 4 Conclusions

We have seen how the polynomial structure of divergences and extended BRS-symmetry determine to all orders in perturbation theory the structure of the counterterms of the axial gauge in the principal value prescription. In contrast to the case of multiplicative renormalization we could not redefine the gauge field order by order. Nevertheless all terms in the renormalized action except those coming from the extension of the BRS-symmetry arise from the original action through a non-multiplicative linear redefinition of the fields and sources.

In the homogeneous gauge, i. e. $\lambda = 0$ in (2), the Ward-identity implies that the counterterms have to be gauge invariant. Therefore in this case the coefficients of $\mathcal{L}_2$ and $\mathcal{L}_3$ vanish and, as in the covariant case, one has multiplicative renormalization, in agreement with early results [2].

A treatment of the light-cone gauge with the same methods as in this article has been given in [14]. Despite the occurrence of nonpolynomial divergences the counterterms can be absorbed by a nonlocal and nonlinear transformation of the fields and sources. The parameters relating these nonlocal counterterms also obey (39).

# References


[1] W. Kummer, Acta Phys. Austr. **14** (1961) 149;
R. L. Arnowitt and S. J. Fickler, Phys. Rev. **127** (1962) 1821;
J. Schwinger, Phys. Rev. **130** (1963) 402;
Y. P. Yao, J. Math. Phys. **5** (1964) 1319.

[2] W. Kummer, Acta Phys. Austr. **41** (1975) 315;
W. Konetschny and W. Kummer, Nucl. Phys. **B100** (1975) 106; **B108** (1976) 397;
J. Frenkel, Phys. Rev. **D13** (1976) 2325;
W. Konetschny, Phys. Rev. **D28** (1983) 354.

[3] S. Mandelstam, Nucl. Phys. **B213** (1983) 149;
G. Leibbrandt, Phys. Rev. **D29** (1984) 1699; **D30** (1984) 2167.

[4] P. Gaigg, M. Kreuzer, M. Schweda, and O. Piguet, J. Math. Phys. **28** (1987) 2781;
M. Kreuzer, G. Pollak, H. Skarke, and M. Schweda, Phys. Lett. **196B** (1987) 557;
P. Gaigg and M. Kreuzer, Phys. Lett. **205B** (1988) 530;
P. Gaigg, M. Kreuzer and G. Pollak, Phys. Rev. **D38** (1988) 2559;
G. Leibbrandt, Nucl. Phys. **B310** (1988) 405.

[5] H. Skarke, J. Math. Phys. **30** (1989) 1270.

[6] W. Kummer, Acta Phys. Austr. Suppl. **XV** (1976) 423.

[7] O. Piguet and K. Sibold, Nucl. Phys. **B248** (1984) 301;
Nucl. Phys. **B253** (1985) 517.

[8] C.N. Yang, R. L. Mills, Phys. Rev. **96** (1954) 191.

[9] C. Becchi, A. Rouet, R. Stora, Phys. Lett. **52B** (1974) 344; Ann. Phys. (NY) **98** (1976) 287;
I.V. Tyutin, Lebedev preprint FIAN No. 39 (1975), unpublished.

[10] P. Gaigg, O. Piguet, A. Rebhan and M. Schweda, Phys. Lett. **175B** (1986) 53;
O. Piguet, M. Schweda and H. Skarke, Phys.Lett. **210B** (1988) 159.

[11] C. Itzykson, J. B. Zuber, *Quantum Field Theory*, McGraw-Hill, New York, 1980.

[12] G.'t Hooft, M. Veltman, Nucl. Phys. **B44** (1972) 189.

[13] W. Kummer, H. Mistelberger, P. Schaller, M. Schweda, Fortschr. Phys. **37** (1989) 261.

[14] H. Skarke and P. Gaigg, Phys. Rev. **D38** (1988) 3205.

# Superaxial Gauges

Wolfgang Kummer

Institut für Theoretische Physik, Technische Universität Wien,

Wiedner Hauptstraße 8-10, A-1040 Wien, Austria

**Abstract:** The analogue of axial gauges where Lorentz invariance is broken are supersymmetric gauge-theories with breaking of supersymmetry. The Wess-Zumino gauge is the most celebrated example. Nevertheless, the full superfield formalism may be retained — just as Lorentz covariance can be made to persist formally in ordinary axial gauges.

## 1 Introduction

In ordinary gauge theories the general axial gauges $n^m A_m = 0$ are characterized by the introduction of a fixed Lorentz vector $n$. Apart from the appearance of $n$ in all quantities describing the theory, Lorentz covariance is formally retained. Analogously, for supersymmetric gauge theories [1] the Wess-Zumino gauge may be included in general 'superaxial' gauges [2,3,4] within the superfield formalism. Moreover consistent renormalization avoiding spurious new IR divergencies of individual supergraphs even *requires* noncovariant supergauges [5].
In supersymmetry the ordinary gauge field $v_m$ becomes a component of the hermitian superfield [6,7] $V = V^+$

$$\begin{aligned} V = & C + \theta\chi + \bar{\theta}\bar{\chi} + \frac{1}{2}\theta^2 M + \frac{1}{2}\bar{\theta}^2\bar{M} + \theta \not{v} \bar{\theta} + \\ & + \frac{1}{2}\theta^2\bar{\theta}\bar{\lambda} + \frac{1}{2}\bar{\theta}^2\theta\lambda + \frac{1}{4}\theta^2\bar{\theta}^2\mathcal{D} \end{aligned} \qquad (1)$$

which contains powers of the Lorentz 2-spinor Grassmann-variables $\theta$ and $\bar{\theta}$. The supergraph IR problem [5] is a consequence of the (mass-) dimension zero of $V$ which in turn follows from the dimension 1 of the vector-field $v_m$ and from the dimension $-\frac{1}{2}$ for $\theta$. The scalar component of $V$ has dimension zero too and hence, for purely dimensional reasons, the $C$−propagator must behave like $1/k^4$. Thus a new type of IR divergence appears. Fixing the gauge in a way which does not break supersymmetry will not provide a solution of this problem. Sufficient for such a solution is the choice of the Wess-Zumino gauge [1] which eliminates all fields except $v, \lambda$ and $\mathcal{D}$. Such a gauge clearly breaks supersymmetry. Nevertheless, in order to be able to still use the powerful modern versions of supergraphs [8] it is desirable to extend the latter method to such gauges. This approach we call 'superaxial gauges'.

## 2 Noncovariant gauge-fixing in supersymmetric Yang-Mills theories

The supersymmetric Young-Mills Lagrangian [6,7] reads

$$L_{YM} = -\frac{1}{256g^2}\int d^6x_+\, Tr\left[\bar{D}^2(e^{-gV}D^\alpha e^{gV})\bar{D}^2(e^{-gV}D_\alpha e^{gV})\right] + h.c. \tag{2}$$

with the 'chiral' gauge-invariance ($\bar{D}_{\dot\alpha}\Lambda_+ = 0,\ V \to t^c_{\underline{ab}}V_c$)

$$(e^{V'})_{\underline{ab}} = (e^{-ig\Lambda^\dagger})_{\underline{ac}}(e^V)_{\underline{cd}}(e^{ig\Lambda})_{\underline{db}} \quad . \tag{3}$$

For an infinitesimal change $\delta\Lambda_+, V' = V + \delta V$ one arrives at

$$\delta V = i\mathcal{L}_{\frac{gV}{2}}\left[(\delta\Lambda + \delta\Lambda^\dagger) + \coth(\mathcal{L}_{\frac{gV}{2}})(\delta\Lambda - \delta\Lambda^\dagger)\right] ,$$

$$\mathcal{L}_A B := [A, B] \tag{4}$$

or

$$\delta V_a = R^b_a(V)\delta\Lambda_b + \bar{R}^b_a(V)\delta\bar{\Lambda}_b \quad . \tag{4'}$$

The gauge function $\Lambda_+$ has 8 degrees of freedom, allowing as many conditions for the components in $V$. E.g. in the Wess-Zumino gauge these conditions are

$$C = \chi = M = f^m v_m = 0 , \tag{5}$$

where $f$ is a fourvector which determines the (homogeneous) gauge of $v(f = \partial$ for the Landau-gauge, $f = n$ for the fixed fourvector in the homogenous axial gauge etc.). This gauge clearly breaks the supersymmetry. The (homogeneous) covariant supergauge is given by

$$D^2 V = \bar{D}^2 V = 0 \ , \tag{6}$$

yielding for the component field $C$ the (spurious, but awkward) infrared singularity of covariant supergauges referred to above.
For the superaxial case the general gauge breaking Lagrangian [2]

$$L_{gb} = \int d^6x_+\, B_a K_+ V_a \ + h.c. = \int d^8x\,(B_a + \bar{B}_a)(K_- + \bar{K}_-)V_a \tag{7}$$

is considered. It implies the gauge conditions

$$\bar{D}^2 K V = D^2 \bar{K} V = 0 \tag{8}$$

and hence by a standard argument the Faddeev-Popov Lagrangian ($u'$ and $u$ are chiral anti-commuting superfields)

$$\begin{aligned} L_{FP} &= \int d^6 x_+ \, u'_a K_+ (\delta V_a)|_{\delta\Lambda \to u} + h.c. = \\ &= \int d^8 x \, (u'_a K - \bar{u}'_a \bar{K})(R^a_b u_b + \bar{R}^a_b \bar{u}_b) \end{aligned} \tag{9}$$

(cf. (4)). The total (bare) Lagrangian consisting of (2), (7) and (9) depends on the breaking of supersymmetry only through the explicit appearance of $K$. This is the analogue of the dependence of a fixed direction $n$ in the axial gauge of ordinary gauge theory [9], referred to above. In the case of supersymmetry with $K$ the situation is more complicated, because (anti-)commuting $K$ with the algebra of $D$-s and $\bar{D}$-s, an enlarged algebra is produced whose elements must be considered as well.
The subsequent derivations are greatly facilitated by a supersymmetric generalization of the elegant compact notation of deWitt [10] in which integrations and summations are denoted by a common index. Writing all fields (Yang-Mills-fields and matter-fields) as components of one 'vector'

$$\phi^I = (V^i, \phi^{\underline{\rho}}) = (V^i, \phi^{\underline{\alpha}}, \phi^{\underline{\dot{\alpha}}}) \tag{10}$$

with

$$\delta\phi^I = R^{I\rho}(\phi)\delta\lambda^\rho \tag{11}$$

and with the definitions $(\bar{D}^2 K, D^2 \bar{K}) \to K^{\rho i}$ the total Lagrangian formally looks like the one in ordinary gauge theory

$$L = L_{inv} + L_{gb} + L_{FP} = L_{inv}(\phi) + B^\rho K^{\rho I} \phi^I + u'^\rho K^{\rho I} R^{I\sigma}(\phi) u_\sigma \ , \tag{12}$$

the only main difference being the nonpolynomial dependence $R(V)$. This notation is also well suited for the evaluation of the $V-$propagator. The quadratic part of (2) with (7) is

$$L_{quad} = \frac{\rho}{2} \phi^I (\Box P_T)_{IJ} \phi^J + B^\rho K^{\rho I} \phi^I \ . \tag{13}$$

We take the propagator to be the negative inverse of $\Gamma$, the 'matrix' in (13):

$$\Delta = \begin{pmatrix} \Delta_{VV} & \Delta_{VB} \\ \Delta_{BV} & \Delta_{BB} \end{pmatrix} \tag{14}$$

We now make the assumption that any admissible $K$ must be such that $(K P_L K^T)$ has an inverse. In this case $\Delta\Gamma = -1$ yields

$$\Delta_{BB} = 0 \tag{15}$$

(no propagation of $B$ !) and

$$\Delta_{VB} = P_L \mathcal{K}^T U \tag{16}$$

with a matrix

$$U = U^T = -(\mathcal{K} P_L \mathcal{K}^T)^{-1} \tag{17}$$

and

$$\Delta_{VV} = -(\rho\Box)^{-1}(P_T + P_T\mathcal{K}^T U \mathcal{K} P_L + P_L \mathcal{K}^T U \mathcal{K} P_T + P_L \mathcal{K}^T U \mathcal{K} P_T \mathcal{K}^T U \mathcal{K} P_L)\,. \tag{18}$$

Since the inversion (16) for $U$ is relatively simple in most practical cases, the full knowledge of the algebra, enlarged by $K$, is not required.
We now turn to a discussion of different choices for $K$. The simplest one is to make $K$ a scalar superfield depending on $\theta$ and $\bar{\theta}$ alone. E.g.

$$K = 1 - \frac{1}{2}\mu^2\theta^2\bar{\theta}^2 \tag{19}$$

has the property that (8) is the same as the covariant supergauge (6) with a modified equation of motion for the $\mathcal{D}$-field. The $C$-field in this case propagates with mass $\mu$ and any mass-term for $C$ is an artefact of gauge-fixing, a result also obtained in [5]. A 'local' ansatz for K like (19) is not sufficient to produce a gauge like (5). We therefore consider the most general 'bilocal' $\hat{K} = \hat{K}(\theta, \bar{\theta}, \theta', \bar{\theta}')$. It is straightforward but lengthy to write down this expression which in fact is just an $N = 2$ superfield. An alternative way to write $K$ uses the derivative operator $K = K(\theta, \bar{\theta}, \partial_\alpha, \partial_{\dot{\alpha}})$. It clearly has the same number of components and is, in fact, simply related to $K$:

$$\hat{K} = K\,\delta^8(x - x') \tag{20}$$

Historically, the first noncovariant supergauge was the '$N^2$-gauge' [2]. Defining

$$N_\alpha = \partial_\alpha - i(\not{n}\bar{\theta})_\alpha \quad , \qquad \bar{N}_{\dot{\alpha}} = \bar{\partial}_{\dot{\alpha}} + i(\not{n}\theta)_{\dot{\alpha}} \tag{21}$$

and $K = N^2$, the algebra of covariant derivatives is enlarged in a very transparent manner. The superpropagator is rather complicated for this gauge, but the $C$-field propagates with a mass $n^2$ and $n^m v_m = 0$ (axial gauge). The advantage of the 'FD-gauge' [3]

$$K = F^\alpha D_\alpha \quad , \qquad F_\alpha = m_\alpha - i(\not{f}\bar{\theta})_\alpha \tag{22}$$

is the simple set of new projection operators.
There are infinitely many choices of $K$ yielding the Wess-Zumino gauge (5) which differ in the sector of the auxiliary field $B$ only. A very simple choice for $K$ is [4]

$$K = \bar{\theta}^2\left[1 + i\partial^2\bar{\partial}^2(\theta\not{f}\bar{\theta})\right] \ . \tag{23}$$

Johanson [11] has proposed another $K$ of a general chiral type.
Supergraph calculations in this approach are perfectly feasible, although the algebra may be

complicated so that computer programs are necessary [12]. E.g. the divergent part of the 1-loop selfenergy of matter is of the form ($\varepsilon = 4 - d$)

$$\Sigma \propto \frac{g^2}{\varepsilon} \int d^8p\, \Phi(-p,\theta)\, X\, \Phi(p,\theta) \ , \tag{24}$$

where $X_{(FD)} = 1, X_{(W.Z.)} = (\theta \not{p} \bar{\theta})$. For the self-energy of $V$ we refer to [12].

# 3 BRS-Invariance, Identities, Remarks on Renormalization and Gauge-Independence

The similarity of (12) to the Lagrangian of ordinary gauge theories immediately allows the introduction of a BRS-transformation [13] with a special gauge transformation involving the Faddeev-Popov field and the anticommuting quantity $\delta\lambda$:

$$\begin{aligned} \delta\phi^I &= R^{I\rho} u_\rho \delta\lambda =: s^I \delta\lambda \\ \delta u_\rho &= \frac{1}{2} g f^{\rho\sigma\tau} u_\sigma u_\tau \delta\lambda =: -s_\rho \delta\lambda \\ \delta u'^\rho &= B^\rho \delta\lambda \\ \delta B &= 0 \end{aligned} \tag{25}$$

$f^{\alpha\beta\gamma}$ are the structure constants in our supercompact notation. The invariance of $L_0 = L_{inv} + L_{gb} + L_{FP}$ with respect to (25) follows from the validity of the 'group relation'

$$\frac{\delta R^{I\sigma}}{\delta\phi^J} R^{J\tau} - (\rho \leftrightarrow \sigma) = -g R^{I\rho} f^{\rho\sigma\tau} \ . \tag{26}$$

Quantization proceeds through a generating functional with a path-integral of the fields

$$\begin{aligned} W &= \int [d\phi dB du du'] e^{i(L + L_s)} \ , \\ L &= L_0 + k^I s^I + l_\rho s_\rho \ , \end{aligned} \tag{27}$$

where the sources are contained in

$$L_s = j^I \phi^I + J^\rho B^\rho + \bar{j}^\sigma u_\sigma + j'_\rho u'^\rho \ . \tag{28}$$

As usual, additional sources are introduced for the composite operators in (25). The latter are BRS-invariant by themselves; therefore, only $\mathcal{L}_s$ breaks the invariance (25). A change of variables (25) in (27) does not change $W$:

$$(j^I \frac{\delta}{\delta k^I} - \bar{j}^\rho \frac{\delta}{\delta l_\rho} + j'^\rho \frac{\delta}{\delta J^\rho}) W = 0 \tag{29}$$

This Slavnov-Taylor identity does not depend explicitly on $\mathcal{K}$. The second important identity obtains for the change $B \to B + \delta B$ in (27):

$$(J^\rho + K^{\rho I} \frac{\delta}{\delta j^I})W = 0 \tag{30}$$

and a third one for the transformation $u' \to u' + \delta u'$ :

$$(j^\rho - K^{\rho I} \frac{\delta}{\delta k^I})W = 0 \tag{31}$$

E.g. from (30) the 'transversality' of all Green's functions with respect to $K$ follows immediately :

$$K^{\rho I_1} \frac{\delta^n W}{\delta j^{I_1} \ldots \delta j^{I_n}} \bigg|_{all\ j=0} = 0 \tag{32}$$

A Legendre-transformation

$$Z_c = -i ln W$$

$$\Gamma(\phi, B, u, u', k, l) = Z_c - j^I \phi^I - \ldots \qquad \phi^I = \frac{\delta Z_c}{\delta j^I} \quad etc. \tag{33}$$

which, however, does not involve $k$ and $l$ leads to the standard Lee-identity, a quadratic identity in the functional $\Gamma$ of one-particle irreducible vertex functions:

$$\frac{\delta \hat{\Gamma}}{\delta \phi^I} \frac{\delta \hat{\Gamma}}{\delta \eta^I} + \frac{\delta \hat{\Gamma}}{\delta u_\rho} \frac{\delta \hat{\Gamma}}{\delta l_\rho} = 0$$

$$\hat{\Gamma} =: \Gamma - L_{gb} \tag{34}$$

$$\eta^I = k^I + u'^\rho K^{\rho I}$$

In order to make contact with previous proofs of renormalization [5] a crucial observation is that any $K$ with bosonic gauge parameters $N$ (bar-indices mean no supercompact notation) may be written as [14]

$$K = \sum_{\bar{i}\bar{j}} N_{\bar{i}\bar{j}} G_{\bar{i}}(D, \bar{D}) \theta^2 \bar{\theta}^2 \hat{G}_{\bar{j}}(D, \bar{D}) \ . \tag{35}$$

Replacing formally

$$\theta^2\bar{\theta}^2 \rightarrow 2\omega \; + \theta^2\bar{\theta}^2 \tag{36}$$

with an external gauge-singlet constant 'field' $w$, the Ward-identity for an arbitrary supersymmetric quantity $A$ ($W_\alpha$ is the generator)

$$W_\alpha A = \sum_{\Phi} \int Q_\alpha \Phi^I \frac{\delta}{\delta \Phi^I} A \tag{37}$$

may be extended to the present broken case:

$$\tilde{W}_\alpha L = \{ \sum_{\Phi \in \{\phi, \ldots, l, w\}} \int Q_\alpha \Phi^I \frac{\delta}{\delta \Phi^I} + \theta_\alpha \bar{\theta}^2 \frac{\partial}{\partial w} \} L = 0 . \tag{38}$$

If, for a 'physical quantity' $w$-independence can be proved, supersymmetry follows.
The inclusion of gauge-parameters $N$ in the BRS-transformations (25) turns out to be very convenient [5]:

$$\delta N_{ij} = M_{ij} \delta\lambda \qquad \delta \; M_{ij} = 0 \tag{39}$$

Further gauge parameters $p$ appear in supersymmetric gauge theories because the dimensionless $V$ may be redefined $V \rightarrow F(V,p)$. Writing $sp = z, sz = 0$ allows their inclusion in the extended BRS-transformation as well.
Similarly also $w$ in (36) acquires a corresponding BRS-transformation $v$. Both terms may be summarized by $N(M)$ in (39). An action which fulfills (25) and (39) is

$$L' = L - M_{ij} u'^{\rho} \frac{\partial K^{\rho I}}{\partial N_{ij}} \phi^I . \tag{40}$$

The corresponding generalization of (34) contains further terms:

$$\frac{\delta \hat{\Gamma}}{\delta \phi^I} \frac{\delta \hat{\Gamma}}{\delta \eta^I} + \frac{\delta \hat{\Gamma}}{\delta u_\rho} \frac{\delta \hat{\Gamma}}{\delta l_\rho} + M_{ij} \frac{\partial \hat{\Gamma}}{\partial N_{ij}} = 0 \tag{41}$$

Here $\hat{\Gamma}$ is defined like $L$ subtracting out $L_{gb}$. Then renormalization can be reduced to the solution of the cohomology problem of a nilpotent 'extended' BRS-operator $\mathcal{B}_{\hat{L}'}$ even in the absence of a supersymmetric regularization scheme [4]. It can be shown that all gauge-dependent counter-terms are necessarily of the type $\mathcal{B}_{\hat{L}'}\Delta$, i.e. they may be absorbed by local counter-terms $-\Delta$ to the action. The cohomology for the rest has been solved (under certain technical assumptions so as to exclude anomalies) in ref. [5].
The proof of gauge-independence refers to an $S-$matrix-element, consisting of an amputated Green's function, decorated by 'polarization vectors' and renormalization factors at external legs. The definition of polarization vectors (physical sources, asymptotic fields) is not without problems in supersymmetric theories [15]. Nevertheless, at least to the degree of rigour achieved in ordinary gauge theories, the proof can be carried through. The 'extended' BRS transformation again is able to exhibit its superiority to other techniques. As a by-product also the covariance of the S-matrix with respect to (globally) supersymmetric transformations can be shown [15]. More details can be found in the comprehensive review [16].

## 4. Conclusion

The generalization of the concept of axial gauges to superspace has very similar formal advantages. It allows the inclusion of manifestly nonsupersymmetric but 'physical' gauges in a supergraph framework. Of course the problems of ordinary axial gauges are not solved in this manner. Nevertheless, also for supersymmetry formal proofs guarantee renormalization and gauge independence.

## References


1. J. Wess and B. Zumino, Phys.Lett. 49B (1974) 52; J. Wess and B. Zumino, Nucl.Phys. B87 (1974) 1; A. Salam and J. Strathdee, Phys.Lett. 51B (1974) 353; S. Ferrara and B. Zumino, Nucl.Phys. B71 (1974) 403,
2. W. Kummer and M. Schweda, Phys.Lett. 141B (1984) 363
3. T. Kreuzberger, W. Kummer, O. Piguet, A. Rebhan and M. Schweda, Phys.Lett. 167 (1986) 393
4. T. Kreuzberger, W. Kummer, H. Mistelberger, P. Schaller, M. Schweda, Nucl.Phys. B281 (1987) 411
5. O. Piguet and K. Sibold, Nucl.Phys. B248 (1984) 336, ibid. B249 (1985) 396
6. S. Ferrara, J. Wess and B. Zumino, Phys.Lett. 51B (1974) 239; A. Salam and J. Strathdee, Phys.Rev. D11 (1975) 1521
7. S. Ferrara and B. Zumino, Nucl.Phys. B79 (1974) 413; A. Salam and J. Strathdee, Phys.Lett. 51B (1974) 353; B. de Witt and D. Freedman, Phys.Rev. D12 (1975) 2286
8. M.T. Grisaru, W. Siegel and M. Roček, Nucl.Phys. B159 (1979) 429
9. W. Kummer, Acta Phys. Austr. 14 (1961) 149; R.L. Arnowitt and S.I. Fickler, Phys.Rev. 127 (1962) 1821; J. Schwinger, Phys.Rev. 130 (1963) 402; Y.P. Yao, Journal of Math.Phys. 5 (1964) 1319; E.S. Fradkin and I.V. Tyutin, Phys.Rev. D2 (1970) 2841; W. Kummer, Acta Phys. Austr. 41 (1975) 315; W. Konetschny and W. Kummer, Nucl.Phys. B100 (1975) 106
10. B.S. de Witt, Phys.Rev. 162 (1967) 1195
11. A.A. Johanson, Superfields in the noncovariant supergauges, Leningrad prepr. 1985
12. T. Kreuzberger, W. Kummer and M. Schweda, SUSYCAL — A program for symbolic computations in supersymmetric theories, HEPHY-Pub. 521/1989
13. C. Becchi, A. Rouet and R. Stora, Phys.Lett. 52B (1974) 344
14. W. Kummer, H. Mistelberger, P. Schaller and M. Schweda, Zs. f. Phys. 40C (1988) 91
15. W. Kummer, H. Mistelberger, and P. Schaller, Zs.f.Phys. 40C (1988) 103
16. W. Kummer, H. Mistelberger, P. Schaller and M. Schweda, Fortschr.d.Phys. 37 (1989) 261

# The Piguet-Sibold Identities and the Light-Cone Gauge

A. Andraši

Ruder Bošković Institute,

41001 Zagreb, P.O.B. 1016 Croatia, Yugoslavia

Abstract

Using diagrammatic methods, we show new identities, additional to BRST identities, which forbid the unwanted counter-terms for non-covariant gauges.

1. Introduction

The light-cone gauge has become increasingly popular with the revival and intensive research of string theories[1]. However, there are problems about defining not only the light cone, but all axial gauges carefully. We start by considering the planar gauge[2] without introducing any regularization. The propagator in this gauge is given by

$$D^{ab}_{\mu\nu} = \delta_{ab}(k^2+i\varepsilon)^{-1}[\delta_{\mu\nu}-(n\cdot k)^{-1}(k_\mu n_\nu+k_\nu n_\mu)] \quad . \tag{1}$$

The simple observation that applying the operation $\partial/\partial n_\gamma$ to the propagator (1) leaves us with the terms proportional to either $k_\mu$ or $k_\nu$, leads to new identities[3] in their simplest form. The factor $k_\mu$ acting on the vertex gives the usual Ward identity.

To show this in detail, for the self-energy to one-loop order we have

$$\begin{aligned}\frac{\partial}{\partial n_\gamma}\Pi^{ab}_{\mu\nu} = {} & A^{ab}[\delta_{\alpha\beta}(2k-p)_\nu+\delta_{\alpha\nu}(-p-k)_\beta+\delta_{\nu\beta}(2p-k)_\alpha] \\ & \times \frac{1}{k^2 n\cdot k}\left[k_\alpha\left(\frac{k_\gamma n_\rho}{n\cdot k}-\delta_{\rho\gamma}\right)+k_\rho\left(\frac{k_\gamma n_\alpha}{n\cdot k}-\delta_{\alpha\gamma}\right)\right] \\ & \times [\delta_{\eta\mu}(2p-k)_\rho+\delta_{\mu\rho}(-k-p)_\eta+\delta_{\rho\eta}(2k-p)_\mu] \\ & \times \frac{1}{(p-k)^2}\left[\delta_{\eta\beta}-\frac{(p-k)_\eta n_\beta+(p-k)_\beta n_\eta}{n(p-k)}\right] \\ & + \text{the term where } \frac{\partial}{\partial n_\gamma} \text{ acts on the propagator with momentum } (p-k)_\lambda \\ A^{ab} = {} & \frac{1}{2}g^2 C(R)\delta_{ab}\int\frac{d^4k}{(2\pi)^4} \quad .\end{aligned} \tag{2}$$

Then, the simple Ward identity

$$k_\alpha\left[\delta_{\alpha\beta}(2k-p)_\nu+\delta_{\alpha\nu}(-p-k)_\beta+\delta_{\nu\beta}(2p-k)_\alpha\right]$$
$$=\delta_{\nu\beta}p^2-p_\nu p_\beta-\delta_{\nu\beta}(p-k)^2+(p-k)_\nu(p-k)_\beta \tag{3}$$

indicates the tensor structure of the form $(\delta_{\nu\beta}p^2-p_\nu p_\beta)$.

Actually, after evaluating these integrals, we obtain

$$\frac{\partial}{\partial n_\gamma}\Pi_{\mu\nu}=2g^2C(R)\delta_{ab}\Big\{\frac{n_\mu}{n^2}\Big(\delta_{\beta\gamma}-\frac{n_\beta n_\gamma}{n^2}\Big)(\delta_{\nu\beta}p^2-p_\nu p_\beta)$$
$$+\frac{n_\rho}{n^2}\Big(\delta_{\nu\gamma}-\frac{n_\nu n_\gamma}{n^2}\Big)(\delta_{\mu\rho}p^2-p_\mu p_\rho)+\frac{n_\nu}{n^2}\Big(\delta_{\beta\gamma}-\frac{n_\beta n_\gamma}{n^2}\Big)$$
$$\times(\delta_{\mu\beta}p^2-p_\mu p_\beta)+\frac{n_\rho}{n^2}\Big(\delta_{\mu\gamma}-\frac{n_\mu n_\gamma}{n^2}\Big)(\delta_{\nu\rho}p^2-p_\nu p_\rho)\Big\}. \tag{4}$$

A counter-term such as $n_\mu n_\nu F^{\mu\lambda}F_{\nu\lambda}$ which would spoil the renormalization in the planar gauge[4] does not satisfy this identity, as it contains the term $\delta_{\mu\nu}(n\cdot p)^2$ and there is no such term in (4) (i.e. the term $\delta_{\mu\nu}(n\cdot p)p_\gamma$).

In the same way, by inserting the light-cone propagator of Mandelstam or Leibbrandt[5], we can show why there are no counter-terms such as

$$n_\mu n^*_\nu F_{\mu\lambda}F_{\nu\lambda}\quad\text{and}\quad(n_\mu n^*_\nu F_{\mu\nu})^2\ .$$

## 2. The ghost-gluon transition

To derive diagrammatic identities for the planar gauge, let us introduce a new vertex

$$\sim\!\sim\!\sim\ \boxed{\xi}\ \text{-->}\ -\ \Big(\frac{n_\xi n_\alpha}{n^2}-\delta_{\xi\alpha}\Big)\ . \tag{5}$$

It symbolizes the ghost-gluon transition (for the planar gauge we can choose $n^2=1$, $n_\xi\delta n_\xi=0$) and then

$$\delta n_\xi\eta\Big(\frac{n_\xi n_\alpha}{n^2}-\delta_{\xi\alpha}\Big)A_\alpha=-\eta\delta n\cdot A\ . \tag{6}$$

Then the structures $\Big[\Big(\frac{k_\gamma n_\rho}{n\cdot k}\Big)-\delta_{\rho\gamma}\Big]$ which occurred in (2) can be represented as the ghost-gluon transition through the new vertex

$$\frac{1}{n\cdot k}\Big(\delta_{\gamma\alpha}-\frac{n_\gamma n_\alpha}{n^2}\Big)\frac{1}{k^2}\Big(\delta_{\alpha\rho}-\frac{k_\alpha n_\rho+k_\rho n_\alpha}{n\cdot k}\Big)=\frac{1}{n\cdot kk^2}\Big(\delta_{\gamma\rho}-\frac{k_\gamma n_\rho}{n\cdot k}\Big). \tag{7}$$

## 3. The vertex graph

Let us take the three-gluon vertex and apply the $\partial/\partial n_\xi$ operation. It will act on each propagator, but let us follow what happens when it acts on the propagator with momentum $r$:

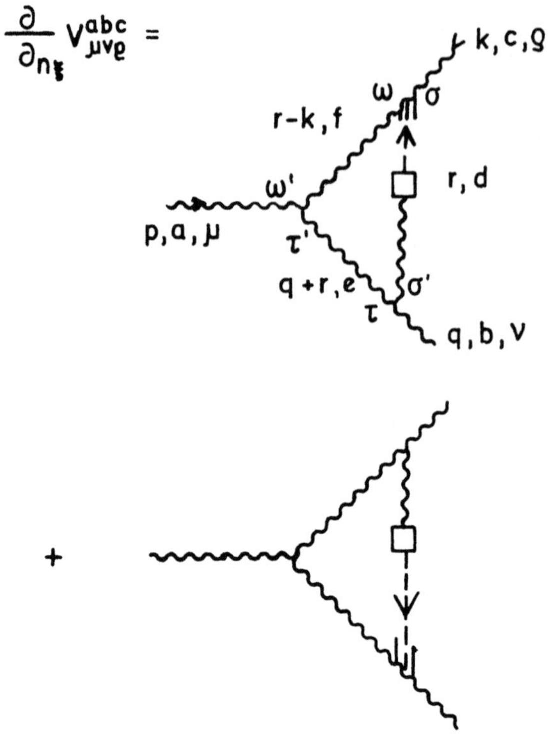


+ (cyclic permutations of propagators)

Fig. 1

Here - → -≡ is the ghost propagator with the factor $r_\sigma$. The factor $r_\sigma$ with the vertex $V_{\sigma\rho\omega}$ gives a Ward identity like the one in (3).

On the mass shell, where $k^2=0$, $e^\rho k_\rho=0$ ($e^\rho$ - the polarization vector), the tensor $(\delta_{\omega\rho}k^2-k_\omega k_\rho)$ vanishes. The other tensor $\delta_{\omega\rho}(r-k)^2 - (r-k)_\omega(r-k)_\rho$ reduces the $(r-k)^2$ denominator, so we are left with the first diagrammatic identity (Fig. 2), where we meet a special ghost-gluon vertex

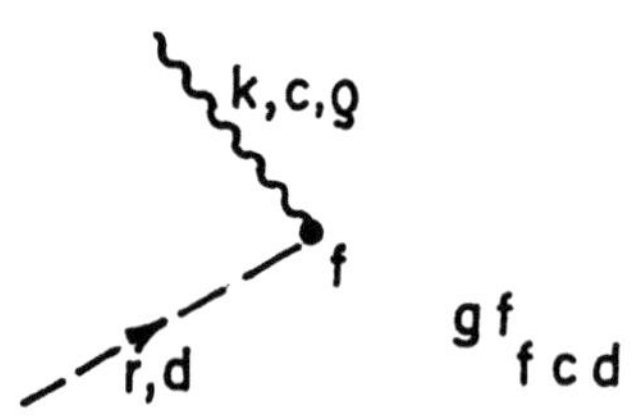


The last diagram in Fig. 2 can be further reduced, until we arrive at the ghost loop shown in Fig. 3.

In this way we have the graph shown in Fig. 4.

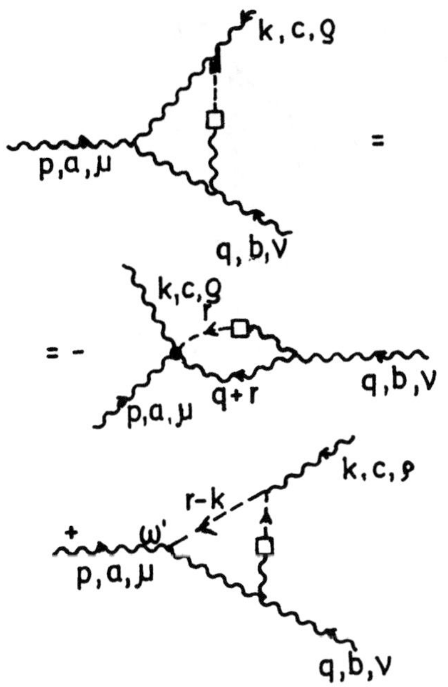
k, c, ρ
p, a, μ
=
q, b, ν
k, c, ρ
r
= -
p, a, μ
q+r
q, b, ν
k, c, ρ
r-k
+
ω'
p, a, μ
q, b, ν

Fig. 2

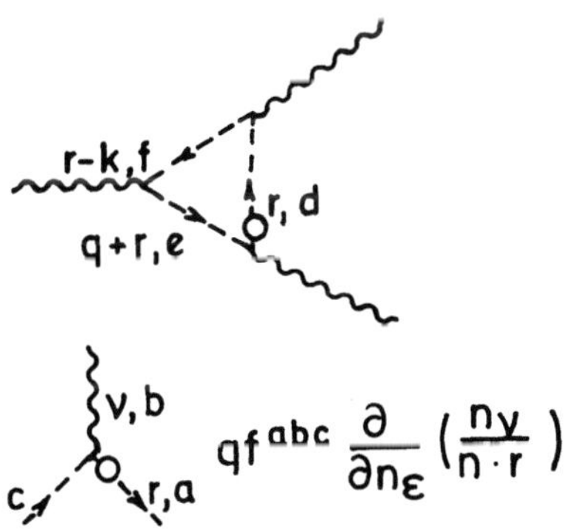
r-k, f
q+r, e
r, d
ν, b
c
r, a
$qf^{abc} \frac{\partial}{\partial n_\varepsilon} \left( \frac{n_\nu}{n \cdot r} \right)$

Fig. 3

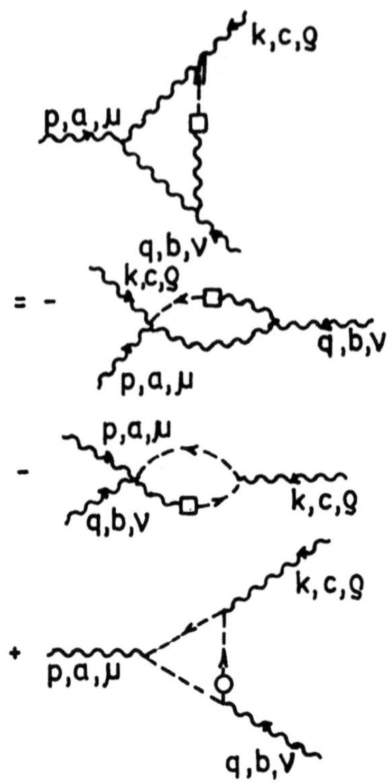


Fig. 4

## 4. Identities for the generating functional

The diagrammatic identities (including the on mass shell terms) can be written as identities for the generating functional:

$$\frac{\delta^3\,\Gamma^{(1)}}{\delta A(1)\,\delta A(2)\,\delta A(3)} = \frac{\delta\Gamma^{(1)}}{\delta A(1)\,\delta A(2)\,\delta\lambda\,\delta u^a_\mu(x)} \times \frac{\delta\Gamma^{(o)}}{\delta A^a_\mu(x)\,\delta A(3)}$$

$$+ \frac{\delta\Gamma^{(1)}}{\delta A(1)\,\delta\lambda\,\delta u_\mu(x)} \times \frac{\delta\Gamma^{(o)}}{\delta A^a_\mu(x)\,\delta A(2)\,\delta A(3)} + \text{permutations of the points } (1,2,3), \qquad (8)$$

where $\delta\Gamma$ is the variation of the generating functional with respect to the gauge parameter $n$, $u_\mu(x)$ is the source coupled to the ghost and the gluon, and $\lambda$ is the strength of the ghost-gluon transition. $\Gamma^{(1)}$ and $\Gamma^{(o)}$ are one-loop and zero-loop generating functionals, respectively.

From Eq. (8) we can deduce the basic equation

$$\delta\Gamma^{(1)} = \int dx\, \frac{\delta^2\Gamma^{(1)}}{\delta\lambda\,\delta u^a(x)} \times \frac{\delta\Gamma^{(o)}}{\delta A^a(x)}\,. \qquad (9)$$

Diagrammatic identities can obviously be derived for any number of external legs. Formally, such identities can be phrased through BRST transformations[3)]:

$$\Delta\Gamma = \Gamma * \Gamma' \, . \tag{10}$$

The change in the generating functional under the change of $n$ (or $n^*$ for the light-cone gauge) has to be given by a BRST convolution $\Gamma * \Gamma'$, where $\Gamma'$ contains a special insertion.

To one-loop order a formal solution to the Piguet-Sibold identity (10) is given by[3c)]

$$\Gamma_1' = \Delta G_1 + G_1 * S' \, , \tag{11}$$

where

$$\Gamma_1 = S * G_1 \tag{12}$$

for some $G_1$, and $S$ is the action.

For the light-cone gauge, we gave an extensive discussion of the divergent parts of $\Gamma_1'$ in Ref. 6). For the two graphs shown there, the complete expression including the finite part of $\Gamma_1'$ is

$$\begin{aligned}
\Gamma_1' \cdot C^{-1} = & -\Delta n^* \cdot p \, \Big\{ \frac{n_\rho n_\alpha}{n \cdot p} \Big[\Gamma\Big(\frac{4-n}{2}\Big) - \ln(-p_+ p_-)\Big]\Big\} \\
& - \frac{\Delta n^* \cdot p}{\underset{\sim}{p}^2} \ln \frac{p_+ p_-}{p^2} \Big[-2p \cdot n \, \delta_{\rho\alpha} - \frac{p^2}{n \cdot p} n_\rho n_\alpha + 2 n_\alpha p_\rho + 2 n_\rho p_\alpha\Big] \\
& + n_\alpha \Delta n_\rho^* \Big[\Gamma\Big(\frac{4-n}{2}\Big) + 2 - \ln(-p_+ p_-) - \frac{p^2}{\underset{\sim}{p}^2} \ln \frac{p_+ p_-}{p^2}\Big] \\
& - 2 n_\alpha \frac{\Delta n^* p}{\underset{\sim}{p}^2} \Big(-1 + \frac{p^2}{\underset{\sim}{p}^2} \ln \frac{p_+ p_-}{p^2}\Big)\Big(p_\rho - \frac{1}{2} n_\rho^* p_+ - \frac{1}{2} n_\rho p_-\Big) \, ,
\end{aligned}$$

where

$$C = g^2 C(R) \, \delta_{mn} \, i\pi^{n/2} (2\pi)^n \, . \tag{13}$$

We can check that it satisfies the Piguet-Sibold identity

$$\Delta\Pi_{\mu\nu}(p) = (p_\mu p_\lambda - \delta_{\lambda\mu} p^2)\Gamma'_{\lambda\nu} + (p_\nu p_\lambda - \delta_{\lambda\nu} p^2)\Gamma'_{\mu\lambda} \, ,$$

where the complete $\Pi_{\mu\nu}$ to one-loop order was calculated in Ref. 7).

Acknowledgment

This research was done in collaboration with Prof. J.C. Taylor.

References

[1] J. Scherk and J.H. Schwarz, Gen. Rel. Grav. 6 (1975) 537.
[2] Yu.L. Dokshitsev, D.J. D'yakonov and S.J. Troyan, Phys. Rep. 58 (1980) 269;
[3] a) O. Piguet and K. Sibold, Nucl. Phys. B253 (1985) 517;
b) P. Gaigg, O. Piguet, A. Rebhan and M. Schweda, Phys. Lett. B175 (1986) 53;
c) A. Andraši and J.C. Taylor, DAMTP 86-11 preprint (1986) unpublished.
[4] A. Andraši and J.C. Taylor, Nucl. Phys. B192 (1981) 283.
[5] S. Mandelstam, Nucl. Phys. B213 (1983) 149;
G. Leibbrandt, Phys. Rev. D29 (1984) 1699.
[6] A. Andraši and J.C. Taylor, Nucl. Phys. B302 (1988) 123.
[7] A. Andraši and J.C. Taylor, Nucl. Phys. B310 (1988) 222.

# BRS Techniques Applied to the Temporal-Planar Gauge with the Vienna Prescription: The Dependence on the Axial Vectors[1]

E. Bagan and C. P. Martin

Department of Applied Mathematics and Theoretical Physics,
University of Cambridge, Silver Street,
Cambridge, CB3 9EW, United Kingdom

## 1 Introduction

Notwithstanding a gauge theory in an axial gauge is supposed to be ghost free, BRS symmetry provides powerful tools for studying its properties. In this gauges, unlike in the Lorentz covariant gauges, ghost-loops do not contribute to any Feynman diagram. However, the remaining diagrams are much more involved than its Lorentz covariant counterparts.

In this report we will show an instance in which BRS symmetry proved to be very useful. The example has to do with the temporal planar gauge (for a review of the temporal gauge see P.V. Landshoff in these proceedings ) as defined by the Vienna prescription (see G. Pollak these Proceedings) We will study the dependence of Green's functions on the axial vectors. In particular, we shall compute the changes in the one-loop UV divergent part of the propagator induced by changes of the axial vectors. The equations these changes satisfy, impose additional constraints on the counterterms. This constraints are welcome since locality of counterterms is lost in our case.

The interested reader will find further details in reference [1].

---

[1] Talk presented by C. P. Martin

## 2 A BRS invariant action and some of its consequences

The aim of this section is to show that if one regards the following path integral

$$Z[J^a,\omega^a,\bar{\omega}^a] \;=\; C^{-1}\int [dA^a\, d\phi^a\, d\bar{\phi}^a] e^{iS+i\int d^{2\omega}x\,(J^a_\mu A^{a\mu}+\bar{\phi}^a\omega^a+\bar{\omega}^a\phi^a)} \tag{1}$$

($S$ defined below) as a formal device to generate the axial-temporal gauge perturbation theory and BRS identities, one can control perturbatively the Green's functions' dependence on the axial vectors.

The action $S$ is defined to be

$$S=\int d^{2\omega}x\Big[-\frac{1}{4}F^a_{\mu\nu}(x)F^{\mu\nu}_a(x)+\frac{1}{2\alpha}N{\cdot}A^a(x)\frac{\partial^2}{N^2}N{\cdot}A^a(x) \\ -\bar{\phi}^a(x)\frac{\partial^2}{N^2}N{\cdot}D^{ab}\phi^b(x)\Big]\;, \tag{2}$$

where $F^a_{\mu\nu}$ denotes the field-strength tensor. The operator $N$ is given by

$$N_\mu \equiv n_\mu + i\eta\frac{\partial_\mu}{n'{\cdot}\partial\,\Box}\quad ;\quad \eta>0\;,$$

where $n$ and $n'$ are fixed vectors satisfying

$$n^2>0\;,\qquad n{\cdot}n'=0\;\Rightarrow\; n'^2<0\;.$$

From equation (1), one can show formally that the correlation functions of gauge-invariant operators are independent of $n$ nor $n'$.

The reader could be puzzled about a complex action occurring in the path integral. Actually, large values of the field could provide infinite contributions. coming from the $e^{iS}$-factor. Let us stress that we consider $Z[J^a,\omega^a,\bar{\omega}^a]$ as defined in perturbation theory. So, the only requirement is the kinetic term to be invertible. The infinite value coming from the free path integral with the sources set to zero is absorbed into the constant $C$. The unitarity of the theory relies on the BRS invariance of the formalism.

We shall be modest as far as computations are concerned. We shall consider the simplest case and show that in the limit $\eta\to 0$ the vacuum polarisation tensor is the one obtained with the Vienna prescription [2]. Besides, its dependence on $n$ and $n'$ can be examined by using the BRS symmetry of $S$.

Before continuing a warning. In doing the computations the limit $\eta\to 0$ must be taken at the end. This seems to be irrelevant in some cases (for instance, when computing Green's functions of the gauge field $< A^{a1}_{\mu 1}\ldots\ldots A^{an}_{\mu n} >$ ) because the integrals have a smooth limit. However, it is absolutely essential in other cases.

The nonlocal operators in $N$ are defined as follows. $\Box^{-1}$ is defined by the Feynman prescription. $1/n'{\cdot}\partial$ is defined in such a way that its Fourier transform fulfils

$$\frac{1}{[n'{\cdot}q]_{\mathrm{P}}}\to\kappa^{-1/2}\frac{1}{[n'{\cdot}q]_{\mathrm{P}}}\quad\text{under }\; q\to\kappa^{1/2}q\;;\;\;\eta\to\kappa\eta\;.$$

The action $S$ is invariant under the following BRS transformation

$$sA^a_\mu = D^{ab}_\mu \phi^b \quad , s\phi^a = -\frac{1}{2} g f^{abc} \phi^b \phi^c \quad , s\bar{\phi}^a = \frac{1}{\alpha} N \cdot A^a \ . \tag{3}$$

Let us consider a function ( or functional), $\hat{O}(A^a_\mu)$ of $A^a_\mu$ and its derivatives. $\hat{O}(A^a_\mu)$ does not depend on $n$ or $n'$ and it does not involve ghosts. We will be interested in the variations of

$$< \hat{O}(A^a_\mu) > = C^{-1} \int [dA^a \, d\phi^a \, d\bar{\phi}^a] \ \hat{O}(A^a_\mu) \ e^{iS} \tag{4}$$

under changes either of $n$ or $n'$. In studying such changes we shall follow Collins' approach [3].

Any variation of $n$ ($n'$) gives rise to a variation of $N_\mu$:

$$\delta_v N_\alpha = \begin{cases} \delta n_\mu & \text{for } v_\mu = n_\mu \\ -i\eta \frac{\partial_\mu}{(n\partial)^2 \Box} \delta n' \cdot \partial & \text{for } v_\mu, = n'_\mu \ . \end{cases} \tag{5}$$

Since $N_\mu$ is the only place in $S$ in which those changes occur, one obtains the following formal equation:

$$\delta_v \left\langle \hat{O}(A) \right\rangle = i \left\langle \hat{O}(A) \ \delta_v S \right\rangle \tag{6}$$

where $\delta_v S$ can be recast into the form

$$\delta_v S = \int d^{2\omega} z \left\{ s \left\{ \bar{\phi}^a \frac{\partial^2}{N^2} \ \delta_v N_\alpha \left[ g^{\alpha\beta} - \frac{N^\alpha N^\beta}{N^2} \right] A^a_\beta \right\} + \frac{N^\alpha}{N^2} \bar{\phi}^a \frac{\partial^2}{N^2} \delta_v N_\alpha N \cdot D^{ab} \phi^b \right\} \tag{7}$$

by trivial BRS algebraic computations.

The last term in equation (7) always yields tadpole-like contributions. Hence, it can be discarded in the framework of dimensional regularisation. Therefore, equation (6) reads

$$\delta_v \langle \hat{O}(A) \rangle = -i \left\langle \int d^{2\omega} z \, \delta_v N_\alpha \, A^a_\beta(z) \left[ g^{\alpha\beta} - \frac{N^\alpha N^\beta}{N^2} \right] \frac{\partial^2}{N^2} \bar{\phi}^a(z) s \left[ \hat{O}(A) \right] \right\rangle \tag{8}$$

In getting equation (8) one has to bear in mind that $s < \ldots\ldots >= 0$.

When $\hat{O}(A)$ is a gauge invariant operator, equation (8) implies

$$\delta_v < \hat{O}(A) > \ = \ 0 \ . \tag{9}$$

Therefore, at least formally, the Green's functions of gauge invariant operators do not change under changes of the axial vectors. This is a consistency requirement for equation (1) to be meaningful.

One could ask the value of $\frac{d}{d\eta} < \hat{O}(A) >$. By using the above techniques one gets

$$\frac{d}{d\eta} \langle \hat{O}(A) \rangle = i \left\langle \int d^{2\omega} z \, \frac{\partial_\alpha}{n' \cdot \partial \Box} A^a_\beta(z) \left[ g^{\alpha\beta} - \frac{N^\alpha N^\beta}{N^2} \right] \frac{\partial^2}{N^2} \bar{\phi}^a(z) s \left[ \hat{O}(A) \right] \right\rangle . \tag{10}$$

Hence, $\frac{d}{d\eta} < \hat{O}(A) >= 0$ when $\hat{O}(A)$ is a gauge invariant operator.

Despite their appealing consequences, equations (8) and (9) are formal ones and involve singular operators. So, it is advisable to check them by means of an explicit computation. We shall study the variations, in the limit $\eta \to 0$, of the UV divergent part of the propagator under changes of $n$ and $n'$.

## 3 The Vacuum Polarisation Tensor and its Dependence on the Axial Vectors

To evaluate the gluon vacuum polarisation in the limit $\eta \to 0$ we need the usual Feynman rules for the three-gluon vertex and for the propagator of the gauge field $iG^{ab}_{\mu\nu}(k)$:

$$iG^{ab}_{\mu\nu}(k) = -i\frac{\delta^{ab}}{k^2+i\epsilon}\left[g_{\mu\nu} - \frac{(k_\mu n_\nu + k_\nu n_\mu)k{\cdot}n'}{(k{\cdot}nk{\cdot}n'+i\eta)} + (1+\alpha)\frac{n^2 k_\mu k_\nu (k{\cdot}n')^2}{(k{\cdot}nk{\cdot}n'+i\eta)^2}\right] + o(\eta)k_\mu k_\nu \tag{11}$$

If the term $o(\eta)$ does not contribute in the limit $\eta \to 0$ we will obtain the Vienna prescription. In our computations this term does not give rise to any new contribution. By calculating in this way we make sure that in order to satisfy the BRS identities no additional $o(\eta)$ terms in the propagator are necessary.

The UV divergent contribution to the vacuum polarisation tensor, $\Pi^{ab}_{\mu\nu}$ has been computed by different groups [1],[2]. In our case, the computations have been carried out for $\eta \neq 0$. All the integrals needed either are $o(\eta^0)$ or $o(\eta^1)$ (see ref. [1]). In the limit $\eta \to 0$ one gets

$$\begin{aligned}
\Pi^{ab}_{\mu\nu} = &- g^2 C(G)\delta^{ab}\frac{i\pi^2}{\epsilon}\Bigg\{ \left[\frac{11}{3} - \frac{2p{\cdot}\hat{n}'}{p{\cdot}\hat{n}}\right](p^2 g_{\mu\nu} - p_\mu p_\nu) \\
&\quad + \left[1 + \frac{(p{\cdot}\hat{n}')^2}{(p{\cdot}\hat{n})^2}\right]\left\{-p{\cdot}\hat{n}' p_{(\mu}\hat{n}_{\nu)} + p{\cdot}\hat{n}\, p_{(\mu}\hat{n}'_{\nu)}\right\} \\
&\quad - p^2\left[1 - \frac{(p{\cdot}\hat{n}')^2}{(p{\cdot}\hat{n})^2}\right]\hat{n}_{(\mu}\hat{n}'_{\nu)} + 2p^2\frac{p{\cdot}\hat{n}'}{p{\cdot}\hat{n}}\left(\hat{n}_\mu\hat{n}_\nu - \hat{n}'_\mu\hat{n}'_\nu\right) \\
&+\alpha\Bigg\{2\left(p^2 g_{\mu\nu} - p_\mu p_\nu\right) \\
&\quad + p{\cdot}\hat{n}\left[2 + 3\frac{p{\cdot}\hat{n}'}{p{\cdot}\hat{n}} - \frac{(p{\cdot}\hat{n}')^3}{(p{\cdot}\hat{n})^3}\right]p_{(\mu}\hat{n}_{\nu)} \\
&\quad - p^2\left[1 - \frac{(p{\cdot}\hat{n}')^2}{(p{\cdot}\hat{n})^2}\right]\hat{n}_{(\mu}\hat{n}'_{\nu)} - 4p^2\left[1 + \frac{p{\cdot}\hat{n}'}{p{\cdot}\hat{n}}\right]\hat{n}_\mu\hat{n}_\nu\Bigg\} \\
&+(1+\alpha)^2\Bigg\{\frac{p^2 p{\cdot}\hat{n}'}{(p{\cdot}\hat{n})^3}\left(p^2 g_{\mu\nu} - p_\mu p_\nu\right) \\
&\quad - \left[\frac{p{\cdot}\hat{n}'}{p{\cdot}\hat{n}} - \frac{(p{\cdot}\hat{n}')^3}{(p{\cdot}\hat{n})^3}\right]p_\mu p_\nu + \frac{p^2 p{\cdot}\hat{n}'}{(p{\cdot}\hat{n})^2}p_{(\mu}\hat{n}_{\nu)} \\
&\quad - \frac{p^2(p{\cdot}\hat{n}')^2}{(p{\cdot}\hat{n})^3}p_{(\mu}\hat{n}'_{\nu)} - \frac{(p^2)^2 p{\cdot}\hat{n}'}{(p{\cdot}\hat{n})^3}\left(\hat{n}_\mu\hat{n}_\nu - \hat{n}'_\mu\hat{n}'_\nu\right)\Bigg\}\Bigg\},
\end{aligned} \tag{12}$$

where $\hat{n}_\mu \equiv n_\mu/\sqrt{n^2}$, $\hat{n}'_\mu \equiv n'_\mu/\sqrt{-n'^2}$ and symmetrisation of the indices in parenthesis is understood.

Now, let us consider $\hat{O}(A) = A^a_\mu(x)A^a_\nu(y)$ . With this operator equation (8) reads:

$$i\delta G^{ab}_{\mu\nu}(p) = i\,\delta n_\alpha \left\langle \int d^{2\omega} z A^c_\beta(z) \left[ g^{\alpha\beta} - \frac{N^\alpha N^\beta}{N^2} \right] \frac{\partial^2}{N^2} \overline{\phi}^c(z) s \left[ \widehat{A}^a_\mu(p) A^b_\nu(0) \right] \right\rangle , \tag{13a}$$

$$i\delta G^{ab}_{\mu\nu}(p) = \eta\,\delta n'_\alpha \left\langle \int d^{2\omega} z A^c_\beta(z) \left[ g^{\beta\gamma} - \frac{N^\beta N^\gamma}{N^2} \right] \frac{-\partial^\alpha \partial_\gamma}{(n'\cdot\partial)^2 N^2} \overline{\phi}^c(z) s \left[ \widehat{A}^a_\mu(p) A^b_\nu(0) \right] \right\rangle . \tag{13b}$$

We have checked that those identities hold as far as one-loop UV divergent parts are concerned. We have considered $n^2 > 0$, $n\cdot n' = 0$ and, therefore, $\delta n$ and $\delta n'$ satisfying $n'\cdot\delta n = 0$ , $n\cdot\delta n' = 0$. The one-loop contribution to the r.h.s. of the above equations stems from the diagrams in figure 1.

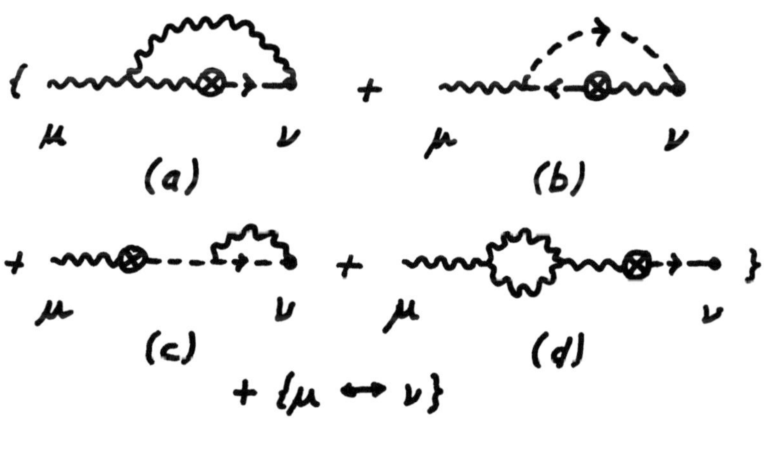


Fig. 1.

The operators $sA^a_\mu$ $(= D^{ab}_\mu \phi^b)$ and $i\delta_v N_\alpha A^c_\beta [g^{\alpha\beta} - N^\alpha N^\beta/N^2]\partial^2/N^2 \overline{\phi}^c$ are represented by • and ⊗ respectively. The dotted line stands for the ghost propagator and the remaining vertices and lines stand for the ones provided by equation (2).

From those diagrams one can work out the following UV divergent contributions to $\delta_n \Pi^{ab}_{\mu\nu}$ and $\delta_{n'} \Pi^{ab}_{\mu\nu}$:

$$
\begin{aligned}
\delta_n \Pi^{ab}_{\mu\nu} = -g^2 C(G)\delta^{ab}\frac{i\pi^2}{\epsilon}\Bigg\{ & \left[2(p{\cdot}\hat{n})^2 - 3p^2(1+\alpha)^2\right]\frac{p{\cdot}\hat{n}'p^2}{(p{\cdot}\hat{n})^4}\wp_\alpha g_{\mu\nu} \\
& + \left\{\left[(p{\cdot}\hat{n})^2 - 3(p{\cdot}\hat{n}')^2 + 3p^2\right](1+\alpha)^2 - 2(p{\cdot}\hat{n})^2\right\}\frac{p{\cdot}\hat{n}'}{(p{\cdot}\hat{n})^4}\wp_\alpha p_\mu p_\nu \\
& + 2\left\{-p{\cdot}\hat{n}'p^2(1+\alpha)^2 + \left[(p{\cdot}\hat{n}')^3 + (p{\cdot}\hat{n})^3\right]\alpha + (p{\cdot}\hat{n}')^3\right\}\frac{1}{(p{\cdot}\hat{n})^3}\wp_\alpha p_{(\mu}\hat{n}_{\nu)} \\
& + \left[3p^2(1+\alpha)^2 + 4(p{\cdot}\hat{n})^2\alpha - 2(p{\cdot}\hat{n})^2\right]\frac{p{\cdot}\hat{n}'p^2}{(p{\cdot}\hat{n})^4}\wp_\alpha \hat{n}_\mu \hat{n}_\nu \\
& + \left\{3(p{\cdot}\hat{n}')^2 p^2(1+\alpha)^2 + (p{\cdot}\hat{n})^2\left[(p{\cdot}\hat{n})^2 - (p{\cdot}\hat{n}')^2\right]\right\}\frac{1}{(p{\cdot}\hat{n})^4}\wp_\alpha p_{(\mu}\hat{n}'_{\nu)} \\
& - 2(\alpha+1)\frac{(p{\cdot}\hat{n}')^2 p^2}{(p{\cdot}\hat{n})^3}\wp_\alpha \hat{n}_{(\mu}\hat{n}'_{\nu)} - \left[3p^2(1+\alpha)^2 - 2(p{\cdot}\hat{n})^2\right]\frac{p{\cdot}\hat{n}'p^2}{(p{\cdot}\hat{n})^4}\wp_\alpha \hat{n}'_\mu \hat{n}'_\nu \\
& + \left\{p{\cdot}\hat{n}'p^2(1+\alpha)^2 + \left[3(p{\cdot}\hat{n})^2 p{\cdot}\hat{n}' - (p{\cdot}\hat{n}')^3 + 2(p{\cdot}\hat{n})^3\right]\alpha - p{\cdot}\hat{n}'\left[(p{\cdot}\hat{n})^2 + (p{\cdot}\hat{n}')^2\right]\right\} \\
& \qquad \times \frac{1}{(p{\cdot}\hat{n})^2}\left[g_{\alpha(\mu}p_{\nu)} - \hat{n}_\alpha p_{(\mu}\hat{n}_{\nu)}\right] \\
& + (\alpha+1)\left[(p{\cdot}\hat{n}')^2 - (p{\cdot}\hat{n})^2\right]\frac{p^2}{(p{\cdot}\hat{n})^2}\left[g_{\alpha(\mu}\hat{n}'_{\nu)} - \hat{n}_\alpha \hat{n}_{(\mu}\hat{n}'_{\nu)}\right] \\
& - \left\{p{\cdot}\hat{n}'p^2(1+\alpha)^2 + 4\alpha(p{\cdot}\hat{n}' + p{\cdot}\hat{n})(p{\cdot}\hat{n})^2 - 2(p{\cdot}\hat{n})^2 p{\cdot}\hat{n}'\right\} \\
& \qquad \times \frac{p^2}{(p{\cdot}\hat{n})^3}\left[g_{\alpha(\mu}\hat{n}_{\nu)} - 2\hat{n}_\alpha \hat{n}_\mu \hat{n}_\nu\right] + \hat{n}'_\alpha \mathcal{Q}_{\mu\nu}\Bigg\}\frac{\delta n_\alpha}{\sqrt{n^2}}\,,
\end{aligned}
\tag{14}
$$

$$
\begin{aligned}
\delta_{n'} \Pi^{ab}_{\mu\nu} = -g^2 C(G)\delta^{ab}\frac{i\pi^2}{\epsilon}\Bigg\{ & \left\{-2p^2(p{\cdot}\hat{n})^2 + (1+\alpha)^2(p^2)^2\right\}\wp'_\alpha g_{\mu\nu} \\
& + \left\{2(p{\cdot}\hat{n})^2 + (1+\alpha)^2\left[3(p{\cdot}\hat{n}')^2 - (p{\cdot}\hat{n})^2 - p^2\right]\right\}\wp'_\alpha p_\mu p_\nu \\
& - \left\{3(p{\cdot}\hat{n}')^2 p{\cdot}\hat{n} + (p{\cdot}\hat{n})^3 + 3\alpha\left[(p{\cdot}\hat{n}')^2 p{\cdot}\hat{n} - (p{\cdot}\hat{n})^3\right] - (1+\alpha)^2 p^2 p{\cdot}\hat{n}\right\}\wp'_\alpha p_{(\mu}\hat{n}_{\nu)} \\
& + \left\{2p^2(p{\cdot}\hat{n})^2 - 4\alpha p^2(p{\cdot}\hat{n})^2 - (1+\alpha)^2(p^2)^2\right\}\wp'_\alpha \hat{n}_\mu \hat{n}_\nu \\
& + \left\{2p{\cdot}\hat{n}'(p{\cdot}\hat{n})^2 - 2(1+\alpha)^2 p^2 p{\cdot}\hat{n}'\right\}\wp'_\alpha p_{(\mu}\hat{n}'_{\nu)} \\
& + 2(1+\alpha)p^2 p{\cdot}\hat{n}' p{\cdot}\hat{n}\wp'_\alpha \hat{n}_{(\mu}\hat{n}'_{\nu)} - \left\{2p^2(p{\cdot}\hat{n})^2 - (1+\alpha)^2(p^2)^2\right\}\wp'_\alpha \hat{n}'_\mu \hat{n}'_\nu \\
& + \left\{(p{\cdot}\hat{n}')^2(p{\cdot}\hat{n})^2 + (p{\cdot}\hat{n})^4 - (1+\alpha)^2 p^2(p{\cdot}\hat{n}')^2\right\}\left[g_{\alpha(\mu}p_{\nu)} + \hat{n}'_\alpha p_{(\mu}\hat{n}'_{\nu)}\right] \\
& + (1+\alpha)\left\{p^2(p{\cdot}\hat{n}')^2 p{\cdot}\hat{n} - p^2(p{\cdot}\hat{n})^3\right\}\left[g_{\alpha(\mu}\hat{n}_{\nu)} + \hat{n}'_\alpha \hat{n}_{(\mu}\hat{n}'{}_{\nu)}\right] \\
& - \left\{2p^2 p{\cdot}\hat{n}'(p{\cdot}\hat{n})^2 - (1+\alpha)^2(p^2)^2 p{\cdot}\hat{n}'\right\}\left[g_{\alpha(\mu}\hat{n}'_{\nu)} + 2\hat{n}'_\alpha \hat{n}'_\mu \hat{n}'_\nu\right] \\
& + \hat{n}_\alpha \mathcal{J}_{\mu\nu}\Bigg\}\frac{\delta n'_\alpha}{\sqrt{-n'^2}\,(p{\cdot}\hat{n})^3}\,,
\end{aligned}
\tag{15}
$$

where $\wp_\alpha \equiv p_\alpha - p{\cdot}\hat{n}\hat{n}_\alpha$ and $\wp'_\alpha \equiv p_\alpha + p{\cdot}\hat{n}'\hat{n}'_\alpha$. The tensor $\mathcal{Q}_{\mu\nu}$ is irrelevant because $\hat{n}'{\cdot}\delta n = 0$. The tensor $\mathcal{J}_{\mu\nu}$ does not matter either since $\hat{n}{\cdot}\delta n' = 0$.

By trivial algebra, one can show that these results can be obtained from equation (12). Hence, both equations (13a) and (13b) hold for the UV contributions and in the limit $\eta \to 0$.

We want to point out that the terms in $\delta_n \Pi^{ab}_{\mu\nu}$ come from integrals whose UV divergent part has a well defined limit as $\eta \to 0$. In fact they could have been obtained by applying the Vienna prescription to the formal temporal gauge Feynman diagrams. On the contrary, the contributions to $\delta_{n'} \Pi^{ab}_{\mu\nu}$ come from integrals having a simple pole at $\eta = 0$. This pole cancels against the $\eta$ in front of the r.h.s. of equation (13b), giving a finite result as $\eta \to 0$ .

Those integrals show up since the operator $1/(n'\cdot\partial)^2$ ( $-1/(q\cdot n')^2$ in momentum space) appearing in the variation $\delta_{n'}$ of $N_\mu$ cancels the numerator of

$$\frac{(q\cdot n')^2}{(q\cdot n q\cdot n' + i\eta)^2}$$

This last quotient is the Vienna prescription for $1/(q\cdot n)^2$. Hence, one obtains ill-defined integrals at $\eta = 0$. For instance, the relevant contribution $o((1+\alpha)^0)$ from the first diagram in Figure 1 is

$$\begin{aligned}
\text{Fig. 1a} &= -ig^2 C(G)\delta^{ab}\eta\,(p^2 g_{\tau\nu} - p_\tau p_\nu) \int_{-\infty}^{\infty} dq_n \int_{-\infty}^{\infty} dq_{n'} \frac{1}{(q_n q_{n'} + i\eta)^2} \\
&\times \int d^{2\omega-2}\widetilde{q}\, \frac{\widetilde{q}\cdot\delta n'}{\widetilde{q}^2[(\widetilde{q}-\widetilde{p})^2 + (p\cdot\hat{n})^2 - (p\cdot\hat{n}')^2]} \\
&\times \frac{\hat{n}^\lambda}{\sqrt{-n'^2}} \left[(p_\mu - 2\widetilde{q}_\mu)\, g_{\lambda\sigma} + (\widetilde{q}_\lambda - 2p_\lambda)\, g_{\mu\sigma} + (p_\sigma + \widetilde{q}_\sigma)\, g_{\mu\lambda}\right] \\
&\times \left\{ g^{\sigma\tau} + \frac{(\widetilde{q}^\sigma - p^\sigma)\hat{n}^\tau + (\widetilde{q}^\tau - p^\tau)\hat{n}^\sigma}{p\cdot\hat{n}} \right\} + O(\alpha+1)\,,
\end{aligned} \tag{16}$$

where we have expressed $q_\alpha$ in the following way

$$q_\alpha \equiv \widetilde{q}_\alpha + q\cdot\hat{n}\hat{n}_\alpha - q\cdot\hat{n}'\hat{n}'_\alpha \equiv \widetilde{q}_\alpha + \frac{q_n}{n^2} n_\alpha + \frac{q_{n'}}{n'^2} n'_\alpha$$

and similarly $g_{\alpha\beta} \equiv \widetilde{g}_{\alpha\beta} + \hat{n}_\alpha\hat{n}_\beta - \hat{n}'_\alpha\hat{n}'_\beta$. $\sim$ means orthogonality to both $n$ and $n'$.

The $\eta$ in front of the integrals cancels against

$$\int_{-\infty}^{\infty} dq_n \int_{-\infty}^{\infty} dq_{n'} \frac{1}{(q_n q_{n'} + i\eta)^2} = -\frac{2\pi}{\eta}\,.$$

Therefore, the limit $\eta \to 0$ of the whole expression exists.

The other singular integral one would need is the following

$$\int_{-\infty}^{\infty} dq_n \int_{-\infty}^{\infty} dq_{n'} \frac{q_n q_{n'}}{(q_n q_{n'} + i\eta)^3} = -\frac{\pi}{\eta}\,.$$

Finally, let us say that to our knowledge, for arbitrary $\alpha$, there is no other way of studying the dependence of the Green's functions on $n'$ than starting with the action

(2). Other methods [4] have been proposed for homogeneous axial gauges, i.e., for $\alpha = 0$, involving likely finite integrals at $\eta = 0$. It would be nice to see whether these methods work for arbitrary values of $\alpha$. These techniques could prove to be more useful than ours for studying the formal properties of the effective action.

### Acknowledgments

We would like to thank Prof. J. C. Taylor for his advice. We are grateful to the Spanish Ministry of Education and Science for the financial support. E.B. also acknowledges the financial support of the British Council. Finally, we thank Prof. Kummer and the organising team of the Vienna Workshop for their help and the pleasant stay we had in Vienna.

## References

[1] E. Bagan and C. P. Martin, Phys. Lett. **223B** (1989) 187; E. Bagan and C. P. Martin, preprint, DAMTP 89-12 (to appear in Int. Journal of Mod. Phys).

[2] P. Gaigg, M. Kreuzer and G. Pollak, Phys. Rev. **D38** (1988) 2559.

[3] J. C. Collins, *"Renormalization"*, Cambridge University Press.

[4] M. Schweda, Phys. Lett. **226B** (1989) 313.

# Comments on the Proof of the Renormalized Action Principle

H. Balasin

Institut für theoretische Physik, Technische Universität Wien,
Wiedner Hauptstr. 8–10, A–1040, Austria

**Abstract:** The action principle holds a central position in renormalized field theory, since it controls the variation of renormalized Green's functions with respect to field transformations (important examples are the Ward-Takahashi and the Slavnov-Taylor-identities). Therefore an investigation of its proof will also shed some light on the prerequisites of its applicability.
Since we are working with renormalized quantities a well-fit framework for the proof of the action principle is provided by the BPHZ-scheme of renormalization.

## 1 The Renormalized Action Principle – Statement of the Problem

### 1.1 Introduction:

Formally the statement of the action principle is given by:

$$i\sum_k < T\phi_{i_1}(x_1)\cdots\Lambda_{i_k}(x_k)\cdots\phi_{i_n}(x_n) >=< T\delta S[\phi]\phi(x_1)\cdots\phi(x_n) > \tag{1}$$

$\phi_i$ ...quantum fields of the theory under consideration

$\Lambda_i$ ...variation of the i-th quantum field

$\delta S[\phi]$ ...variation of the action functional

However, all the objects in the above relation are ill-defined quantities since they are expectation values of products of operator-valued distributions. Therefore it is necessary to define these objects via a suitable renormalization procedure. But now the question arises wether the equality sign still holds. In the following our task will be to prove eq.(1) by introducing appropriate field products.

### 1.2 The BPHZ-Scheme:

Let me give a brief introduction to the renormalization scheme which will be used in the following. It is important to point out that we are working strictly in four dimensions and that no intermediate regularization procedure will be needed.
The BPHZ-scheme [1,2,3] assumes that the so-called unrenormalized Feynman-integrand associated with a given diagram $\Gamma$ is obtained via the usual Feynman rules. Since the Green's functions under consideration may also contain insertions of composite operators, there will be additional vertices in the corresponding diagrams. With the help of Zimmermann's forest formula, which in turn is an explicit solution of Bogoliubov's recursion relation, one obtains the renormalized integrand $R_\Gamma$:

$$R_\Gamma(p,k) = \sum_{U \in \mathcal{U}(\Gamma)} \prod_{\gamma \in U} (-t^\gamma) I_\Gamma(U) \tag{2}$$

The sum in eq.(2) ranges over all forests $U$ of the diagram $\Gamma$. These are sets of nonoverlapping 1PI-subdiagrams of $\Gamma$, i.e. $\gamma, \gamma' \in U \Rightarrow \gamma \subseteq \gamma'$ or $\gamma' \subseteq \gamma$ or $\gamma \cap \gamma' = \emptyset$. $\mathcal{U}(\Gamma)$ denotes the set of all $\Gamma$-forests. $t^\gamma$ is the Taylor operator acting on the external momenta of $\gamma$ which expands a given function of these variables into a Taylor polynomial of the order $\delta(\gamma)$. $\delta(\gamma)$ denotes the degree-function which is a generalization of the usual degree of divergence of a diagram. I will comment on this in the next section. Finally $I_\Gamma(U)$ stands for the unrenormalized integrand with a forest-appropriate momentum choice in order to faciliate the application of the Taylor operators. The contribution of the diagram $\Gamma$ to the Green's function is obtained by integrating over the internal momenta $k$.

### 1.3 Renormalized Composite Operators – Vertices

#### 1.3.1 Vertices:

I will now introduce the notion of a vertex, which is a generalization of the normalproducts [2,4] invented by Zimmermann in order to renormalize composite operators.

**Definition 1:** A simple vertex $v$ is a pair $[f^v, \alpha^v]$, where $f^v$ is an N-tuplet of arbitrary order derivatives of the quantum fields under consideration. These will be referred to as the N "legs" of the vertex. $\alpha^v$ is a map assigning to every set of legs represented by their position in the N-tuplet a positive integer, subject to the following restrictions:

$$\alpha(S) \leq \alpha(R) \text{ if } S \subseteq R \tag{5}$$

$$\alpha(S) = 0 \text{ if } N = 0, 1$$

(N will also be referred to as the order of $v$)

The degree function $\delta(\gamma)$ for an arbitrary 1PI-diagram $\gamma$ is now given by

$$\delta(\gamma) = 4 - \sum_{L\in\mathcal{E}(\gamma)} d_{\alpha(L)} + \sum_{v\in\mathcal{V}(\gamma)} (d_v + \alpha^v(\gamma) - 4) \qquad (6)$$

$\mathcal{E}(\gamma)\ldots$external lines of $\gamma$
$\mathcal{V}(\gamma)\ldots$vertices of $\gamma$
$d_{\alpha(L)}\ldots$dimension of the field associated with L
$\alpha^v(\gamma) := \alpha(\{i|f_i \in \gamma\})$

At this point we make contact with perturbation theory and the BPHZ-scheme. Eq.(6) is interpreted as the degree function corresponding to a Green's function of the free theory containing the field monomials $\{v\}$.
The important point noticed by Zimmermann is that the forest formula retains its convergence properties even in the case of oversubtractions, i.e. it is possible to use Taylor operators with a degree of subtraction larger than the one needed for convergence. Since in both cases the infinities are removed their difference appears in finite counterterms only.
Green's functions containing operator insertions and which are renormalized via the BPHZ-procedure with the above degree-function will be called Green's functions with vertex insertions. They will turn out to be the appropriately renormalized objects for the proof of the action principle.

### 1.3.2 Substitution of Vertices:

In order to define the substitution of a given field $\psi$ in a vertex $w$ by a vertex $v$, it is necessary to the define the field part $h^{(j)}$ and the subtraction function $\gamma^{(j)}$ resulting from the replacement of the j-th occurrence of $\psi$ in $f^w$ by all elements of $f^v$. The corresponding set of vertices $[h^{(j)}, \gamma^{(j)}]$ is summed over and defines the substitution of vertices $(\frac{\delta w}{\delta\psi}|v)$.

**Definition 2:** Let $w$ and $v$ be order-M and order-N vertices respectively, and let $\psi$ denote an arbitrary-order derivative of $\phi$; then the substitution of $\psi$ in $w$ by $v$ is given by:

$$(\frac{\delta w}{\delta \psi}|v) = \sum_{j\in U^w_\psi} [h^{(j)}, \gamma^{(j)}]$$

$$U^w_\psi = \{j | f^w_j = \psi\}$$

$$h^{(j)} = (f^w_1, ..., f^w_{j-1}, f^v_1, ..., f^v_N, f^w_{j+1}, ..., f^w_M) = \\ = (h^{(j)}_1, ..., h^{(j)}_{M+N-1}) \qquad (5a)$$

To define $\gamma^{(j)}$ we split every $S \subseteq I_{N+M-1}$ into a v-leg- and w-leg-part $S = S^j_v \cup S^j_w$

$$S^j_v = \{i | i \in S \quad \hat{h}^{(j)}_i \in f^v\}$$
$$S^j_w = \{i | i \in S \quad \hat{h}^{(j)}_i \in f^w\} \qquad (5b)$$
$$\gamma^{(j)}(S) := \begin{cases} \alpha^v(\tilde{S}^j_v) & \text{if } S = S^j_v \\ \alpha^w(\tilde{S}^j_w) & \text{if } S = S^j_w \\ \alpha^v(\tilde{S}^j_v) + \alpha^w(\tilde{S}^j_w \cup \{j\}) & \text{otherwise} \end{cases}$$

$\tilde{S}^j_v, \tilde{S}^j_w$ are the corresponding subsets of $I_N$ and $I_M$ respectively

## 2 Proof of the Action Principle:

The dynamical setting for the proof of the action principle is given by the action $S[\phi]$ that may be split into a free part $S_0$ and an interaction part $S_I$. The corresponding Lagrangeans are $\mathcal{L}_0$ and $\mathcal{L}_I$ respectively.
We will now set out for the proof of

$$< T(w(x)\frac{\delta S}{\delta\phi(x)})v^1(x^1)...v^N(x^N) >= i\sum_k < Tv^1(x^1)...(w(x)\frac{\delta v^k(x^k)}{\delta\phi(x)})...v^N(x^N) > \quad (6)$$

which in turn implies that eq.(1). ($v^1, ...v^N$ are arbitrary composit operators)
A formal proof of eq.(6) is given by:

$$\begin{aligned} &< T(eqn_0(\phi(x))w(x))v^1(x^1)...v^N(x^N) >= \\ &\sum_{n=0}^{\infty}\frac{i^n}{n!} < T(eqn_0(\phi(x))w(x))v^1(x^1)...v^N(x^N)(S_I)^n >_0= \\ &\sum_{n=1}^{\infty}\frac{i^{n-1}}{(n-1)!} < T(w(x)\frac{\delta S_I}{\delta\phi(x)})v^1(x^1)...v^N(x^N)(S_I)^{n-1} >_0 + \\ &-i\sum_{k=1}^{N}\sum_{n=0}^{\infty}\frac{i^n}{n!} < Tv^1(x^1)...(w(x)\frac{\delta v^k(x^k)}{\delta\phi(x)})...v^N(x^N)(S_I)^n >_0= \\ &=< T(w(x)\frac{\delta S_I}{\delta\phi(x)})v^1(x^1)...v^N(x^N) > -i\sum_k < Tv^1(x^1)...(w(x)\frac{\delta v^k(x^k)}{\delta\phi(x)})...v^N(x^N) > \end{aligned} \quad (7)$$

The formal proof uses essentially the fact that the Wick- contraction of the free equation of motion, represented by the differential operator $eqn_0$, yields a delta-function.

$$< -\frac{\delta S_0}{\delta\phi}\phi >=< eqn_0(\phi)\phi >= eqn_0 < \phi\phi >= -i\delta \quad (8)$$

The above derivation made use of the formal Gell-Mann–Low formula and the Wick-theorem.
Graphically this corresponds to an identity between unrenormalized integrands. In order to take renormalization into account we have to consider the forest structure of the diagrams.
Nevertheless let the above derivation be some kind of guideline for the renormalized proof. We begin with the renormalized contribution of a (connected) graph $\Gamma$ to the first line of eq.(7). In the following $v^1, ..., v^N$ are replaced by arbitrary vertices and $(eqn_0(\phi)w)$ by a sum of vertices, the field-part of each corresponding to a product of an arbitrary summand in $eqn_0(\phi)$ with w, together with an appropriately chosen subtraction function. Figure 1 represents a typical graph $\Gamma$.

$r$ is the graphical counterpart of $(eqn_0(\phi)w)$. The blob in figure 1 represents contractions with monomials of $\mathcal{L}_{int}$.
In the following we will not distinguish between the $v^i$'s and the monomials of $\mathcal{L}_{int}$ but we will use a uniform notation $u, r, s, t, ...$ for the vertices of $\Gamma$.

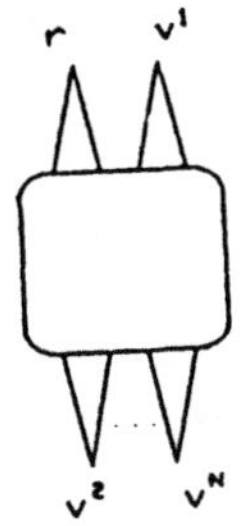


**Fig. 1.** Graph $\Gamma$ with the external vertices $r, v^1, ..., v^N$.

The pairing of the equation of motion-leg of $r$ with an arbitrary leg of one of the remaining vertices, say $u$, is shown in figure 2. $L$ represents the corresponding line. However, the propagator associated with $L$ (after a reshuffling of momentum factors) is essentially one. This fact motivates the interpretation of the unrenormalized integrand of $\Gamma$ as coming from the diagram $\hat{\Gamma}$, which is obtained from $\Gamma$ by shrinking of the line $L$ and joining $u$ and $r$ into a new vertex $\widehat{ur}$. Unfortunately this does not carry over to the renormalized integrands, since in general the forest structures of $\Gamma$ and $\hat{\Gamma}$ are different: Shrinking of $L$ in all diagrams of a given $\Gamma$-forest does not necessarily produce a $\hat{\Gamma}$-forest.
Nevertheless, a more sophisticated identification is possible. Therefore we have at first to get rid of those diagrams $\Gamma$ and those $\Gamma$-forests that have no counterpart after shrinking of $L$.

**Fig. 2.** Pairing of the equation of motion-leg of $r$ with a leg of an arbitrary vertex $u$ of $\Gamma$.

i) If a $\Gamma$-forest contains a diagram $\mu$ such that $u, r \in \mu$ but $L \notin \mu$ then the shrinking of $L$ does not affect $\mu$. However,

**Lemma 1** *The $\Gamma$-forests containing $\mu$ do not contribute to the forest formula*

To illustrate the typical techniques, we will show how lemma 1 is proven.

*Proof.* Let $U$ be a $\mu$-forest. Then there exists a maximal element $\hat{\mu} \supseteq \mu \wedge \hat{\mu} \in U$ which is of the same type as $\mu$ (i.e. $u, r \in \hat{\mu}$ but $L \notin \hat{\mu}$). Let $\mu_1 = \hat{\mu} \cup \{L\}$. The contribution of both forests, the one containing $\mu_1$ and the one which does not, vanishes due to the identity $(1 - t^{\mu_1})(-t^{\hat{\mu}}) = 0$.

ii) If $\Gamma$ is such that there is another line $L_1$ connecting $u$ and $r$ the contribution of the shrunken diagram does not exist in the BPHZ-framework. However,

**Lemma 2** *$R_\Gamma = 0$, if there is another line $L_1$ connecting $u$ and $r$.*

The proof of lemma 2 rests on the fact that the forests may be partioned into two classes. One which allows subdiagrams with only one line joining $u$ and $r$ and one which does not. The contribution of the first class vanishes by an argument analogous to the one in the proof of lemma 1. The second class may further be partitioned into pairs of forests that differ from each other by the one-loop diagram formed from $L$ and $L_1$. The contribution of such a pair can be shown to vanish.
The remaining $\Gamma$-forests allow a decomposition

$$U = \{\gamma_{1i}\} \cup \{\rho_m\} \cup \{\gamma_{2j}\} \tag{9}$$

where the $\gamma_1$'s contain $u$ but not $r$, the $\gamma_2$'s contain $r$ but not $u$ and the $\rho_m$'s contain either both or none. In the following these $\Gamma$-forests will be denoted by $\mathcal{U}_0(\Gamma)$.
We will now define a mapping $f$ which has the following properties:

$$\begin{gathered} f : \mathcal{U}_0(\Gamma) \to P(\mathcal{U}(\hat{\Gamma})) \qquad P(X)\text{...power set of the set } X \\ i) f(U) \cap f(U') = \emptyset \text{ if } U \neq U' \\ ii) \hat{U} \in \mathcal{U}(\hat{\Gamma}) \Rightarrow \exists U \in \mathcal{U}_0(\Gamma) : \hat{U} \in f(U) \end{gathered} \tag{10}$$

The construction of $f$ is due to Breitenlohner and Lowenstein [4,5] and is closely related to the Taylor-operator identity:

$$\begin{aligned} t_x^{\delta_1} t_y^{\delta_2} &= t_{x,y}^{\delta_1+\delta_2}(t_x^{\delta_1} t_y^{\delta_2}) = \\ &= (-t_{x,y}^{\delta_1+\delta_2})(1 - t_x^{\delta_1} - t_y^{\delta_2}) \end{aligned} \tag{11}$$

which may be seen from figure 3, where the abscissa depicts the expansion in $x$ and the ordinate the expansion in $y$.

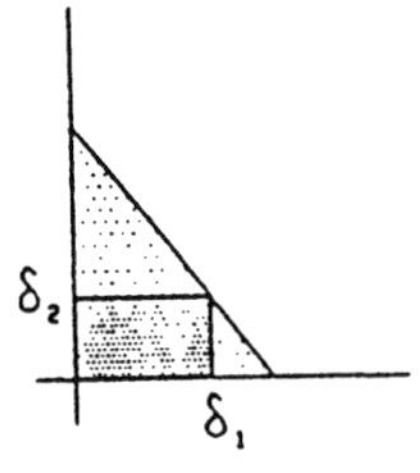


**Fig. 3.** Taylor operator identity

The construction of $f$ is such that

$$\prod_{\gamma \in U} (-t^\gamma) I_\Gamma = \sum_{\hat{U} \in f(U)} \prod_{\hat{\gamma} \in \hat{U}} (-t^{\hat{\gamma}}) I_{\hat{\Gamma}} \tag{12}$$

holds. The subtraction function of the r.h.s. is calculated by assigning to $\widehat{ur}$ the vertex $(\frac{\delta u}{\delta \psi}|w)$.

The properties of $f$ imply that:

$$\sum_{U \in \mathcal{U}_0(\Gamma)} \prod_{\gamma \in U} t^{\gamma} I_{\Gamma} = \sum_{\hat{U} \in \mathcal{U}(\hat{\Gamma})} \prod_{\hat{\gamma} \in U} t^{\hat{\gamma}} I_{\hat{\Gamma}} \tag{13}$$

Summation over all graphs gives eq.(6). This completes the proof.

## 4 Conclusion:

Since the only prerequisites needed were a power-counting renormalizable theory without infrared divergences, and since the main arguments of the proof were based on graphical properties (i.e. forest-structure and the Taylor identity) we see that this proof is highly model-independent. Even the incorporation of infrared singularities within the framework of BPHZL-renormalization [6] seems to leave the above properties intact. However, further work on this subject is in progress. It will also cover Yang-Mills theories in axial-type gauges and supersymmetric models within the supergraph framework [7]. The question is still open for the Leibbrandt-Mandelstam [8] prescription and supersymmetric gauge theories in noncovariant gauges. Finally I want to apologize for the sketchy form of the presentation. For a more detailed version the reader is referred to [9].

## References

[1] W.Zimmermann, 1970 Brandeis lectures;
Lectures on Elementary Particles and Quantum Field Theory,
eds. S.Deser, M.Grisaru and H.Pendleton, MIT Press, Cambridge Mass., 1971.

[2] W.Zimmermann, Commun. Math. Phys. **15** (1969) 208.

[3] J.H.Lowenstein, Renormalization theory (1975 Erice lectures)
eds. G.Velo and A.Wightman, Reidel (1976).

[4] Y.-M.P.Lam, Phys. Rev. **D6** (1972) 2145.

[5] P.Breitenlohner, D.Maison, Commun. Math. Phys. **52** (1977) 11.

[6] T.E.Clark, J.H.Lowenstein, Nucl. Phys. **B113** (1976) 109.

[7] O.Piguet, K.Sibold, Renormalized Supersymmetry,
Progress in Physics, Birkhäuser (1986).

[8] S.Mandelstam, Nucl. Phys. **B123** (1983) 149;
G.Leibbrandt, Phys. Rev. **D29** (1984) 1699; D30 (1984) 2167;
P.Gaigg et al., J. Math. Phys. **28** (1987) 2781;
P.Gaigg, M.Kreuzer, Phys. Lett. **205B** (1988) 530.

[9] H.Balasin, M.Stierle, The Action Principle Revisited,
TU-preprint, 1989.

# On-Shell Compatibility of Clashing Gauges in Quark Condensate Contributions to Standard Model Selfenergies

Victor Elias

Department of Applied Mathematics, University of Western Ontario, London, Ontario N5A 6B9 Canada

In quantum field theory, our ability to perform mass renormalizations that make physical sense is contingent upon such renormalizations proceeding order-by-order in a gauge invariant manner. Such gauge invariance is possible only if unrenormalized fermion self energies are themselves gauge-parameter independent when evaluated on the fermion mass shell. This property is well known for unbroken-symmetry field theories, such as QED, and by extension, QCD [1],[2].

It is not so generally well known that self energies exhibit rigorous on-shell gauge parameter independence in perturbative quantum field theories in which fermion masses arise from spontaneous symmetry breaking. For example, in standard spontaneously broken $SU(2) \times U(1)$ theory [3], the Z's gauge parameter cancels entirely from the summation of the three graphs of Figure 1, involving the Z's scalar partner $\chi_3$ and the physical Higgs field $\phi$, as well as from the two graphs of Figure 2, involving the Faddeev-Popov ghost $G_Z$ [2],[4].

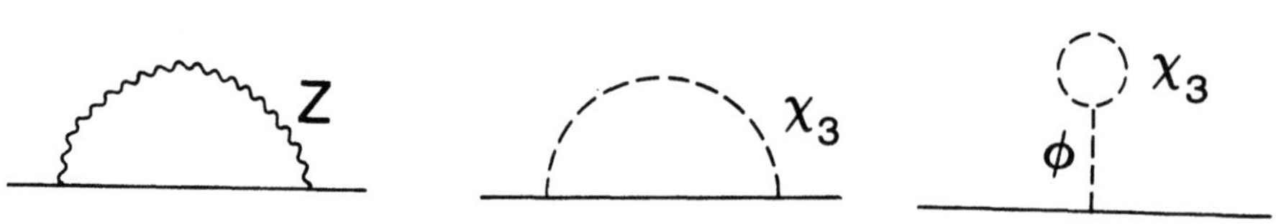


Figure 1

Figure 2

Similar cancellation of the W's gauge parameter occurs on shell for arbitrary K-M mixing [2].

Such gauge independence of a spontaneously broken theory may be disrupted if the vacuum expectation value (VEV) $\langle\Phi\rangle$ is no longer the only symmetry breaking order parameter of the theory. Indeed, a clear difference between SU(2) X U(1) electroweak theory considered in isolation and the SU(2) X U(1) electroweak sector of the $SU(3)_c$ X SU(2) X U(1) standard model is that the vacuum of the latter case admits the occurrence of additional chiral-noninvariant order parameters, the vacuum condensates of QCD.

As in QCD sum rule physics [5], suppose we let nonperturbative QCD order parameters couple to QCD Green's functions via residual normal ordered terms in the Wick expansion [6]. For example, we denote schematically by $g\bar{\psi}B\psi$ the interaction of a quark field $\psi$ with a boson B. Then the order-$g^2$ term of the vacuum expectation value of the time ordered product of Heisenberg quark fields (i.e, the fermion two-point function) considered in the interaction picture,

$$iS^{(2)}(p) = \frac{g^2}{2}\int d^4x\int d^4y\int d^4z \langle 0|T\psi(x)\bar{\psi}(y)B(y)\psi(y)\bar{\psi}(z)B(z)\psi(z)\bar{\psi}(0)|0\rangle e^{ipx} \tag{1}$$

has a Wick expansion containing both the usual completely contracted perturbative (P) self energy contribution

$$iS_P^{(2)}(p) = \frac{g^2}{2}\int d^4x\int d^4y\int d^4z\; S^{(0)}(x-y)\; D(y-z)S^{(0)}(y-z)\; S^{(0)}(z)e^{ipx}, \tag{2}$$

$[S^{(0)}(x-y) = \langle 0|T\psi(x)\bar{\psi}(y)|0\rangle]$ as well as an incompletely-contracted contribution

$$iS_{NP}^{(2)}(p) = \frac{g^2}{2}\int d^4x\int d^4y\int d^4z S^{(0)}(x-y)\; D(y-z)\langle 0|:\psi(y)\bar{\psi}(z):|0\rangle S^{(0)}(z)\; e^{ipx}. \tag{3}$$

In purely perturbative contexts, the VEV of normal ordered fields is zero. The nonlocal normal ordered expectation value (in (3)) is the vehicle by which vacuum condensates can enter the theory, **provided** $|0\rangle$ has nonperturbative (NP) content. Let us consider the momentum space content of (3), using $S^{(0)}(p) = \int d^4w\, e^{ipw}S^{(0)}(w)$. If we rewrite $e^{ipx} = e^{ip(x-y)}e^{ip(y-z)}e^{ipz}$ and note that a trivial transformation allows replacement of $\int d^4x\int d^4y\int d^4z$ with $\int d^4(x-y)\int d^4(y-z)\int d^4z$, find that (3) becomes [7], [8]

$$iS_{NP}^{(2)}(p) = \frac{g^2}{2}S^{(0)}(p)\left\{\int d^4(y-z)\; D(y-z)\langle 0|:\psi(y)\bar{\psi}(z):|0\rangle e^{ip(y-z)}\right\}S^{(0)}(p). \tag{4}$$

The integral in (y-z) is a source of gauge parameter dependence. D(y-z), the perturbative boson propagator, contains a perturbative gauge parameter if that boson is either a gauge boson of $SU(3)_c$ X SU(2) X U(1) or a scalar partner $(\chi^{\pm},\chi_3)$ of the gauge bosons in the spontaneously broken sector. However, the nonperturbative content of the remaining VEV in (4) is also intrinsically gauge dependent; if we expand $\psi(y)$ and $\overline{\psi}(z)$ as a Taylor series about y=0, z=0 [$\psi(y) = \psi(0) + y^{\mu}(\partial_{\mu}\psi(0)) + \ldots$], the Taylor series coefficients involve conventional (noncovariant) derivatives. To express $<0|:\psi(y)\overline{\psi}(z):|0>$ into explicitly covariant form, one may employ a separate fixed-point Foch-Schwinger (FS) gauge condition [9] for the nonperturbative portion of (4): $x^{\tau}A_{\tau}(x) = 0$, where $A^{\tau}(x)$ is any $SU(3)_c$ gluon field. In this gauge, every factor of $\partial_{\mu}$ in the Taylor series coefficients described above can be replaced with the $SU(3)_c$ covariant-derivative $D_{\mu}(0)$, a result which can be extended to the replacement of any number of ordinary derivatives with their $SU(3)_c$ covariant derivative analogues [10]. Consequently we see that application of gauge condition to the nonperturbative VEV in (4) yields explicitly $SU(3)_c$-covariant Taylor series coefficients (n and r are Dirac indices):

$$<0|:\overline{\psi}_r(z)\psi_n(y):|0> = - <0|:\psi_n(y)\overline{\psi}_r(z):|0>$$

$$= <0|:\overline{\psi}_r(0)\psi_r(0):|> + y^{\mu}<0|:\overline{\psi}_r(0)\overrightarrow{D}_{\mu}(0)\psi_n(0):|0>$$

$$+ z^{\mu}<0|:\overline{\psi}_r(0)\overleftarrow{D}_{\mu}(0)\psi_n(0):|> + \frac{1}{2} y^{\mu}y^{\nu}<0|:\overline{\psi}_r(0)\overrightarrow{D}_{\mu}(0)\overrightarrow{D}_{\nu}(0)\psi_n(0):|0>$$

$$+ \frac{1}{2} z^{\mu}z^{\nu}<0|:\overline{\psi}_r(0)\overleftarrow{D}_{\mu}(0)\overleftarrow{D}_{\nu}(0)\psi_n(0):|0> + y^{\mu}z^{\nu}<0|:\overline{\psi}_r(0)\overleftarrow{D}_{\nu}(0)\overrightarrow{D}_{\mu}(0)\psi_n(0):|0>$$

$$+ \ldots \quad . \tag{5}$$

The covariantized Taylor series coefficients $<0|:\overline{\psi}_r(0)D_{\mu}(0)D_{\nu}(0)\ldots\psi_n(0):|0>$ are specific to the FS gauge. This gauge condition necessarily clashes with any perturbative choice of gauge used to specify D(y-z)in (4). Since on shell amplitudes are necessarily gauge independent, we might still hope that the clashing gauges in (4), factorized into perturbative [D(y-z)] and nonperturbative [$<0|:\psi(y)\overline{\psi}(z):|0>$] contributions, will yield a gauge independent amplitude on shell, provided **both** contributions are constrained to appropriate mass-shell values.

For the nonperturbative contribution, the Taylor series coefficients $<0|:\overline{\psi}_r(0)D_{\mu}(0)D_{\nu}(0)\ldots\psi_n(0):|0>$ can be evaluated "on shell" through use of the QCD field equations at x=0 [6]:

$$\not{D}(x)\psi(x) = -im\ \psi(x) \ , \tag{6}$$

$$[D_{\mu}(x), D_{\nu}(x)] = -ig_s(\lambda^a/2)F^a_{\mu\nu}(x) \ , \tag{7}$$

$$[D_{\mu}(x), -i\lambda^a F_a^{\mu\nu}(x)] = -ig_s\lambda^b\sum_q \overline{\psi}_q(x)\lambda^b\gamma^{\nu}\psi_q(x) \ . \tag{8}$$

These equations can be used to replace every $D_\mu(0)$ from the Taylor series coefficients with the dimensionally appropriate product of quark masses (m), field strengths ($F_{\mu\nu}(0)$) and fermion currents ($\sum_q \bar{\psi}_q(0)\lambda^b\gamma^\nu\psi_q(0)$). For example, the leading three terms in (5) are found from straightforward tensor algebra to be [6], [8]

$$<0|:\bar{\psi}_i^\alpha(0)\psi_j^\beta(0):|0> = (\delta^{\alpha\beta}/3)(\delta_{ij}/4)<\bar{q}q> , \tag{9}$$

$$<0|:\bar{\psi}_i^\alpha(0)D_\rho^{\beta\tau}(0)\psi_j^\tau(0):|0> = (-im(\gamma_\rho)_{ji}\delta^{\alpha\beta}/48)<\bar{q}q> , \tag{10}$$

$$<0|:\bar{\psi}_i^\alpha(0)D_\mu^{\rho\tau}(0)D_\nu^{\tau\beta}(0)\psi_j^\beta(0):|0> = (-m^2\delta^{\alpha\rho}\delta_{ij}/96)<\bar{q}q> + \left[g_s[-3i\delta_{ij}g_{\mu\nu} - (\sigma_{\mu\nu})_{ji}]\delta^{\alpha\rho}/96\right]<\bar{q}F\cdot\sigma q> , \tag{11}$$

where the actual condensates are defined to be the following local Lorentz scalar and colour scalar vacuum expectation values:
$<\bar{q}q> \equiv <0|:\bar{\psi}_\ell^\alpha(0)\psi_\ell^\alpha(0):|0>$, $<\bar{q}F\cdot\sigma q> \equiv \frac{1}{2}<0|:\bar{\psi}_k^\alpha(0)\lambda^a_{\alpha\beta}(\sigma_{\rho\tau})_{k\ell}F_a^{\rho\tau}(0)\psi_\ell^\beta(0):|0>$.

It is important to note that the on-shell use of (6)-(8) to facilitate the removal of "D(0)'s" from the covariantized Taylor series coefficients in (5) is necessarily coupled to the introduction either of a mass [eq. (6)], or of additional fields accompanied by the QCD coupling constant $g_s$ [eqs. (7), (8)]. These additional fields necessarily project onto higher dimensional condensates, as in (11). Thus we see that higher dimensional condensates are necessarily accompanied by higher powers of $g_s$ in the Taylor series coefficients of (5), or alternatively, that the ordering of QCD-vacuum condensate contributions from lowest dimension upwards is consistent with a perturbative ordering of such contributions in the QCD coupling constant $g_s$. One can therefore use (6)-(8) to transform (5) into the following "on-shell" expression for the normal ordered VEV in (4) [8]:

$$-<0|:\psi_j^\sigma(y)\bar{\psi}_i^\rho(z):|0> = \delta^{\rho\sigma}<\bar{q}q>\sum_{n=0}^{\infty}\left[-im\left(\gamma\cdot(z-y)\right)\right]^n C_n + g_s<\bar{q}F\cdot\sigma q>\left\{\left[-i\delta_{ji}(z-y)^2/64 + \sigma_{ji}^{\mu\nu}z_\mu y_\nu\right] + m\left[\gamma_{ji}^\mu(z-y)_\mu(z-y)^2/384\right] + \mathcal{O}(m^2)\right\} + g_s^2<\bar{q}q\bar{q}q>\{...\} + \mathcal{O}(g_s^3) , \tag{12}$$

$$C_n^{-1} \equiv 3\left(\frac{n}{2}\right)!\left(\frac{n}{2}+1\right)!4^{n/2+1}\left[n \text{ even}\right];\ C_n^{-1} \equiv 6\left(\frac{n-1}{2}\right)!\left(\frac{n+3}{2}\right)!4^{n/2+1}\left[n \text{ odd}\right]. \tag{13}$$

To illustrate how (12) can be incorporated into (4) in order to yield gauge independent on-shell contributions to the quark propagator, we consider first the $O(\alpha_s)$ contribution of the quark condensate to the quark propagator, corresponding to having D(y-z) in (4) represent a gluon propagator in arbitrary (gauge parameter $\xi$) covariant gauge. The quark condensate projection of the self energy corresponding to the curly-bracketed term in (4), is evaluated via (12) [7], [8]:

$$\Sigma_{<\bar{q}q>}(p) = -\frac{4g_s^2}{3} <\bar{q}q>\gamma^\mu \sum_{j=0}^{\infty} c_j \left(-m\,\gamma\cdot\frac{\partial}{\partial p}\right)^j \left[\frac{-g_{\mu\nu}}{p^2} + \frac{(1-\xi)p_\mu p_\nu}{(p^2)^2}\right]\gamma^\nu$$
$$= \frac{g_s^2 <\bar{q}q>}{9p^2}\left[(3+\xi) - \xi m\not{p}/p^2\right] . \qquad (14)$$

It is interesting to note that all $j \geq 2$ contributions in the intermediate line of (14) vanish, leaving only the lead and next-to-lead contributions of the final line [11]. The perturbative gauge parameter dependence of (14) vanishes once we sandwich (14) between on-shell spinors through use of (6) [$\not{p}\psi(p) = m\psi(p)$]:

$$\Sigma_{<\bar{q}q>}(p) = g_s^2 <\bar{q}q>/3m^2 \qquad (15)$$

regardless of the value of the covariant-gauge parameter $\xi$.

This result, in which a general covariant gauge was employed for D(y-z) in (4), can be obtained from noncovariant gauges as well. Had we used the light-cone gauge expression for the gluon propagator D(y-z), we would have found that [12]

$$\Sigma_{<\bar{q}q>}(p) = -\frac{4g_s^2}{3} <\bar{q}q>\gamma^\mu \sum_{j=0}^{\infty} c_j \left(-m\,\gamma\cdot\frac{\partial}{\partial p}\right)^j \left[-\frac{g_{\mu\nu}}{p^2} + \frac{p_\mu n_\nu + p_\nu n_\mu}{p^2\, p\cdot n}\right]\gamma^\nu . \qquad (16)$$

To evaluate (16), we note the light-cone gauge condition $n^2 = 0$ implies that powers of $n\cdot p$ in the denominator of the gluon propagator cannot be increased though successive differentiations of $1/(p^2 n\cdot p)$:

$$\left(\frac{\partial^2}{\partial p^2}\right)^N \frac{1}{p^2 n\cdot p} = \frac{4^N (N!)^2}{p^{2N+2}(n\cdot p)} . \qquad (17)$$

Upon substitution of (17) into (16), we find after summation of an infinite series over j that [12]

$$\Sigma_{<\bar{q}q>}(p) = \frac{2g_s^2}{9} <\bar{q}q>\left\{\frac{1}{p^2} - \frac{m\gamma\cdot p}{2p^4} + \frac{\gamma\cdot n}{mn\cdot p}\left[1 + \left(\frac{p^2}{m^2} - 1\right)\ln\left(1 - \frac{m^2}{p^2}\right)\right]\right\} . \qquad (18)$$

It is important to realize that (14) and (18) yield different asymptotic structure in the large-p limit; such asymptotic structure is therefore dependent on the choice of gauge. Nevertheless, one can take (18) onto the quark mass shell by sandwiching it between on shell quark spinors, for which $\bar{\psi}(p)\gamma^\mu\psi(p) = (p^\mu/m)\bar{\psi}(p)\psi(p)$. Substitution of $(p\cdot n)/m$ for $\gamma\cdot n$ in (18), and further use of the on shell substitution $\not{p}\psi(p) = m\psi(p)$ leads immediately to the recovery of (15) [12].

The on shell validity of (15) has also been confirmed using plane wave methods for the evaluation of $<0|:\psi(x)\bar{\psi}(y):|0>$, and using either generalized-covariant [13] or generalized-axial [14] gauges for the gluon propagator. We thus see that there is no clash resulting from the use of the FS gauge for the nonperturbative content of $\Sigma_{<\bar{q}q>}$ on shell, and the use of either covariant or noncovariant gauges for its perturbative content. Such clashing is avoided because the FS gauge expression (5) is independently taken "on shell" through use of (6)-(8). Equation (12), the net result of this process, is therefore to be taken seriously (i.e., gauge independently) as the QCD-vacuum contribution to the vacuum expectation value of the normal-ordered product of quark and antiquark fields [15].

A provocative consequence of the above approach is the possibility of a distinction between the vacuum of purely perturbative SU(2) X U(1) and the vacuum of $SU(3)_c$ X SU(2) X U(1), as the latter quark condensing vacuum allows VEV's, such as (5), to enter the Wick expansions of perturbative amplitudes. In the presence of such vacuum-condensate-sensitive quantities (which would be expected to couple to hadronic weak interactions as well), the quark propagator would be expected to obtain new condensate-sensitive contributions from Z exchanges (fig. 3) and Z scalar partner exchanges (fig. 4) analogous to those of fig. 1. Focusing on the down quark (d) and the new order parameter $<\bar{q}q>$, one can obtain the gauge dependent ($\xi$- dependent) portion of the self-energies corresponding to figs. 3 and 4 by substituting the $<\bar{q}q>$-component of (12) into the appropriate electroweak analog of (4)[2]:

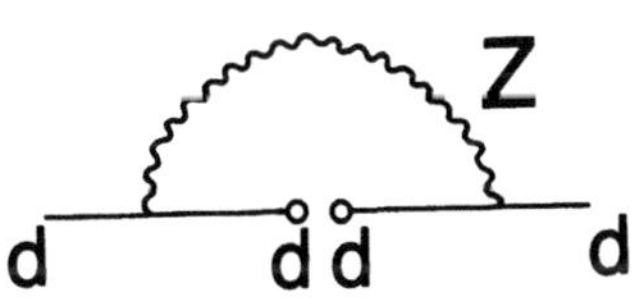

Figure 3

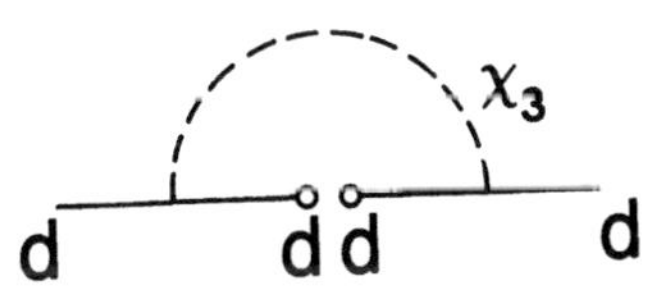

Figuro 4

$$\Sigma_\xi^{(3)}(p) = \frac{e^2 M_Z^4 <\bar{q}q>}{4M_W^2(M_Z^2 - M_W^2)}\bar{d}(p)\left[\gamma^\mu \frac{(\lambda+\gamma_5)}{2}\sum_{j=0}^{\infty} c_j\left(-m_d\gamma\cdot\frac{\partial}{\partial p}\right)^j\left(\frac{p_\mu p_\nu/M_Z^2}{p^2-\xi M_Z^2}\right)\gamma^\nu\frac{(\lambda+\gamma_5)}{2}\right]d(p), \quad (19)$$

$$\Sigma_\xi^{(4)}(p) = \frac{e^2 (m_d^y)^2 M_Z^2 <\bar{q}q>}{4M_W^2(M_Z^2 - M_W^2)}\bar{d}(p)\left[\gamma_5\sum_{j=0}^{\infty} c_j\left(-m_d\gamma\cdot\frac{\partial}{\partial p}\right)^j\left(\frac{-1}{p^2-\xi M_Z^2}\right)\gamma_5\right]d(p) \ . \quad (20)$$

[$\lambda \equiv -1 + 4(M_Z^2 - M_W^2)/3M_Z^2$.] In (20) we make a distinction between $m_d$, the d quark mass shell arising ultimately from (6), and $m_d^y$, the d quark mass entering the Yukawa interaction that characterizes the $\chi_3$ vertices of fig. 4. After some algebraic manipulation, one finds on shell that [2]:

$$\Sigma_\xi^{(3)}(p)+\Sigma_\xi^{(4)}(p) = \frac{e^2<\bar{d}d>m_d^2}{48\xi M_Z^2(M_Z^2 - M_W^2)}\sum_{\ell=0}^{\infty}\left(\frac{m_d^2}{\xi M_Z^2}\right)^\ell \frac{(2\ell+4)!}{(\ell+2)!(\ell+3)!}\left[\frac{m_d^2 - (m_d^y)^2}{2m_d^2}\right]\bar{d}(p)d(p), \quad (21)$$

a result consistent with gauge parameter independence provided the d quark mass $m_d^Y$ entering the Yukawa interaction continues to be identified with the mass shell value $m_d$ appropriate to (6). This result is not peculiar to Z and $\chi_3$-exchanges, but is upheld for W and $\chi^{\pm}$ exchanges as well for arbitrary Cabibbo (KM) mixing and arbitrary values of the $<\bar{c}c>/<\bar{u}u>$ ratio [2].

Thus it appears that SU(2) X U(1) gauge parameter independence is upheld on shell **even in the presence of a $<\bar{q}q>$ condensing $SU(3)_c$ X SU(2) X U(1) vacuum,** provided the quark propagator mass shell $m_d$ continues to be identified with the quark mass $m_d^Y$ generated through spontaneous symmetry breaking. The mass $m_d^Y$, however, is generated entirely through the order parameter $<\Phi>$, while the additional order parameter $<\bar{q}q>$ is expected to contribute via gluon exchanges to the on shell quark mass (in the absence of explicit mechanisms for confinement) analogously to its contribution to the nucleon mass [8],[16]. In other words, $m_d^Y$ and $m_d$ should **not** coincide. This conflict with SU(2) X U(1) gauge parameter independence is resolved upon the realization that $<\bar{q}q>$ **explicitly** (as opposed to spontaneously) breaks SU(2) X U(1) local symmetry. Except under special circumstances [$m_d^Y = m_d$], the simultaneous presence of $<\Phi>$ and $<\bar{q}q>$ leads to a clashing of symmetry breaking mechanisms that will undo the hidden SU(2) X U(1) symmetry of the spontaneously broken theory, as manifested in the distinct Z- and W-gauge-parameter independence of on-shell electroweak amplitudes. Indeed, if we wish to uphold SU(2) X U(1) gauge parameter independence and simultaneously allow both $<\Phi>$ and $<\bar{q}q>$ to contribute to quark masses, we must modify the vertices of perturbative electroweak theory to retain consistency with SU(2) X U(1) Ward-identity constraints. For example, if we schematically denote the $d\bar{d}Z$ vertex arising from spontaneously broken SU(2) X U(1) theory as $\Gamma^{0\mu}_{d\bar{d}Z} = \gamma^{\mu}(a-b\gamma_5)$, in which case the $d\bar{d}\chi_3$ vertex is [3] $\Gamma^0_{d\bar{d}\chi_3} = -2im_d^Y b\gamma_5/M_Z$, we find that the incorporation of a $<\bar{q}q>$- sensitive Dirac scalar self-energy in the quark propagator [$S(p) = \{m_d^Y - \not{p} - \Sigma_{<\bar{q}q>}(p^2)\}^{-1}$] is possible only if one modifies the former two vertices to retain consistency with the Ward identity [17]

$$M_Z\Gamma_{d\bar{d}\chi_3}(p;p-q,q) = i(p-q)_{\mu}\Gamma^{\mu}_{d\bar{d}Z}(p;p-q,q) - i(a+b\gamma_5)S^{-1}(q) + iS^{-1}(p)(a-b\gamma_5) \qquad (22)$$

[Momenta before and after the semicolon are inbound and outbound, respectively]. This identity, which is upheld by the usual SU(2) X U(1) Feynman rules [i.e, $\Sigma_{<\bar{q}q>} = 0$], can be derived by requiring SU(2) x U(1) BRST invariance of the Green's function $<\psi(x)\bar{G}^Z(y)\bar{\psi}(z)>$, where $\bar{G}^Z$ is the Z's conjugate Faddeev-Popov ghost field [17]. Indeed, if $\Sigma_{<\bar{q}q>}$ is a constant, (24) is upheld quite trivially through replacement of $m_d^Y$ with $m_d^Y - \Sigma_{<\bar{q}q>} = m_d$ in the vertex $\Gamma_{d\bar{d}\chi_3}$. However, the more general problem of external momentum dependent self energy contributions to S(p), which is the subject of present research, requires simultaneous consistency with a number of additional Ward identities [17].

References

[1] V. Elias, Phys. Rev. **D21** (1980) 1113; R. Tarrach, Nucl.Phys. B183 (1981) 384.

[2] M. R. Ahmady, V. Elias, R. R. Mendel, M. D. Scadron, and T.Steele, Phys. Rev. **D39** (1989) 2764.

[3] S. Weinberg, Phys. Rev. Lett. **19** (1967) 1264; Abdus Salam, *in Elementary Particle Theory: Relativistic Groups and Analyticity (Nobel Symposium No. 8)*, N. Svartholm, ed. (Almqvist and Wiksell, Stockholm, 1968) 367. See K. Aoki, Z. Hioki, R. Kawabe, M. Konuma, and T. Muta, Suppl. Prog. Theor. Phys. **73** (1982) 1, for Feynman rules of spontaneously broken SU92) X U(1) expressed entirely in terms of particle masses, KM mixing angles, and the electromagnetic coupling constant e.

[4] R. Delbourgo and M. D. Scadron, Nuovo Cimento Lett. **44** (1985) 193.

[5] M. Shifman, A. Vainshtein, and V. Zakharov, Nucl. Phys. **B147** (1979) 385 and 448.

[6] P. Pascual and R. Tarrach, in *QCD: Renormalization for the Practitioner [Lecture Notes in Physics, Vol. 194]* H. Araki, J. Ehlers, K. Hepp, J. Kippenhahn, H. A. Weidenmuller, and J. Zittartz, editors (Springer-Verlag, Berlin, 1984) 9 and 168-184.

[7] V. Elias, M. D. Scadron, and R. Tarrach, Phys. Lett. **B162** (1985) 176.

[8] V. Elias, T. G. Steele, and M. D. Scadron, Phys. Rev. **D38** (1988) 1584.

[9] V. A. Fock, Phys. Z. Sowjetunion **12** (1937) 404; J. Schwinger, *Particles and Fields* (Addison-Wesley, New York, 1970); M. A. Shifman, Nucl. Phys. **B173** (1980) 13; C. Cronstroem, Phys. Lett. **B90** (1980) 267; M. Dubikov and A. Smilga, Nucl. Phys. **B185** (1981) 109.

[10] Successive differentiations of $x \cdot A = 0$ evaluated at $x = 0$ are proportional to $D_{(\mu}(0)D_{\nu}(0)\ldots D_{\sigma)}(0) - \partial_\mu\partial_\nu\ldots\partial_\sigma$, [V. Elias, Can. J. Phys. **64** (1986) 595.]

[11] V. Elias, T. Steele, M. D. Scadron, and R. Tarrach, Phys. Rev. **D34** (1986) 3537.

[12] T. Steele and V. Elias, Phys. Rev. **D40** (1989) 2669.

[13] L. J. Reinders and K. Stam, Phys. Lett **B180** (1986) 125.

[14] As reported by M. Lavelle at this meeting [Noncovariant Gauges, Vienna, Austria, September, 1989] based on the mass-shell extrapolation of eq. (3.9a) of M. Lavelle and M. Schaden, Phys. Lett. **B217** (1989) 551.

[15] This expression was recently derived in F. J. Yndurain, Z. Phys. **C12** (1989) 653, without reference to the FS gauge.

[16] B. L. Ioffe, Nucl. Phys. **B188** (1981) 317.

[17] M. R. Ahmady, V. Elias, and R. R. Mendel, research in progress.

# Beyond Perturbation Theory in the Axial Gauge[1]

Martin Lavelle and Martin Schaden

Institut für Theoretische Physik, Universität Regensburg, Universitätsstraße 31, D–8400 Regensburg, FRG

**Abstract:** We present a path integral formulation of a regulated axial gauge. The Slavnov-Taylor identity is derived, Gauss' law shown to apply and the Gribov ambiguity discussed. We show how to introduce nonperturbative quarks in these gauges, and calculate the $<: \bar{\psi}\psi :>$ contributions to the quark and gluon propagators to lowest order in the coupling. Results are also given in the light-cone and Coulomb gauges for massless quarks. The gluon condensate in the axial gauge is discussed and the necessity of operator mixing is shown.

## 1 Introduction

The notorious difficulty with the longitudinal singularity in axial gauges [1] has recently been explained by Cheng and Tsai [2] as a consequence of inconsistently regularising this singularity. In this picture, they showed that there is a great arbitrariness in the choice of regularisation scheme [2]. Hence, in perturbation theory, the question now [3] is how best to regulate these gauges so as to preserve the various advantages of unregulated axial gauges (ghost–free, no longitudinal gluons).

Here we will investigate the inclusion of *nonperturbative* effects in the shape of condensates into (non-light-like) axial gauges. The impressive success of QCD sum-rules [4] points to the significance of the nonperturbative structure of the vacuum. This is echoed in phenomenological models [5] and in attempts to relate the condensates to, e.g., the bag model [6], nonrelativistic quark models [7], the Skyrme model [8] and the Pomeron [9].

Until now it has been popular to calculate the coefficients of various condensates in Fock–Schwinger gauge [10]; however, beyond oneloop, or indeed anywhere where there are perturbative gluons, this is practically impossible due to the untractable nature of the gluon propagator in this gauge [11]. Other gauges must be used. Indeed axial gauges

[1] Talk presented by M. Lavelle

– if one (in some regularisation?) can indeed show that their Hilbert spaces are positive definite – may be particularly interesting [12].

In Sect. 2 the path integral formalism for an (arbitrary) regularisation of a (non–light-like) axial gauge is presented. The Slavnov-Taylor-identity is derived, and the applicability of Gauss' law for this regulated theory is shown. The quark condensate, $<:\bar{\psi}\psi:>$, is discussed in Sect. 3; a method is found which allows one to evaluate, in axial gauges, the contribution of $<:\bar{\psi}\psi:>$ to any Greens' function. Results are also presented, for massless quarks, in the Coulomb and light-cone gauges. The gluon condensate of dimension 4, $<:G^2:>$, is investigated in Sect. 4. A comparison is made with results in covariant gauges, and it is shown that there is operator mixing. Finally in Sect. 5, we give some conclusions.

## 2 The Path Integral Formalism for a Regulated Axial Gauge

As discussed above there are many ways of regulating the axial gauge [2]; here we arbitrarily choose a case [13] which has been shown by Cheng and Tsai to correctly yield the Wilson loop to order $g^4$. Our regulated gluon propagator is

$$D^{ab(0)}_{\mu\nu}(k) = \frac{-\delta^{ab}}{k^2+i\epsilon'}\left[g_{\mu\nu} - \frac{(k_\mu\eta_\nu + k_\nu\eta_\mu)k.\eta - \eta^2 k_\mu k_\nu}{(k.\eta)^2 - \epsilon^2\eta^2}\right], \tag{1}$$

and the corresponding ghost-gluon vertex (with outgoing ghost propagator included) is

$$G^{abc(0)}_\mu(k) = \frac{-gf^{abc}}{k^2+i\epsilon'}\,\frac{k^2 k.\eta\eta_\mu - \epsilon^2\eta^2 k_\mu}{(k.\eta)^2-\epsilon^2\eta^2}, \tag{2}$$

where $\epsilon^2$ is a regularisation parameter. This set of Feynman rules may be viewed as a *flow gauge* [14], which for $\epsilon^2 \to 0$ becomes the unregulated axial gauge, and for $\epsilon^2 \to \infty$ the Feynman gauge. We stress that the apparent decoupling of ghosts in the $\epsilon^2 \to 0$ limit is spurious, and that ghosts do *not* decouple [2].

This set of Feynman rules may be derived [13] from the gauge fixing term

$$\mathcal{L}_{Fix} = -\frac{1}{2}\phi^a\left(\Box+\epsilon^2\right)\phi^a - \frac{1}{2}\left(\partial^\mu + \frac{\partial.\eta\Box\eta^\mu}{\epsilon^2\eta^2}\right)A^a_\mu\frac{\epsilon^2}{\Box+\epsilon^2}\left(\partial^\nu+\frac{\partial.\eta\Box\eta^\nu}{\epsilon^2\eta^2}\right)A^a_\nu, \tag{3}$$

whence one, by the Faddeev-Popov procedure, obtains the gauge fixed Lagrangian [13]

$$\begin{aligned}\mathcal{L}_G = &-\frac{1}{4}G^{a^2}_{\mu\nu} - \frac{1}{2}\phi^a\left(\Box+\epsilon^2\right)\phi^a + i\epsilon\phi^a\left(\partial^\mu+\frac{\partial.\eta\,\Box\,\eta^\mu}{\epsilon^2\eta^2}\right)A^a_\mu\\ &+\bar{c}^a\left(\partial^\mu+\frac{\partial.\eta\,\Box\,\eta^\mu}{\epsilon^2\eta^2}\right)\left(\delta^{ab}\partial_\mu - g\,f^{abc}A^c_\mu\right)c^b,\end{aligned} \tag{4}$$

where the $\phi^a$ are auxiliary fields. Note that this equivalence is straightforwardly seen to fit into the more general formalism of Andraši and Taylor [15].

The Lagrangian (2.4) has a BRS-invariance [13], which we may use to derive the Slavnov-Taylor identity. Due to the presence of ghost fields the derivation runs essentially parallel to that in covariant gauges [16], and one obtains [13]

$$\left(k^\mu - \frac{k.\eta k^2 \eta^\mu}{\epsilon^2 \eta^2}\right)\left(k^\nu - \frac{k.\eta k^2 \eta^\nu}{\epsilon^2 \eta^2}\right) D^{ab}_{\mu\nu}(k) = \frac{(k^2 - \epsilon^2)}{\epsilon^2}\delta^{ab}, \tag{5}$$

where $D^{ab}_{\mu\nu}$ is the full gluon propagator. In the limit $\epsilon^2 \to 0$, this leads to

$$\eta^\mu \eta^\nu D^{ab}_{\mu\nu} = 0. \tag{6}$$

For the polarisation we obtain the constraint

$$k^\mu k^\nu \Pi^{ab}_{\mu\nu}(k) = O(\alpha_s^2). \tag{7}$$

This identity is less restrictive than that usually quoted for the axial gauge [17], $k^\mu \Pi_{\mu\nu} = 0$, due to the presence of the axial vector $\eta_\mu$, which allows us to build a tensor orthogonal to $k^\mu k^\nu$ but not, in contradistinction to covariant gauges, to $k^\mu$ or $k^\nu$ singly.

We note also that this regulated theory obeys Gauss' law – this is most easily shown by generalising the method of Steiner [18] – and hence is a theory of physical states [19]. It is not clear to us however, if (unlike naive axial gauges [20]) this theory suffers from a Gribov ambiguity [21]: although various regulators of the axial gauge (e.g. Ref. 18) can be seen to be unique (one follows the proof of Chan and Halpern [14] the case chosen here does not fall into this category, and indeed, if one tries brute force, one obtains a third order partial differential equation that we are, as yet, unable to solve.

## 3 The Quark Condensate in the Axial Gauge

The nonperturbative part of a Green's function is defined as the difference of the full Green's function and the perturbative part. In an axial gauge the (nonperturbative) propagator may be written as [13]

$$S^{nonpert}(k) = S(k) - S^{pert}(k) = \not k \tilde{A}(k^2, (k.\eta)^2) + \tilde{B}(k^2, (k.\eta)^2) + \frac{\not\eta \eta.k}{\eta^2}\tilde{C}(k^2, (k.\eta)^2), \tag{8}$$

where we have dropped explicit indices, and $\tilde{A}$, $\tilde{B}$ and $\tilde{C}$ are unknown functions that depend only on $k^2$ and $(k.\eta)^2/\eta^2$. With the aid of the renormalisation point, $\mu^2$, we divide momentum space into perturbative and nonperturbative parts, and identify [13]

$$\int_{\text{"soft"}} i\frac{d^4k}{(2\pi)^4} m\tilde{B} = \int_{\text{"soft"}} i\frac{d^4k}{(2\pi)^4}\left(k^2\tilde{A} + \frac{(\eta.k)^2}{\eta^2}\tilde{C}\right)$$
$$= -\frac{m <:\bar{\psi}\psi:>}{4N_c} + O(g)\,, \tag{9}$$

where we work in 4 dimensions, with $N_f$ light quark flavours and an $SU(N_c)$ gauge group. Note that because of the axial vector, $\eta_\mu$, *soft* momenta here are such that $k^2 > -\mu^2$ and $(k.\eta)^2 < \mu^2$. Unlike the case of covariant gauges [22], we have not specified all the possible moments of the quark propagator which can contribute to $<:\bar{\psi}\psi:>$. A further constraint is required.

We now calculate the contributions from $<:\bar{\psi}\psi:>$ to the quark and gluon selfenergies to lowest order in $\alpha_s$ in the deep Euclidean region, $p^2 < -\mu^2$, $(p.\eta)^2 > \mu^2$, where the second bound is again a consequence of our gauge. Expanding the loop integral in powers of $1/p^2$ and $1/(p.\eta)^2$ and using (3.2), we obtain for the quark selfenergy

$$\Sigma^{<:\bar{\psi}\psi:>}(p) = \frac{4\pi\alpha_s C_f}{p^2}\Bigg[\left(-2 - \frac{\eta^2p^2}{(p.\eta)^2}\right)\int_{\text{"soft"}} i\frac{d^4k}{(2\pi)^4}\tilde{B}$$
$$+\left(-\frac{4\not{p}}{p^2} + \frac{4\not{\eta}}{p.\eta} - \frac{4\eta^2\not{p}}{(p.\eta)^2} - \frac{2\eta^2p^2\not{\eta}}{(p.\eta)^3}\right)\frac{1}{3}\int_{\text{"soft"}} i\frac{d^4k}{(2\pi)^4}\left(k^2 - \frac{(k.\eta)^2}{\eta^2}\right)\tilde{A}$$
$$+\left(-\frac{4p.\eta\not{\eta}}{p^2\eta^2} - \frac{2\eta^2\not{p}}{(p.\eta)^2}\right)\int_{\text{"soft"}} i\frac{d^4k}{(2\pi)^4}\left(\left(\frac{4}{3}\tilde{A} + \tilde{C}\right)\frac{(k.\eta)^2}{\eta^2} - \frac{\tilde{A}k^2}{3}\right)\Bigg]\,, \tag{10}$$

and for the gluon polarisation

$$\Pi_{\mu\nu}^{<:\bar{\psi}\psi:>}(p) = \frac{16\pi\alpha_s N_f}{p^2}\Bigg[\frac{m <:\bar{\psi}\psi:>}{2N_c} g^{(2)}_{T\mu\nu}(p)$$
$$+\int_{\text{"soft"}} i\frac{d^4k}{(2\pi)^4}\tilde{A}\left(k^2 - \frac{(k.\eta)^2}{\eta^2}\right)\left(\frac{8}{3}g^{(2)}_{T\mu\nu}(p) - \frac{4}{3}g^{(1)}_{T\mu\nu}(p)\right)\Bigg]\,,$$
$$g^{(1)}_{T\mu\nu}(p) := g_{\mu\nu} - \frac{p_\mu p_\nu}{p^2}\;,\quad g^{(2)}_{T\mu\nu}(p) := \frac{g_{\mu\nu}(p.\eta)^2}{p^2\eta^2} - \frac{(p_\mu\eta_\nu + p_\nu\eta_\mu)p.\eta}{p^2\eta^2} + \frac{\eta_\mu\eta_\nu}{\eta^2}\,, \tag{11}$$

which satisfies the Slavnov-Taylor-identity. For massive quarks, we see that neither of these expressions can be directly expressed in terms of the condensate. We look for our further constraint in the gauge invariance of the effective potential [23]. To lowest order in the coupling the leading divergence of the effective potential from $<:\bar{\psi}\psi:>$ may be calculated either from the quark selfenergy or the gluon polarisation [22]. From (11) we obtain

$$V^{Fig.3}(<:\bar{\psi}\psi:>) = \frac{6N_f(N_c^2-1)\pi\alpha_s m <:\bar{\psi}\psi:>}{N_c} \int_{"hard"} -i\frac{d^4p}{(2\pi)^4}\frac{1}{p^4}, \tag{12}$$

where we used

$$\int_{"hard"} -i\frac{d^4p}{(2\pi)^4}\frac{1}{p^4} = \int_{"hard"} -i\frac{d^4p}{(2\pi)^4}\frac{4(p.\eta)^2}{p^6} = \int_{"hard"} -i\frac{d^4p}{(2\pi)^4}\frac{8(p.\eta)^4}{p^8}, \tag{13}$$

which is true up to finite terms. The undetermined moment in (11) has cancelled, and the result is exactly that found in covariant gauges [22].

In the calculation of the potential from the quark self energy new power divergent integrals appear

$$\int_{"hard"} -i\frac{d^4p}{(2\pi)^4}\frac{\eta^2}{p^2(p.\eta)^2}, \quad \int_{"hard"} -i\frac{d^4p}{(2\pi)^4}\frac{\eta^4}{(p.\eta)^4}, \tag{14}$$

which cannot contribute to the (logarithmically divergent) potential. Demanding that the coefficient of these gauge artefacts is zero implies

$$\int_{"soft"} i\frac{d^4k}{(2\pi)^4}\left(k^2 - \frac{(k.\eta)^2}{\eta^2}\right)\tilde{A} = -\frac{3m <:\bar{\psi}\psi:>}{8N_c}, \tag{15}$$

which is the extra constraint we needed. The logarithmically divergent part of the potential can now also be found from the quark selfenergy, and the result is once more that of the covariant case, this gives us confidence in our approach.

The final results for the nonperturbative quark contributions to the quark selfenergy and the gluon polarisation in terms of $<:\bar{\psi}\psi:>$ are

$$\Sigma^{<:\bar{\psi}\psi:>} = \frac{\pi\alpha_s C_f <:\bar{\psi}\psi:>}{N_c p^2}\left[2 + \frac{\eta^2p^2}{(p.\eta)^2} + m\left(-\frac{2\not{p}}{p^2} + \not{\eta}\left(\frac{4p.\eta}{p^2\eta^2} + \frac{2}{p.\eta} - \frac{\eta^2p^2}{(p.\eta)^3}\right)\right)\right], \tag{16}$$

$$\Pi_{\mu\nu}^{<:\bar{\psi}\psi:>} = \frac{8\pi\alpha_s N_f m <:\bar{\psi}\psi:>}{p^2N_c}\left(g_{T\mu\nu}^{(1)}(p) - g_{T\mu\nu}^{(2)}(p)\right). \tag{17}$$

The method found above can now be used to find $<:\bar{\psi}\psi:>$ contributions to any Green's function in the deep Euclidean region. The result for the quark self energy may also be used to test the claimed [24] gauge invariance of the on shell value of $\Sigma^{<:\bar{\psi}\psi:>}$, using the on shell spinor relation, $\bar{u}\gamma_\mu u = \frac{p_\mu}{m}\bar{u}u$, and averaging over angles one obtains the usual result [24]

$$\Sigma^{<:\bar{\psi}\psi:>}_{p^2=m^2,\not{p}=m} = \frac{3\pi\alpha_s C_f <:\bar{\psi}\psi:>}{N_c p^2}. \tag{18}$$

Although one could question the applicability of (16) at low momenta, this is certainly a nontrivial check of the formal gauge invariance of the quark condensate contribution to the on shell selfenergy.

Finally we present the results of parallel calculations for massless quarks in the Coulomb and light-cone gauges. From similar calculations one finds in the light-cone gauge

$$\Sigma^{<:\bar{\psi}\psi:>} = \frac{2\pi\alpha_s C_f <:\bar{\psi}\psi:>}{N_c p^2}, \tag{19}$$

and in the Coulomb gauge

$$\Sigma^{<:\bar{\psi}\psi:>} = \frac{3\pi\alpha_s C_f <:\bar{\psi}\psi:>}{N_c p^2}\left[3 - \frac{p_0^2}{p^2}\right], \tag{20}$$

The last result in the instantaneous approximation, $p_0 = 0$ is that of Ref. 25, and coincides with that in the Landau gauge (as one would expect since Coulomb gauge in this approximation is equivalent to the Landau gauge). The extension to massive quarks of calculations in these gauges is currently under investigation.

## 4 First Steps with the Gluon Condensate in the Axial Gauge

The (abelian part of the) gluon condensate is defined by a second moment of the propagator

$$< G_{\mu\nu}G_{\rho\sigma} >= \int_{\text{"soft"}} i\frac{d^4k}{(2\pi)^4}\left(\tilde{D}_{\nu\sigma}k_\mu k_\nu + \tilde{D}_{\mu\rho}k_\nu k_\sigma - \tilde{D}_{\nu\rho}k_\mu k_\sigma - \tilde{D}_{\mu\sigma}k_\nu k_\rho\right) + O(g), \tag{21}$$

where

$$\tilde{D}_{\mu\nu}(k) := D^{nonpert}_{\mu\nu}(k) = D_{\mu\nu}(k) - D^{pert}_{\mu\nu}(k). \tag{22}$$

We parameterise these moments by

$$\int_{\text{"soft"}} i\frac{d^4k}{(2\pi)^4}\,\tilde{D}_{\mu\nu}(k)\,k_\alpha k_\beta = \left(g_{\mu\nu} - \frac{\eta_\mu\eta_\nu}{\eta^2}\right)\left(a g_{\alpha\beta} + b\frac{\eta_\alpha\eta_\beta}{\eta^2}\right)$$
$$+ c\left(\left(g_{\mu\alpha} - \frac{\eta_\mu\eta_\alpha}{\eta^2}\right)\left(g_{\nu\beta} - \frac{\eta_\nu\eta_\beta}{\eta^2}\right)\right.$$
$$\left. + \left(g_{\nu\alpha} - \frac{\eta_\nu\eta_\alpha}{\eta^2}\right)\left(g_{\mu\beta} - \frac{\eta_\mu\eta_\beta}{\eta^2}\right)\right) \tag{23}$$
$$+ d\left(g_{\mu\alpha}\,g_{\nu\beta} + g_{\nu\alpha}\,g_{\mu\beta} - 2\frac{\eta_\mu\eta_\nu\eta_\alpha\eta_\beta}{\eta^4}\right)$$
$$+ \frac{\eta_\mu\eta_\nu}{\eta^2}\left(e\frac{\eta_\alpha\eta_\beta}{\eta^2} + f g_{\alpha\beta}\right),$$

where $a$, $b$ and $c$ are coefficients of structures that satisfy $\eta_\mu\tilde{D}^{\mu\nu} = 0$, the structure with coefficient $d$ satisfies (6), and the remaining terms do not satisfy the ST-identity. From (20) and (22) we have

$$\frac{<: G^2 :>}{6\,(N_c^2 - 1)} = 3a + b - 2c - 4d + f + O(g)\,. \tag{24}$$

The contribution of this condensate to the quark-selfenergy is easily found to be [13]

$$\Sigma^{<:G^2:>}(p) = -\frac{4\pi\alpha_s C_f}{p^2}\left[\frac{\not{p}}{p^2}((2a + b - 2c - 6d + e + 2f) + \frac{(p.\eta)^2}{p^2\eta^2}(-4b - 8c\right.$$
$$\left. + 8d - 4e)) + \frac{\not{\eta}p.\eta}{p^2\eta^2}((4a + 8b + 8d - 4e - 4f) + \frac{(p.\eta)^2}{p^2\eta^2}(-8b + 16c - 16d + 8e))\right]. \tag{25)}$$

We see that (22) does not suffice to express this in terms of $<: G^2 :>$; however, the corresponding term in the effective potential is so determined [13] and once more the covariant gauge result [22] is obtained.

The nonperturbative part of the gluon selfpolarisation does not directly satisfy the Slavnon-Taylor-identity: the result is complicated [13], and demanding (7) implies

$$3a + b - 2c - 2d + f \equiv 0\,, \tag{26}$$

from which we see that if we only allow for transverse gluons then the Slavnov-Taylor-identity and (26) would necessitate $<: G^2 :>= 0$! A similar effect has been found in covariant gauges [22], and has been explained [26,27] as a result of operator mixing with an equation of motion condensate. Presumably this is also the case here.

## 5 Conclusions

It has been shown how to include <: $\bar{\psi}\psi$ :> effects in the e. The gauge invariance of the pole term in the quark propagator from this condensate (neglecting other effects) was verified. The situation with <: $G^2$ :> is still unclear, but of great interest, the main question being if one can demand $\tilde{D}_{0\mu} = 0$. Finally, although the regulated axial gauges all obey Gauss' law, it is generally not clear if they have a Gribov ambiguity.

### Acknowledgments

We thank Prof. E. Werner for his support and interest. M.J.L thanks Dr. A. Vladikas for collaboration on Gauss' law and the Gribov problem in axial gauges, Prof. V. Elias for a useful discussion on the pole mass, and the organisers for a profitable and enjoyable conference. One of the authors (M. L.) was supported by the BMFT under contract MEP 0233 REB 4).

## References

[1] See, e.g., G. Leibbrandt, Rev. Mod. Phys. **59** (1987) 1067.

[2] H. Cheng & E. C. Tsai, Phys. Rev. **D 36** (1987) 3196.

[3] P.V. Landshoff, this proceedings.

[4] See, e.g., E. V. Shuryak, Phys. Rep. **C115** (1984) 151.

[5] M. J. Lavelle, E. Werner & St. Głazek, in Few Body Systems, Supplementum 2, Ed.s J.-L. Ballot & M.Fabre de la Ripelle.

[6] V. A. Novikov et al, Nucl. Phys. **B191** (1981) 301.

[7] St. Głazek & M. Schaden, Phys. Lett. **198B** (1987) 42.

[8] H.G. Dosch & S. Narison, Phys. Lett. **180B** (1986) 390 ; **184B** (1987) 78.

[9] P. V. Landshoff & O. Nachtmann, Z. Phys. **C31** (1987) 405.

[10] V. A. Novikov et al, Fortschr. d. Physik **32** (1984) 585.

[11] W. Kummer & J. Weiser, Z. Phys. **C31** (1986) 105.

[12] See e.g. G. B. West, Phys. Rev. **D27** (1983) 1878.

[13] M. J. Lavelle, & M. Schaden, Phys. Lett. **217B** (1989) 551.

[14] H. S. Chan & M. B. Halpern, Phys. Rev. **D 33** (1985) 540.

[15] A. Andraši & J. C. Taylor, Nucl. Phys. **B310** (1988) 222.

[16] See e.g. P. Becher, M. Böhm & H. Joos, *Eichtheorien* (Teubner Verlag, Stuttgart, 1983).

[17] D. M. Capper & G. Leibbrandt, Phys. Rev. **D25** (1982) 1002.

[18] F. Steiner, Phys. Lett. **173B** (1986)321.

[19] See, e.g. ,G. C. Rossi & M. Testa, Nucl. Phys. **B163** (1980) 109.

[20] W. I. Weisberger, in *Asymptotic Realms of Physics* ed. A. H. Guth et al. (MIT Press, Cambridge, Mass. 1983).

[21] V. N. Gribov, Nucl. Phys. **B139** (1978) 1.

[22] M. J. Lavelle, & M. Schaden, Phys. Lett. **208B** (1988) 297.

[23] S. Coleman & E. Weinberg, Phys. Rev. **D7** (1973) 1888.

[24] V. Elias & M. Scadron, Phys. Rev. **D30** (1984) 647 ; V. Elias, this proceedings.

[25] R. Alkofer & P. A. Amundsen, Nucl. Phys. **B306** (1988) 305.

[26] E. Bagan & T. G. Steele, Phys. Lett. **226B** (1989) 142.

[27] V. P. Spiridonov, INR Preprint No. 2346 (1989) , (in Russian).

# Cancellation of $1/n\cdot p$-Factors in Light-Cone Gauge Yang-Mills Theory

Mats Ögren

Institute for Theoretical Physics, S-412 96 Göteborg, Sweden

**Abstract:** We discuss the cancellation of poles of type $1/n\cdot p$ in (two-component) light-cone gauge Yang-Mills theory. These poles should not contribute to physical quantities, and we exhibit their cancellation explicitely in a few simple examples.

## 1 Introduction

The light-cone gauge is, as all axial gauges, plagued by the presence of inverse powers of a single component of the momenta. Since these are gauge artifacts, they should be spurious and their regularization should primarily be a technical issue. The problem is to regularize the integrals in such a way that the contributions from these poles cancel when the terms are integrated first, and summed afterwards. Within dimensional regularization, one technical problem is that the analytic continuation to D dimensions is ambiguous, and this gives rise to so-called "surface terms". Usually, any undesirable effects caused by this can be compensated by adding finite local counterterms. However, for the Leibbrandt-Mandelstam [1,2] prescription, it has been shown [3] that certain diagrams involving $1/n\cdot p$-factors suffer from divergent surface terms, and then this cure cannot be taken for granted. The principal value (PV) prescription, on the other hand, gives rise to new divergencies [4]. A straightforward application of this prescription also has the inconvenient property of not being compatible with Ward[1]-identities [2]. Still, the PV prescription has the advantage of not involving a second light-like vector, thus preserving one more Lorentz-symmetry. Accordingly, efforts to rehabilitate this prescription would be most welcome (Some progress has been made in [5,6]). Let us also remind the reader of Cornwall's [4] elegant formula for computing (or defining) integrals with the PV prescription, when the result is convergent.

If the gauge symmetry is anomalous, one can no longer expect the $1/n\cdot p$-factors to be spurious, and the pole prescription may have physical significance. At present, only two-dimensional anomalous theories can be quantized consistently, and for them the Leibbrandt-Mandelstam (LM) prescription is very natural, leading to an ordinary propagator in the denominator. See eg. [7,8] (Note that the Fourier transform of $(1/np)_{\mathrm{LM}}$

[1] -Lee-Slavnov-Taylor-

is proportional to $(1/n{\cdot}x)_{\rm LM}$). In other cases, some kind of regularized PV prescription is often used in two dimensions [9].

In this paper, we shall give the result of explicit calculations for the simplest cases, and verify the cancellation of the $1/n{\cdot}p$-factors in physical quantities. (Similar calculations have been done long ago, but the focus has been on the divergent terms [10,11]).

## 2 Two-component light-cone gauge Yang-Mills theory

There are two ways two obtain the two-component formulation of Yang-Mills theory. The first is to start with the covariant gauge-invariant lagrangian. Then the light-cone gauge is chosen, and the unphysical degrees of freedom are eliminated. The second is to start from free fields built from the creation and annihilation operators for physical states. Compare the approach of Weinberg [12], but in our case the construction of covariant quantum fields is not needed. Then, interaction terms are added to the generators of the Poincaré algebra in such a way that the algebra is preserved [13]. The latter procedure gives an independent motivation for the two-component formalism, but it should be sufficient to compare [11] the complexity of computations in the two formalisms to see its merit. The two-component formalism also has advantages for those supersymmetric theories for which covariant off-shell supersymmetric formulations are not known, or are very complicated.

The easiest way to obtain a two-component formulation of Yang-Mills is to start from a covariant formulation, though. Thus we begin with the standard lagrangian

$$\mathcal{L} = -\frac{1}{4}(F^a_{\mu\nu})^2 + \mathrm{i}\overline{\Psi}\gamma^\mu D_\mu \Psi. \tag{1}$$

The light-cone coordinates are defined using three light-like vectors, $n^\mu$, $n^{*\nu}$ (real), and $\epsilon^\mu$ (complex), satisfying

$$n^\mu n^*_\mu = 1, \qquad \overline{\epsilon}^\mu \epsilon_\mu = 1, \qquad n^\mu \epsilon_\mu = 0, \qquad n^{*\mu}\epsilon_\mu = 0, \tag{2}$$

where $\overline{\epsilon}^\mu$ is the complex conjugate of $\epsilon^\mu$. Then

$$A^+ \equiv n{\cdot}A, \qquad A^- \equiv n^*{\cdot}A, \qquad A \equiv \epsilon{\cdot}A, \qquad \overline{A} \equiv \overline{\epsilon}{\cdot}A, \tag{3}$$

and similarly for other four-vectors. Conversely,

$$A^\mu = n^{*\mu}A^+ + n^\mu A^- + \overline{\epsilon}^\mu A + \epsilon^\mu \overline{A}. \tag{4}$$

For the fermions

$$\Psi^{(+)} \equiv \frac{1}{2}\gamma^-\gamma^+\Psi \qquad \text{and} \qquad \Psi^{(-)} \equiv \frac{1}{2}\gamma^+\gamma^-\Psi. \tag{5}$$

When the lagrangian is expressed in terms of light-cone coordinates, the explicit four-vectors obviously dissappear. Next we chose the light-cone gauge $A^+ = 0$, and shift $A^-$ and $\Psi^-$ to isolated the unphysical degrees of freedom.

$$\begin{aligned} A'^{-} &\equiv A^{-} - \frac{1}{\partial_{-}}(\overline{\partial} A + \partial \overline{A}) + \text{interaction terms}, \\ \Psi'^{(-)} &\equiv \Psi^{(-)} + \frac{1}{2\partial_{-}}(\overline{\partial}\gamma + \partial\overline{\gamma})\gamma^{+}\Psi^{(+)} + \text{interaction terms}. \end{aligned} \tag{6}$$

The primed fields do not interact, and their propagators are just $1/(\partial_-)^2$ and $1/\partial_-$ respectively. They are therefore redundant. Note that they do contribute to expectation values of manifestly Lorentz covariant objects, though. The shifts can only be performed after the prescription for $1/p^+$ has been determined. Within canonical quantization, this was done by Bassetto et al. [14], with the LM prescription as the result. Clearly, the two-component formulation lacks both Ward[1]-identities and the Piguet-Sibold extended BRST-invariance [15]. We would therefore like to be able to use Lorentz-covariance in their place. (N.B. some Lorentz-transformations are nonlinear). Unfortunately, the LM prescription leads to Greens functions that explicitely violate the relations that are obtained from a naive application of the Lorentz-transformations. (For example, the self energy should be a function of the two variables $p^+$ and $p^\mu p_\mu$ only, nevertheless it contains a Spence function with the argument $\left(\frac{p^\mu p_\mu}{p^+p^-}\right)$. Incidentally, unitarity is not quite manifest either [16]).

The final result is the following (effective) lagrangian (cf. [17])

$$\begin{aligned} L = &-\overline{A}^a \Box A^a + \frac{i}{2}\psi^\dagger \frac{\Box}{\partial_-}\psi \\ &+ g\psi^\dagger \tilde{T}^a \left(\frac{\partial}{\partial_-}\overline{A}^a \psi - \overline{A}^a \frac{\partial}{\partial_-}\psi\right) + g\left(\psi^\dagger \frac{\overline{\partial}}{\partial_-}A^a - \frac{\overline{\partial}}{\partial_-}\psi^\dagger A^a\right)\tilde{T}^a\psi - \\ &- 2gf^{abc}\left(\partial_-\overline{A}^a A^b \frac{\overline{\partial}}{\partial_-}A^c + \partial_- A^a \overline{A}^b \frac{\partial}{\partial_-}\overline{A}^c\right) - \\ &- \frac{1}{2}g^2\left(\frac{1}{\partial_-}\psi^\dagger \tilde{T}\psi\right)^2 + ig^2\left(\psi^\dagger \overline{A}^a\right)\tilde{T}^a \frac{1}{\partial_-}\tilde{T}^b\left(\psi A^b\right) + \\ &+ g^2 f^{abc}\left(\overline{A}^a \partial_- A^b - \partial_- \overline{A}^a A^b\right)\frac{1}{\partial_-{}^2}\left(\psi^\dagger \tilde{T}^c \psi\right) - \\ &- 2g^2 f^{abc} f^{ade}\frac{1}{\partial_-}\left(\overline{A}^b \partial_- A^c\right)\frac{1}{\partial_-}\left(A^d \partial_- \overline{A}^e\right). \end{aligned} \tag{7}$$

To simplify the Feynman rules, the left-handed fermions are reversed, and an irrelevant factor $\sqrt{2}$ is dropped. Thus $\psi$ contains two right-handed physical spinors, and

$$\tilde{T}^a = \begin{pmatrix} T^a_{\mathrm{R}} & 0 \\ 0 & -(T^a_{\mathrm{L}})^t \end{pmatrix}, \tag{8}$$

where $T^a_{\mathrm{R}}$ and $T^a_{\mathrm{L}}$ are the generators of the Lie algebra in the representation of the right-handed and left-handed fermions, respectively. (The $\tilde{T}$'s satisfy the Lie algebra: $\left[\tilde{T}^a, \tilde{T}^b\right] = if^{abc}\tilde{T}^c$). In momentum space, the cubic vertices contain the combination $\left(\frac{p}{p^+} - \frac{q}{q^+}\right)$, where $p$ and $q$ are the appropriate incoming momenta. (This combination is dictated by the linearly realized part of the Lorentz-group [13] ). It is therefore convenient to introduce a special notation [18]

$$(p, q) \equiv (pq^+ - qp^+). \tag{9}$$

The helicity basis yields very beautiful Feynman rules, but these are not necessarily the most suitable for computations, since every consistent combination of helicities in a process gives rise to a separate diagram. One could also object that dimensional regularization does not respect the Lorentz symmetry, since the number of degrees of freedom changes with the dimension, and in D dimensions one should really work with a D-2 component formalism. However, the "errors" can be absorbed by finite local counterterms, provided that one can avoid the additional singularities at D=4 that appear from a naive application of the PV prescription [4]. Other regularizations are not very attractive. Pauli-Villars would introduce new vertices and regularization by means of higher covariant derivatives is not practical even in covariant gauges.

## 3 Cancellations

First one should observe that at tree-level, gauge invariance implies that all terms with poles $1/n{\cdot}p$, where $p$ is an internal momentum, has to vanish when the external momenta are put on shell. Second, the quartic vertices contain poles that exactly cancel one power of $1/n{\cdot}p$ from the corresponding "H"-diagrams with two cubic vertices and one propagator. Let us just give the simplest case as example, namely $\psi^\dagger(p-r)\overline{A}(r)A(s)\psi(-p-s)$. Suppressing the $\tilde{T}$-matrices we have

$$\begin{aligned}-ig^2\frac{1}{p^+} + ig\frac{\overline{(r,p)}}{r^+p^+}2i\frac{p^+}{p^2}ig\frac{(s,p)}{s^+p^+} = \\ ig^2\frac{s^2}{s^+p^2} - ig^2\frac{(p+s)^2}{(p^+ + s^+)p^2} - 2ig^2\frac{(p,s)\overline{(p+s,r)}}{r^+s^+(p^+ + s^+)p^2}\end{aligned} \tag{10}$$

The way to collect terms is not unique. Here we have used the identity

$$\overline{(r,p)} = \frac{1}{p^+ + s^+}\left(\overline{(r,p+s)}p^+ - \overline{(p,p+s)}r^+\right), \tag{11}$$

and a special case of the identity

$$(k,l)\overline{(p,q)} + \overline{(k,l)}(p,q) = -(k^\mu l^+ - l^\mu k^+)(p_\mu q^+ - q_\mu p^+). \tag{12}$$

This is one way to avoid a naked $p^-$ in the numerator, but the price is a new spurious pole $1/(p^+ + s^+)$. We have found it quite difficult to collect terms in this way for other than the simplest diagrams. When all diagrams are added, the net effect of these cancellations is that the internal fermion lines are free of $1/p^+$-poles, and that only simple poles survive on internal boson lines. (Actually, we do not have a complete proof of the latter statement beyond one-loop, but there is no doubt that it is true. The problem is to show that the cancellations are not destroyed by multi-loop self-energy like corrections).

Expectation values of manifestly gauge invariant and Lorentz covariant objects like $< \overline{\Psi}(x)\gamma^\mu\Psi(x)\,\overline{\Psi}(y)\gamma^\nu\Psi(y) >$ should obviously exhibit manifest cancellation of all $1/p^+$-poles. It has been checked explicitely for one component of this particular example at two loops, but the result is mainly a test that the Feynman rules are correct.

We have not found a general rule for the remaining cancellations, so we can only give an example. The one-loop self-energies have parts involving the integral

$$L(p) \equiv \int d^4k \frac{1}{k^+ k^2 (k-p)^2}. \tag{13}$$

As we shall see now, the cubic vertices include such parts that exactly cancel half of the corresponding ones from the self-energies. Let us first define

$$I(p) \equiv \int d^4k \frac{1}{k^2 (k-p)^2}, \tag{14}$$

and state our conventions (cf. [19])

$$\begin{aligned} d^{abc} &\equiv tr\{\tilde{T}^a(\tilde{T}^b\tilde{T}^c + \tilde{T}^c\tilde{T}^b)\}, & C\delta^{ab} &\equiv f^{cda} f^{cdb}, \\ \tilde{T}_f \delta^{ab} &\equiv (T_{f\mathrm{L}} + T_{f\mathrm{R}}) \equiv tr\{\tilde{T}^a\tilde{T}^b\}, & C_f \mathbb{1} &\equiv \sum_a \tilde{T}^a\tilde{T}^a. \end{aligned} \tag{15}$$

The one-loop gluon self-energy is

$$\delta^{ab} p^2 \frac{g^2}{(2\pi)^4} \left\{ \left( \frac{2}{3}\tilde{T}_f - \frac{11}{3}C \right) I(p) + 4CL(p) \right\}. \tag{16}$$

The one-loop fermion self-energy is

$$\mathbb{1} \frac{p^2}{2p^+} \frac{g^2}{(2\pi)^4} \left\{ -3C_f I(p) + 4C_f L(p) \right\}. \tag{17}$$

The one-loop contribution to the $\overline{A}^a(p)\overline{A}^b(q)A^c(r)$-vertex is (where $r = -p - q$)

$$\begin{aligned} & 2g f^{abc} r^+ \frac{(p,q)}{p^+ q^+} \frac{ig^2}{(2\pi)^4} \left\{ \left( \frac{11}{3}C - \frac{2}{3}\tilde{T}_f \right) I(r) - 2C\left(L(p) + L(q) + L(r)\right) \right\} + \\ & + 2g f^{abc} \frac{ig^2}{(2\pi)^4} \int d^4k \frac{1}{k^2(k-p)^2(k+q)^2} \times \\ & \times \Bigg[ T_f \left( q^2(k,p) + p^2(k,q) \right) \frac{(k^+ - p^+)^2 + (k^+ + q^+)^2}{p^+ q^+ r^+} - \\ & - Cp^2 \left( \frac{q^+ r^+}{p^+ k^+ (k^+ + q^+)} (k - p, r) + \frac{(k^+ - p^+)^4 + (k^+ + q^+)^4}{p^+ q^+ r^+ (k^+ - p^+)(k^+ + q^+)} (k,q) \right) - \\ & - Cq^2 \left( \frac{p^+ r^+}{q^+ k^+ (k^+ - p^+)} (k + q, r) + \frac{(k^+ - p^+)^4 + (k^+ + q^+)^4}{p^+ q^+ r^+ (k^+ - p^+)(k^+ + q^+)} (k,p) \right) - \\ & - Cr^2 \frac{r^+}{k^+} \left( \frac{q^+(k,p)}{p^+(k^+ + q^+)} + \frac{p^+(k,q)}{q^+(k^+ - p^+)} \right) \Bigg] + \\ & + 2g d^{abc} \frac{ig^2}{(2\pi)^4} \int d^4k \frac{1}{k^2(k-p)^2(k+q)^2} \frac{(2k^+ - p^+ + q^+)}{p^+ q^+} \left( p^2(k,q) + q^2(k,p) \right). \end{aligned} \tag{18}$$

The one-loop contribution to the $\psi^\dagger(q)A^a(r)\psi(p)$-vertex is

$$
\begin{aligned}
& ig\tilde{T}^a \frac{\overline{(r,p)}}{r^+p^+}\frac{ig^2}{(2\pi)^4}\{3C_f I(q) - 2C_f\,(L(p)+L(q)+L(r))\} + \\
& + ig\tilde{T}^a \frac{ig^2}{(2\pi)^4}\int d^4k \frac{1}{k^2(k-p)^2(k+q)^2}\times \\
& \times \Bigg[Cp^2\left(\frac{\overline{(k,q)}}{p^+(k^+-p^+)} - \frac{\overline{(k,r)}}{p^+r^+} + \frac{k^{+2}+(k^++q^+)^2}{k^+q^+p^+r^+}\overline{(k-p,r)}\right) + \\
& + Cq^2\left(-\frac{\overline{(k,p)}}{p^+(k^++q^+)} - \frac{\overline{(k-p,r)}}{k^+r^+} - \frac{\overline{(r,p)}}{r^+p^+}\right) - \\
& - 2C_f p^2\frac{(k^++q^+)^2}{k^+q^+p^+r^+}\overline{(k-p,r)} + 2C_f q^2\frac{\overline{(k-p,r)}}{k^+r^+} - \\
& - 2C_f r^2\left(\frac{k^++q^+}{p^+q^+r^+}\overline{(k,p)} + \frac{\overline{(k+q,p)}}{p^+r^+}\right)\Bigg].
\end{aligned}
\tag{19}
$$

The $AA$-propagator vanish identically, and the $AAA$-vertex equals the finite $1/k^+$ free integral

$$
\frac{8ig^3}{(2\pi)^4}f^{abc}(C-\tilde{T}_f)\frac{1}{p^+q^+r^+}\int d^4k\frac{(k,p)(k,q)(k-p,r)}{k^2(k-p)^2(k+q)^2} \tag{20}
$$

The non-cancelling terms with $1/k^+$-poles in the integrands are proportional to squares of external momenta. This is reassuring, although the cubic interactions vanish on shell anyway, due to conservation laws. There may be better ways to collect terms. Particularly disturbing is the presence of combinations like $(k,p)/k^+$. When these are integrated, the transverse momenta will no longer appear in the combination (9). Observe that in order to reproduce the correct value for the $\beta$-function, it suffices to regularize the $1/k^+$-poles in such a way that the genuinely cubic integrals are finite.

Finally we would like to say a few words about anomalies. It seems likely that anomalies will destroy the cancellation of $1/k^+$-poles at higher loop orders. On the other hand, one might guess that no new divergent terms appear if the LM prescription is used. We have made an attempt to study the two-loop "anomalous" (i.e. $d^{abc}$-containing) contribution to the fermion-gluon cubic vertex, and found it to be finite. However, we have not managed see what happens with the $1/k^+$-poles. The significance, if any, of such terms thus remains to be determined. Of course, it is possible to compute the standard axial anomaly in this formalism. However, depending on how the computation is organized, there might be some difficulty in keeping track of the surface term, since it is necessary to add diagrams in order to cancel the $1/k^+$-poles. A calculation of this kind, using the PV prescription, was performed in [20]. (Note, however, that these authors failed to apply the light-cone gauge projectors $\eta^{\mu\nu} - (p^\mu n^\nu + p^\nu n^\mu)/n{\cdot}p$ at a certain step, so their discussion of the result is somewhat missleading). In this case there is obviously a manifest cancellation of the $1/k^+$-poles, so the result should be prescription independent. However, in [20] the integrations are done before collecting terms, and an erroneous, quite peculiar, extra term is obtained. This suggests that the calculation in question can be used as a test for prospective regularizations of the PV prescription. A good regularization must not yield any extra terms.

# 4 Discussion

What is then a safe and practical way to compute integrals in the light-cone gauge? At present, the unique finiteness properties of the LM prescription makes it the best buy for most integrals, but one should try to collect terms so that no divergent integrals involving $1/k^+$-poles need to be computed. We are not very optimistic about the possibility to find an algorithm that would allow all spurious poles to be explicitely cancelled. In our opinion, the LM prescription is not very good for the computation of finite quantities, so a regularization procedure that would allow the PV prescription to be used is desirable.

# References


[1] S. Mandelstam: Nucl. Phys. **B213** (1983) 149
[2] G. Leibbrandt: Phys. Rev. **D29** (1984) 1699
[3] R. B. Mann, L. V. Tarasov: Mod. Phys. Lett. **A1** (1986) 525
[4] J. M. Cornwall: Phys. Rev. **D10** (1974) 500
[5] C. Becchi: these proceedings.
[6] O. Piguet: these proceedings.
[7] G. Wanders: these proceedings.
[8] A.M. Polyakov: Mod. Phys. Lett. A **2** (1987) 893
[9] G. 't Hooft: Nucl. Phys. **B75** (1974) 461;
S. J. Brodsky, H.-C. Pauli: Phys. Rev. **D32** (1985) 1993,2001;
A. Hadrianath, J. Vary: Phys. Rev. **D37** (1988) 3010;
St. Glazek: Phys. Rev. **D38** (1988) 3277
[10] G. Leibbrandt, S. L. Nyeo: Phys. Lett. **140B**(1984) 417;
M. Dalbosco: Phys. Lett. **163B** (1985) 181
[11] A. Bassetto, M. Dalbosco, R. Soldati: Phys. Lett. **159B** (1985) 311;
H.C. Lee, M.S. Milgram: Phys. Rev. Lett. **55** (1985) 2122, Nucl. Phys. **B268** (1986) 543
[13] A. Bengtsson, I. Bengtsson, L. Brink: Nucl. Phys. **B227** (1983) 31
[12] S. Weinberg: Phys. Rev. **134** (1964) B882, **138** (1965) B988
[14] A. Bassetto, M. Dalbosco, I. Lazzizzera, R. Soldati: Phys. Rev. **D31** (1985) 2012
[15] O. Piguet, K. Sibold: Nucl. Phys. **B253** (1985) 517;
O. Piguet, M. Schweda, H. Skarke: Phys. Lett. **210B** (1988) 159
[16] A. Andraši, J.C. Taylor: Nucl. Phys. **B310** (1988) 222
[17] L. Brink, O. Lindgren, B.E.W. Nilsson: Nucl. Phys. **B212** (1983) 401
[18] A. Smith: Nucl. Phys. **B261**(1985) 285
[19] J. D. Bjorken, S. D. Drell: Relativistic quantum fields (McGraw Hill, 1965)
[20] D.M. Capper, M.J. Litvak: J. Phys. **A18** (1985) 955

# The Gluon Propagator in Two Dimensional Space-Time

G. Wanders

Institut de Physique Théorique, Université de Lausanne, CH - 1015 Lausanne, Switzerland

**Abstract:** The gluon propagator in $1 + 1$ dimensions with $N$ flavors of massless quarks is obtained from its $1/N$ expansion. The construction is based on an exact identity for the vertex generating functional

In two dimensional gauge theories with massless quarks the fermion determinant $\det \not{D} = \exp iW$ is known explicitely [1] and the gluon sector is controled by an effective action:

$$S = S_{\mathrm{YM}} + W, \tag{1}$$

$S_{\mathrm{YM}} = -(1/2)tr(F_{\mu\nu}F^{\mu\nu})$ being the pure Yang-Mills action. All effects of the quark vacuum polarization on the gluons are described by $W$. Starting from the action (1) D. Cangemi, M. Makowka and I have investigated the light-cone gauge gluon propagator in the limit of a large number $N$ of quark flavors [2]. This report gives an outline of our results.

The questions we try to answer originate in the following observations. In the Schwinger model, $QED_2$ with massless fermions, $W$ is bilinear in the gauge potential $A$. It is a gauge invariant mass term (mass $m = e/\sqrt{\pi}$) and the transverse part of $A$ becomes a physical free massive boson. In $QCD_2$, $W$ contains multilinear terms of all orders, the bilinear term being the same as in $QED_2$ (with $m = g/\sqrt{2\pi}$). One may think that the transverse part of the non abelian gauge potential describes a massive self-interacting boson. We want to inquire whether this simple picture is adequate.

It is easy to see that it may well be incorrect. Apart its massive component the gauge field has also a massless one and in the simplest self-energy graph (Fig. 1) one finds contributions of intermediate states containing both components. They produce a branch point at $k^2 = m^2$ in the gluon selfenergy $\pi(k)$. Therefore the Schwinger mass pole at $k^2 = m^2$ is not an isolated singularity; it coincides with a branch point. As a consequence, the gauge field carries no asymptotic massive gluon states having a well defined mass. To get an insight into the status of the massive component of the gauge potential we look for the singularity structure of the gluon propagator at the Schwinger mass.

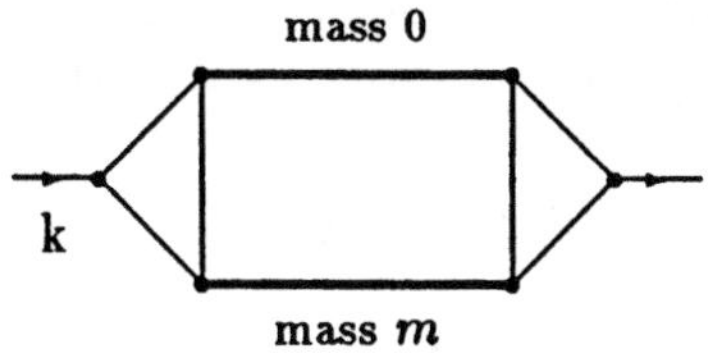


**Fig. 1.** Gluon self-energy diagram with 2 internal gluon lines. The mass 0 contribution to one of the lines together with the mass $m$ contribution of the other one produces a branch point at $k^2 = m^2$.

The functional W[A] is too complicated for an exact treatment; an approximation scheme is needed. The perturbation expansion in the gauge coupling constant $g$ is of no use because we want the Schwinger mass to show up already in zero order. To get an additional parameter we consider a model with $N$ quark flavors: if $W$ is the 1-flavor sum of closed quark loops, it gets replaced by $NW$ in the $N$-flavor case. The following rescaling of coordinates and gauge field:

$$x \rightarrow x' = \sqrt{N}x, \quad A(x) \rightarrow A'(x') = (1/\sqrt{N})A(x) \tag{2}$$

has the effect that $S_{YM}$ too is multiplied by $N$, the 1-flavor $W$ being unchanged. The action (1) is transformed into:

$$S = N(S_{\mathrm{YM}} + W). \tag{3}$$

If $m$ is the 1-flavor Schwinger mass, the $N$-flavor mass is $\sqrt{N}m$. The length scale is changed in (2) in such a way that the numerical value of the $N$-flavor Schwinger mass is reduced to $m$. The form (3) of the effective action allows a $1/N$ expansion if $N$ is large. In this limit, the gluon self-interaction becomes weak with respect to the mass generating mechanism. This does not change the coincidence of the Schwinger mass pole with a branch point and the question raised before remains meaningful.

If we want the implications of (3) beyond first order in $1/N$ we need an efficient tool. We have discovered one in the form of an identity for the gluon vertices generating functional $\Gamma[A]$ in the light-cone gauge $A_- = 0$ ($A_\pm = (1/\sqrt{2})(A_0 \pm A_1)$). To get this identity we perform the following change of variable in the functional integral giving the Green's functions generating functional $Z$:

$$A_+ \rightarrow A_+ + D_+\phi, \quad A_- \rightarrow A_-(=0). \tag{4}$$

$\phi$ is infinitesimal and our identity expresses the insensitivity of $Z$ to this change of variable. Remarkably the explicit form of $W$ is not required. The variation of $S$ under (4) involves $D_+(\delta W/\delta A_+)$. One knows that:

$$D_+ \frac{\delta W}{\delta A_+} = \frac{g}{2} D_+ j^+, \tag{5}$$

where $j$ is the current induced in the vacuum by the external potential $A$. This current is conserved, $D_\mu j^\mu = 0$ and its axial counterpart obeys the anomaly equation $D_\mu j_5^\mu = (g/\pi)F$ ($F \doteq F_{01} = \partial_- A_+$ in the light-cone gauge). As $j_5^\mu = \epsilon^{\mu\nu} j_\nu$, $D_+ j^+$ is proportional to the field strength and:

$$D_+ \frac{\delta W}{\delta A_+} = -\frac{1}{2} m^2 F. \tag{5}$$

In spite of the nonlinearity and nonlocality of $W[A]$, the axial anomaly implies that its variation under (4) is both linear and local in $A_+$.

Standard manipulations lead to the following identity for $\Gamma[A]$, the gluon vertices generating functional:

$$D_+^{ab} \frac{\delta \Gamma}{\delta A_+^b(x)} = -\left([D^\mu D_\mu + m^2]F\right)^a$$

$$-\frac{ig}{N} f^{abc} \partial_-^x \lim_{y \to x} \left(\partial_-^y \Delta^{bc}(x, y, A)\right). \tag{6}$$

$\Delta$ is the (++) component of the gluon propagator in the external potential $A$. Although we are in the light-cone gauge the first term in the right-hand side has been written in covariant form in order to make clear that it involves a covariant Klein-Gordon operator with mass equal to the Schwinger mass.

The identity (6) is not the full quantum equation of motion which would be obtained from an arbitrary translation of $A_+$ in (4). Equation (6) can be obtained from that equation of motion by derivation; this explains why $D_+$ acts on $\delta\Gamma/\delta A_+$ in the left-hand side. To get the gluon vertices from (6) we need the inverse of $\partial_+$; it is uniquely defined once causality is imposed. To see this, observe that $1/k_+ = 2k_-/k^2$ because $k^2 = 2k_- k_+$. Causality requires $1/k^2$ to be interpreted as $1/(k^2 + i\epsilon)$. This agrees with the Leibbrandt-Mandelstam prescription [3] which defines $1/(n \cdot k)$ as $1/[(n \cdot k) + (n^* \cdot k)i\epsilon]$ ($n^2 = n^{*2} = 0$, $(n \cdot n^*) \neq 0$). In 2 space time dimensions $n^*$ is unique; if $n = (1, -1)/\sqrt{2}$, $n^* = (1, 1)/\sqrt{2}$, and:

$$\frac{1}{(n \cdot k) + (n^* \cdot k)i\epsilon} = \frac{1}{k_+ + k_- i\epsilon} = \frac{2k_-}{k^2 + i\epsilon}. \tag{7}$$

With $1/\partial_+$ defined unambiguously, the identity (6) is a substitute of the quantum equation of motion. It leads to a Dyson-Schwinger type equation for the two-point gluon vertex $\Gamma_2$ which can be solved by means of a $1/N$ expansion. We have obtained detailed information on the first and second order terms. They are given by sums of convergent integrals. Their infrared convergence is secured by the Schwinger mass which acts as a regulator. Our results indicate that $\Gamma_2$ has the following structure:

$$\Gamma_2^{ab}(k) = \frac{\delta^{ab}}{2\pi} \frac{(k^+)^2}{k^2 + i\epsilon} \pi(k^2), \tag{8}$$

$$\pi(k^2) = \sum_{p=0}^{\infty} \frac{1}{N^p} \pi_p(k^2). \tag{9}$$

If we work at fixed coupling constant $g$, there is a finite renormalization of the Schwinger mass: $m^2 = g^2/2\pi \to M^2 = m^2 + \delta m^2$ where $\delta m^2$ is a finite function of $m^2$ given by its $1/N$ expansion. The Schwinger mass pole is shifted from $m^2$ to $M^2$ whereas all $\pi_p'$s have a singularity at $k^2 = m^2$. It is clear from Fig. 1 that a change in the mass has to be accompagnied by a corresponding shift of the branch point. Therefore the sum $\pi(k^2)$ of the series (9) has to exhibit a branch point at $k^2 = M^2$ instead of $m^2$ and one can try to identify the higher order graphs which are responsible for this displacement. This is not too easy and it is better to work at fixed renormalized mass $M$. In this approach an order by order consistency check replaces non practicable partial summations. It is the bare Schwinger mass which is now a function of the renormalized mass:

$$m^2 = \sum_{p=0}^{\infty} \frac{1}{N^p} m_p^2 \tag{10}$$

and the $m_p$'s have to be adjusted in such a way that each $\pi_p$ vanishes at $M^2$:

$$\pi_p(M^2) = 0 \quad p = 0, 1, 2, \ldots$$

In zero order $\pi_0(k^2) = k^2 - m_0^2$ and $m_0$ has to be set equal to $M$. A consequence of this is that each $\pi_p$, $p \geq 1$, has now a branch point at $k^2 = M^2$, as it has to be. In first and second order this is in fact the lowest branch point. Surprisingly the gluon self-energy seems to be regular at $k^2 = 0$ in the light-cone gauge.

As $\pi_p$ is complex for $k^2 \geq M^2$ it is not guaranteed that (11) can be fulfilled just by tuning the real parameter $m_p^2$. The real part of $\pi_p$ has to be bounded at $M^2$, independently of the choice of $m_p^2$, and its imaginary part has to vanish at this point. Whether this is true or not depends on the singularity structure at $M^2$ and has to be checked order by order.

The function $\pi_1(k^2)$ has been computed explicitely and its singular part at $M^2$ is:

$$\pi_1^{\text{sing}}(k^2) = 2C(k^2 - M^2)\left(\log \frac{M^2 - k^2 - i\epsilon}{M^2} + \ldots\right) \tag{12}$$

the dots represent less singular terms and $C$ is the coefficient in the second Casimir invariant: $f^{acd} f^{bcd} = C\delta^{ab}$. We see that $\operatorname{Im} \pi_1(M^2) = 0$ and condition (11) can be fulfilled for $p = 1$.

Using the Cutkosky cutting rule we have evaluated the imaginary part of $\pi_2$ in the neighbourhood of $M^2$. This is possible because, according to (7) we have no noncovariant denominators in our integrals. We find:

$$\pi_2^{\text{sing}}(k^2) = 2C^2(k^2 - M^2)\left[\left(\log \frac{M^2 - k^2 - i\epsilon}{M^2}\right)^2 + \ldots\right]. \tag{13}$$

Again, we see that $\pi_2(k^2)$ can be brought to vanish. Our first and second order results are consistent with a pole and a branch point both located at $M^2$, the square of the renormalized Schwinger mass.

If we assume that this result holds for each order, the $(++)$ component $\Delta$ of the gluon propagator has the following form:

$$\Delta^{ab}(k) = \frac{2\delta^{ab}}{\pi} \frac{(k^-)^2}{k^2 + i\epsilon} \frac{1}{k^2 - M^2 + i\epsilon} G(k^2), \tag{14}$$

where $G(k^2)$ has its first branch point at $M^2$. The form of (12) and (13) suggests that the leading logarithmic singularities exponentiate. We are thus led to speculate that the dominant singularity of the propagator at $M^2$ is:

$$\Delta^{ab}_{\rm sing}(k) = -\frac{2\delta^{ab}}{\pi}\,\frac{1}{M^2(1-\alpha)}\,\frac{1}{(M^2-k^2-i\epsilon)^{1+\alpha}} \tag{15}$$

with $\alpha = 2C/N$. The pole and the branch point have merged into an algebraic singularity and the gauge potential has no asymptotic massive 1-particle states. The picture of the gluon as a particle with well defined mass is wrong.

In 2 space-time dimensions a field describing a particle of mass $M$ has a propagator which behaves as $t^{-1/2}\exp(-iMt)$ for large times $t$. If (15) holds the factor $t^{-1/2}$ is replaced by $t^{-1/2(1-2\alpha)}$, the decay of the propagator is slowed down. This might seem suspicious. However one notices that the imaginary part of $\Delta_{\rm sing}$ is positive; with our conventions this means that it is a state with negative $(\text{norm})^2$ which is propagated. Our results suggest that the gluon, which is a massive physical boson in zero order is turned into a non physical object by its self-interaction.

The first order gluon selfenergy has been computed in a family of gauges interpolating the noncovariant light-cone gauge and the covariant Landau gauge [4]. In these gauges $\pi_1$ has a branch point at $k^2 = 0$. In spite of that, $\mathrm{Im}\,\pi_1(M^2) = 0$ and one has the coalescence of a pole and a branch point at $k^2 = M^2$ found in the light-cone gauge. Furthermore the first order bare mass correction $m_1^2$ and the exponent $\alpha$ (which can be read off from $\pi_1$) are the same for all gauges of the family. They seem to be gauge independent parameters specifying physical properties of the gauge field.

## References

[1] A.M. Polyakov, P.B. Wiegmann: Phys. Lett. B **131** 121 (1983); M. Makowka, G. Wanders: Helv. Phys. Acta **59** 1366 (1986)

[2] D. Cangemi, M. Makowka, G. Wanders: University of Lausanne preprint 1989, to appear in Nucl. Phys. B

[3] S. Mandelstam: Nucl. Phys. B **213** 149 (1983); G. Leibbrandt: Rev. Mod. Phys. **59** 1067 (1987)

[4] M. Makowka: private communication

# Introduction

The treatment of QCD at finite temperature quite naturally leads to a preferred Lorentz frame. At the time of the workshop this topic turned out to be the most important "practical" application of noncovariant gauges. Gauge dependent results for the plasma damping constant, which should be an observable quantity, had been the source of a sometimes heated international controversy. However, at the workshop the picture emerged that general proofs of gauge-independence (Kobes-Kunstatter) demand a nonperturbative treatment. Although perturbative methods are valuable for "theoretical" comparisons with lattice calculations (Kajantie), even an "on-shell" perturbative calculation (Rebhan et al.) cannot be expected to and does not yield a gauge-independent result. Further special contributions (d'Emilio, James) confirmed the difficulties of perturbative approaches. A first nonperturbative "prototype" calculation was presented (Rebhan et al).

# IV

# QCD at Finite Temperature

# Hot Gluon Matter and Physical Gauges

K.Kajantie

Department of Theoretical Physics, University of Helsinki,
Siltavuorenpenger 20 C, 00170 Helsinki, Finland,
Bitnet address: KAJANTIE at FINUHCB

## 1 Introduction

QCD at $T = 0$ is an extremely successful theory, convincingly tested by experiment. Factorisation theorems permit one to treat separately the perturbatively calculable hard and incalculable soft parts. The former are computed theoretically, the latter are taken from experiments other than the one being analysed. The situation with QCD at finite $T$ [1] is much less satisfactory. For the first, we have so far no experiments to guide us. For the second, perturbative and nonperturbative phenomena seem to be hard to disentangle. I shall discuss this briefly in Section 2.

The validity of perturbative methods at finite $T$ is ultimately an experimental question. Lacking genuine experiments, in Section 2 I shall attempt to compare one perturbative calculation with "experimental" data from a computer experiment [2]. The quantity studied is the correlation function $\langle L(r)L(0)\rangle$ of two Wilson lines, which is related to the free energy of a system with two heavy test quarks. Although this quantity is not perturbatively calculable to higher orders [3], the agreement of the lowest order calculation and "experimental" data is surprisingly good, especially in view of the uncertainties in the theoretical calculation. This observation may be useful in convincing one of the usefulness of perturbative calculations in thermal QCD.

Encouraged by the above observation, in Section 3 I shall discuss 1-loop calculations in three different physical gauges, the temporal axial gauge TAG:

$$A_0^a = 0, \tag{1.1}$$

the Coulomb gauge CG:

$$\partial_i A_i^a = 0, \tag{1.2}$$

and the covariant background gauge BG with the gauge fixing term

$$\mathcal{L}_{\text{gf}} = -\frac{1}{2\xi}[D^\mu(A)Q_\mu]^2, \tag{1.3}$$

where $A$ is the background field and $Q$ the quantum field propagating only in internal loops. In the standard formulation of finite $T$ field theory, based on including bosonic field configurations periodic in imaginary time $t$, it is actually not possible to impose the condition (1.1) for all $t$. It must be violated at least on one infinitesimal interval $t, t+\delta t$. As a consequence, even at the 1-loop level, one needs a prescription to handle a singularity appearing. With a correct prescription one obtains results in agreement with those in other gauges, if gauge independent quantities are studied. For gauge dependent quantities the results clearly will vary. As an example I shall discuss the question of damping of coloured plasma oscillations. Ideally one would need an explicitly gauge invariant quantity for determining the parameters (mass, width) of plasma oscillations and I shall at the end suggest one for this purpose.

## 2 Finite temperature SU(3) gauge theory

The formulation of finite $T$ field theory involves essentially the compactification of the imaginary time direction to the interval $0 < t < 1/T \equiv \beta$ and the inclusion of periodic field configurations:

$$A_\mu(t,\mathbf{x}) \equiv A_\mu^a(t,\mathbf{x})T^a = A_\mu(t+\beta,\mathbf{x}), \quad \beta = 1/T. \tag{2.1}$$

A particularly elegant and unambiguous formulation of the theory is the lattice one, which replaces $A_\mu(n)$ at the site $n = (t,\mathbf{x})$ by the matrix ($a$ is the lattice constant)

$$U_\mu(n) = \exp(igaA_\mu(n)). \tag{2.2}$$

This is described in great detail in [4].

For a finite $T$ system the rest frame of the system singles out the four-vector $n^\mu = (1,0)$ and this suggests that the temporal axial gauge $n \cdot A = A_0 = 0$ could play a special role. Indeed, $A_0$ does play a special role but the condition $A_0 = 0$ everywhere is incompatible with the periodicity (2.1). This is particularly clear in the lattice formulation. Consider a string

$$L(\mathbf{x}) = U_0(1,\mathbf{x})\ldots U_0(N_t,\mathbf{x}) = P\exp[ig\int_0^\beta dtA_0(t,\mathbf{x})] \tag{2.3}$$

of matrices spanning all the temporal direction at a fixed spatial coordinate $\mathbf{x}$. In view of Eq. (2.2), transforming to TAG would mean gauge transforming all the matrices of this string to unit matrices. However, because of the periodicity (2.1) $L(\mathbf{x})$ is actually a closed loop and all the matrices of a closed loop cannot be transformed to unit matrices because the trace of a loop is invariant and the trace of a unit matrix would be $N_c$. The best one can do is thus to gauge transform all the matrices of the string to unit matrices (or $A_0 = 0$) on all temporal links but one which carries the tr$L$ of the original configuration. One thus almost gets to TAG, the topological condition prevents it only at one infinitesimal line element of the imaginary time direction.

Although $L(\mathbf{x})$ is a closed loop there is a special set of gauge transformations under which its trace is not invariant. To maintain the periodicity (2.1) the gauge transformations have to be periodic but only modulo an element $z = \exp(2\pi ik/3)$ , $k = 1,2,3$ of the center Z(3) of the gauge group. Under gauge transformations along the temporal line at $\mathbf{x}$ then ($g_1$ is one of the gauge transformations along this line)

$$\mathrm{tr}L(\mathbf{x}) \to \mathrm{tr} z g_1 L(\mathbf{x}) g_1^{-1} = z\,\mathrm{tr}L(\mathbf{x}). \tag{2.4}$$

Instead of transforming $A_0(t,\mathbf{x}) \to 0$ everywhere but on one surface $t = t(\mathbf{x})$, one can distribute the deviation of $A_0$ from 0 evenly over all values of $t$. Then one has transformed the field configuration to the static TAG (STAG), $A_0(t,\mathbf{x}) = A_0(\mathbf{x})$. In this gauge

$$L(\mathbf{x}) = \exp[ig\beta A_0(\mathbf{x})] \tag{2.5}$$

and, for instance, in SU(2),

$$\mathrm{tr}L(\mathbf{x}) = 2\cos(\tfrac{1}{2}g\beta|A_0|),\ |A_0|^2 = \sum_1^3 A_0^a(\mathbf{x})A_0^a(\mathbf{x}). \tag{2.6}$$

A very special static configuration is one in which $A_0(\mathbf{x}) = a_0 =$ constant. At $T = 0$ a configuration of this type can be trivially gauge transformed away. As indicated above this is not possible at finite $T$.

The quantity $\langle \mathrm{tr}L(\mathbf{x})\rangle$ is an order parameter for pure SU(3) gauge theory and jumps at $\beta \equiv 6/g^2 = \beta_c$ to some value close to one of the three values $zN_c$ but so that the absolute value is below the maximum value $N_c = 3$. This is in analogy with spin models; only at $T = 0$ (corresponding here to $T = \infty$) the spins are completely parallel and the expectation value of the total spin maximal. In view of its relation to $A_0$ this jump can also be described in terms of $A_0$, but in a gauge dependent way. E.g., in SU(2) Eq.(2.6) explicitly relates $\langle \mathrm{tr}L(\mathbf{x})\rangle$ and $\langle \cos|\tfrac{1}{2}g\beta A_0|\rangle$. As long as $\langle \mathrm{tr}L(\mathbf{x})\rangle < N_c$ the expectation value of $\langle |A_0|\rangle > 0$, but the magnitude of this expectation value is gauge dependent. This has led to speculations that the expectation value of the quantum operator develops a nonzero value, $< A_0^a >= a_0^a$. On the lattice this has been studied in the static temporal axial gauge [5] and in the Landau gauge [6].

If there is a condensate, usual high $T$ perturbation theory, which is based on expanding around $A_0 = 0$, has problems. In more physical terms, the relevant scales of the plasma are $1/T$ (distance between particles), $1/gT$ (electric screening length) and $1/g^2T$ (magnetic screening length). Since for a QCD plasma $g$ is never small (even at the GUT scale $g \approx 0.6$), the scales are degenerate and there is no small plasma parameter to expand with. However, the validity of perturbation theory is ultimately an empirical question. To test it, I shall next compare a theoretical perturbative prediction with "experiment". In spite of the doubts concerning calculability in high orders, the lowest order prediction is seen to fit the "data" surprisingly well.

## 3 Perturbation theory and finite $T$

At $T = 0$ we know that perturbation theory works simply since we have experimental data to verify the predictions with. Of course, soft or large-distance physics can never be entirely decoupled from the verification (e.g., a quark or gluon jet will fragment into hadrons), and higher order corrections seem to be large so that the accuracy of the comparison is at best on the 10% level and does not even remotely approach that one is used to in QED. For $T > 0$ there is so far no experimental data. However, to the extent QCD is the correct theory and to the extent lattice Monte Carlo computations are able to produce correct numbers, it should be as good to compare with "experimental" data obtained with lattice computations.

The quantity I shall consider is the correlator of two (traces of) Wilson lines (2.3), computed perturbatively in [7-8]. The result is

$$\begin{aligned}\frac{<\mathrm{tr}L(r)\mathrm{tr}L(0)>}{<\mathrm{tr}L>^2} &= \frac{1}{9}e^{-V_1/T}+\frac{8}{9}e^{-V_8/T} \\ &\equiv e^{-V/T}=1+\frac{V_1^2}{16T^2}+\dots \\ &=1+\frac{1}{(12\pi Tr)^2}g^4e^{-2gTr}+\dots,\end{aligned} \tag{3.1}$$

where $V_1$ and $V_8$ are the singlet and octet potentials. In perturbation theory the lowest order contribution comes from the exchange of two gluons so that $V_1 = -8V_8$ and the correlator contains the square of a Yukawa potential. This arises from the Fourier transform of the dressed propagator:

$$D_{00}(0,r) \sim \int d^3p \frac{e^{i\mathbf{p}\cdot\mathbf{x}}}{\mathbf{p}^2+\Pi_{00}(0,\mathbf{p})}, \tag{3.2}$$

which has been regulated by the static limit of the gluon polarisation tensor with $\Pi_{00}(0,0) = g^2T^2$ (see Eq. (4.2)). In this sense this result is not entirely perturbative; it contains a dressed propagator.

The "experimental" data used to test (3.1) has been obtained by the Bielefeld group [2]. They measured the correlation for pure SU(3) gauge fields on a $4\cdot 12^3$ lattice at three different temperatures, $T/T_c = 1.2, 2.3, 14$ in systems of spatial sizes $(0.2\ \mathrm{fm})^3$, $(0.1\ \mathrm{fm})^3$, $(0.013\ \mathrm{fm})^3$, respectively. Distance is naturally measured in units of lattice spacing $a$. Now $N_t = 4$ so that $a = 1/4T$ and

$$\frac{r}{a} = 4Tr.$$

Since $N_s = 12$, the correlator can be measured for $r/a = 1, \dots, 6$.

To give an idea of the raw data, for $T = 14T_c$ one finds that $< \mathrm{tr}L > = 1.44292$ and that the l.h.s. of (3.1) is 1.0217, 1.0040, ... for $r/a = 1, 2$ .... . Thus indeed $V_1/T$ is small and to get a meaningful result a very good statistical accuracy is needed.

The comparison of theory and "experiment" is shown in Fig. 1. The interpretation of this comparison has been discussed in [2] with the rather reserved conclusion that "perturbative relations are at best recovered for very high temperatures". I find the agreement between lowest order perturbation theory and "experiment" surprisingly good and at least as good as one is accustomed to in $T = 0$ QCD. Even for $T = 0$ it has often happened that simple predictions often work well in spite of obvious sources of corrections, scaling sets in early, higher twists do not matter, etc.

In view of the good agreement observed in Fig. 1 it would be very interesting to continue the "experiments" at higher $T$ in order to verify whether the observed rapid decrease of the correlator at fixed $Tr$ as a function of $T$ slows down and turns to a slower decrease caused by a $T$ dependence of $g$.

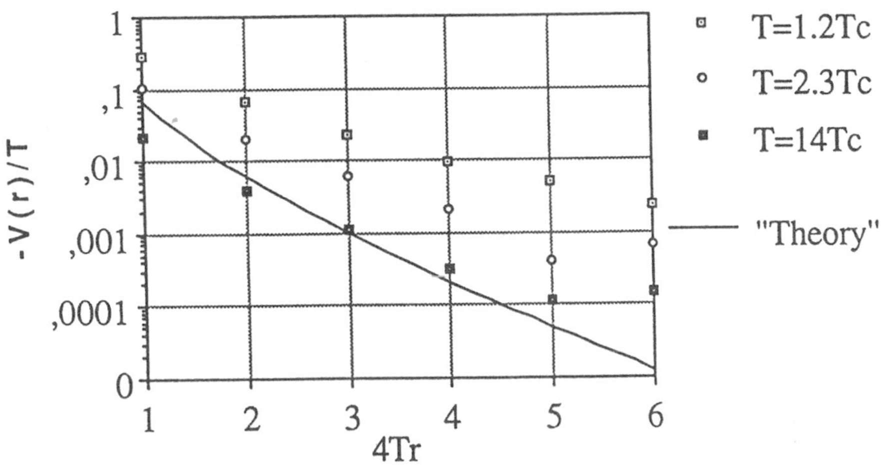


Fig. 1. Data [2] for the correlator in Eq. (3.1) compared with the perturbative prediction for $-V(r)/T$ in the RHS of Eq. (3.1). The prediction is plotted for $g^2 = 4$ ($\alpha_s = 0.3$).

## 4 One-loop results

In this section, I shall present and discuss some results for the gluon self energy in the physical gauges (1.1-3), obtained in [9-13]. The reasons for using these gauges are for TAG that for a thermal system it is singled out by the rest frame 4-vector and that a simple Hamiltonian formalism is available for linear response studies, for CG that it is the gauge of choice for the study of intractions between charges and for BG that it is particularly elegant from a theoretical point of view. For simplicity, the results will mainly be quoted in the two limits:

$$\begin{aligned} \text{Static}: &\quad \Pi(0, \mathbf{k}) \\ \text{Plasmon}: &\quad \Pi(k_0, \mathbf{0}). \end{aligned}$$

At finite $T$ one must separate the time-time, space-time and space-space components of the gluon polarisation tensor:

$$\Pi_{\mu\nu}(k_0, \mathbf{k}) = \begin{pmatrix} \Pi_{00} & \hat{k}_j \Pi_{0j} \hat{k}_i \\ \hat{k}_j \Pi_{0j} \hat{k}_i & \Pi_T(\delta_{ij} - \hat{k}_i \hat{k}_j) + \Pi_L \hat{k}_i \hat{k}_j \end{pmatrix}. \tag{4.1}$$

If transversality $k^\mu \Pi_{\mu\nu} = 0$ holds, which is the case only in TAG or BG at any $\xi$, only two of the four scalars in (4.1) are independent. These are most naturally taken to be $\Pi_T$ and $\Pi_L$. Then, for instance,

$$\begin{aligned} k_0^2 \Pi_{00} &= \mathbf{k}^2 \Pi_L, \\ \Pi_{ii} &= 2\Pi_T + \Pi_L. \end{aligned} \tag{4.2}$$

One-loop expressions are derived by using standard Feynman rules. The propagator in TAG is:

$$D_{\mu\nu}(k_0, \mathbf{k}) = \begin{pmatrix} 0 & 0 \\ 0 & (\delta_{ij} - \hat{k}_i \hat{k}_j)/k^2 + \hat{k}_i \hat{k}_j / k_0^2 \end{pmatrix} \tag{4.3}$$

Here the regularisation of the longitudinal (L) $1/k_0^2$ pole is the central question in the application of TAG also at finite $T$, where $k_0 = 2\pi i T n$; the transverse (T) pole is the usual one. In the CG one, instead, has a momentum divergence:

$$D_{\mu\nu}(k_0, \mathbf{k}) = \begin{pmatrix} 1/\mathbf{k}^2 & 0 \\ 0 & (\delta_{ij} - \hat{k}_i \hat{k}_j)/k^2 \end{pmatrix}, \tag{4.4}$$

which, e.g., gives rise to the non-transversality of $\Pi_{\mu\nu}$ in this gauge. Both singularities are regulated in various interpolating gauges. For instance, in the STAG-CG-Feynman gauge interpolating gauge [11,14] ($a^2$ is a parameter)

$$D_{\mu\nu}(k_0, \mathbf{k}) = \begin{pmatrix} 1/(k_0^2/a^2 + \mathbf{k}^2) & 0 \\ 0 & (\delta_{ij} - \hat{k}_i \hat{k}_j)/k^2 + \hat{k}_i \hat{k}_j / (k_0^2 + a^2 \mathbf{k}^2) \end{pmatrix}, \tag{4.5}$$

but then one looses the very simple structure of the propagator in (4.3). In the covariant BG with gauge parameter $\xi$ the propagator is the standard covariant gauge propagator.

In TAG the 1-loop result for $\Pi_{00}(k_0, \mathbf{k})$ is

$$\Pi_{00}(k_0, \mathbf{k})/g^2 N_c = \tfrac{1}{2} \int \frac{d^3p}{(2\pi)^3} T \sum_n \{ \frac{4}{p^2} + \frac{2}{p_0^2} - (p_0 - q_0)^2 [\frac{1+c^2}{p^2 q^2} + \frac{1-c^2}{p^2 q_0^2} + \frac{1-c^2}{q^2 p_0^2} + \frac{c^2}{p_0^2 q_0^2}]\}, \tag{4.6}$$

where $k + p + q = 0$ and $c \equiv \hat{p} \cdot \hat{q}$. The T and L poles of the loop diagram with a 4-gluon vertex and the TT, TL and LL poles of the loop diagram with two 3-vertices are clearly distinguishable in (4.6).

The prescription to treat the $n = 0$ term in the $1/p_0^2, p_0 = 2\pi i T n$ terms applied in [9-13] is to associate the pole at $p_0 = 0$ with the vacuum part. Thus L and LL terms in (4.6) (the 2nd and last terms in the summand) do not at all contribute to the finite $T$ part and and the L poles of the TL terms only contribute as the residues of the usual poles at $p_0 = \pm|\mathbf{p}|$. Formally, the prescription can be motivated by noting that there are two small parameters in the problem:

$$\frac{1}{p^2 + i\epsilon}, \quad \frac{1}{p_0^2} \equiv \tfrac{1}{2} [\frac{1}{(p_0 + i\eta)^2} + \frac{1}{(p_0 - i\eta)^2}] \tag{4.7}$$

and taking $\epsilon \to 0$ first, then $\eta \to 0$. The pole at $p_0 = 0$ is then effectively associated with the vacuum $T = 0$ part. In this way one at least obtains all the results also obtained in the covariant gauge, various contradictory results [15-16] are based on neglecting the ordering of the limits in (4.7).

In the static limit $k_0 = 0$ one obtains both in TAG and CG

$$\Pi_{00}(0,\mathbf{k})/g^2N_c = -\frac{1}{3}T^2 + \frac{1}{4}kT + \frac{11}{48\pi^2}k^2\ln\frac{k^2}{T^2} + \ldots; \tag{4.8}$$

in BG with any $\xi$ the coefficient of the $kT$ term is 1/2 instead of 1/4. The fact that one obtains the same coefficients for the $T^2$ and $kT$ terms in TAG and CG (and also in the usual covariant gauge) can be understood by the fact that these terms also give the $g^3$ and $g^4\ln g$ terms of the perturbative expansion of the pressure [17]. In BG the polarisation tensor has the background field on the external legs while the quantum field propagates in the loop. The pressure does not have diagrams of this type and thus the coefficient of the $kT$ term can be different. Further, the coefficient of the $k^2\ln k$ is in these physical gauges determined by the QCD beta function.

Similarly, for $\Pi_{ii}(0,\mathbf{k})$ one obtains

$$\begin{aligned}\Pi_{ii}(0,\mathbf{k})/g^2N_c &= -\frac{5}{8}kT - \frac{11}{24\pi^2}k^2\ln\frac{k^2}{T^2} + \ldots \quad \text{TAG},\\ &= -\frac{21+6\xi+\xi^2}{32}kT - \frac{11}{24\pi^2}k^2\ln\frac{k^2}{T^2} + \ldots \quad \text{BG, any } \xi.\end{aligned} \tag{4.9}$$

The simple form of $\Pi_{ii}$ in TAG arises from a rather complicated expression through the application of the prescription described above [11] and to me its correctness is shown by the fact that one indeed also gets the beta function coefficient in the $k^2\ln k$ term here.

According to (4.9) the static spatial propagator has the form

$$D_{ij}(0,\mathbf{k}) = \frac{-\delta_{ij} + \hat{k}_i\hat{k}_j}{\mathbf{k}^2 - ag^2N_cT|\mathbf{k}| + g^2O(\mathbf{k}^2\ln\mathbf{k}^2)},$$

where $a$ is a gauge dependent constant. The propagator thus has a singularity at $k < g^2T$ which means that the theoretical treatment is insufficient at these scales. If one instead of $< A_i(x)A_j(y) >$ studies the correlator $< B_i(x)B_j(y) >$ of colour magnetic fields in TAG, the constant $a$ actually vanishes as a result of complicated cancellations [13]. There could exist some so far unknown dynamical reason for this cancellation.

Consider then the plasmon limit $k \ll k_0 \ll T$. In this limit $\Pi_{ij}(k_0,0) = \Pi_T\delta_{ij}$ so that $\Pi_T = \Pi_L$,

$$\Pi_{00}(k_0,\mathbf{k}\to 0) = \Pi_L(k_0,0)\frac{\mathbf{k}^2}{k_0^2} = \Pi_{ii}(k_0,0)\frac{\mathbf{k}^2}{3k_0^2},$$

and

$$\begin{aligned}\Pi_L(k_0,0)/g^2N_c &= \frac{1}{3\pi^2}\int_0^\infty dp\, p\, n(p)\left[2 - \frac{k_0^2}{4p^2-k_0^2} + \frac{k_0^2(3p^2-k_0^2)}{(p^2-k_0^2)^2}\right] \quad \text{TAG}\\ &= \frac{1}{9}T^2 + (-\frac{i}{12\pi} + c)\omega T + O(\omega^2\ln\omega) \quad \text{TAG}\\ &= \frac{1}{9}T^2 + (\frac{[-1+12+(1-\xi)^2/4]i}{12\pi} + c')\omega T + \ldots \quad \text{BG}\\ &= \frac{T\omega^2}{16k} + \text{TAG}, \qquad \text{CG}\end{aligned}$$

where $c, c'$ are real constants, $n(p) = 1/[\exp(p/T) - 1]$ and the analytic continuation to real frequency is performed by writing

$$k_0 = \omega + i\epsilon.$$

In the expression for TAG the second term in the bracket is the TT and the third the TL term; with the prescription above only the TT term contributes to the imaginary part giving a positive damping constant or stable plasma oscillations. In BG the imaginary part is related to the beta function coefficient and one obtains unstable plasma oscillations [18]. In CG one actually obtains a term diverging when $\mathbf{k} \to 0$, due to the momentum divergence in $D_{00}$. This term is also the root of the violation of the transversality condition $k^\mu \Pi_{\mu\nu} = 0$ in CG. When considering plasma oscillations it is eliminated by considering the colour electric field correlator instead of the propagator [10].

Above as well as in calculations in other gauges one obtains the same result for the plasmon mass

$$m_{\rm pl} = \frac{1}{\sqrt{3}} gT\,,$$

but a gauge dependent result for the damping constant. This is not surprising, as the quantity one starts from is not gauge independent.

An explicitly gauge invariant way of defining the screening mass is to study the correlations in spatial distance (3.1) of the static gauge invariant quantity tr$L(\mathbf{x})$. I suggest that a similarly gauge invariant determination of the properties of the plasmon could be achieved by studying the correlations in imaginary time, summed over $\mathbf{x}$ to obtain the $\mathbf{k} = 0$ component, of the components of the energy momentum tensor. This operator is obviously a singlet and thus couples to a pair of plasmons. It also couples singly to singlet glueball states. In the spectral representation of, say, the imaginary time $\tau, 0 < \tau < \beta$ correlator of the component $T_{12}$:

$$< T_{12}(\tau) T_{12}(0) >= \int_{-\infty}^{\infty} \frac{d\omega}{\pi} \rho(\omega) \frac{e^{-\tau\omega}}{1 - e^{-\beta\omega}},$$

the spectral function $\rho(\omega)$,

$$\rho(-\omega) = -\rho(\omega),$$

thus for positive $\omega$ has a discontinuity starting at $2m_{\rm pl}$, in addition to possible poles corresponding to glueball states. This correlator has already been studied on the lattice [19] (of size $4 * 8^3$, near $T = T_c$) for the determination of the viscosity of QCD matter, which is dependent of the behaviour of $\rho(\omega)$ near $\omega = 0$. Separating a cut starting at $2m_{\rm pl}$ ($\approx 2gT/\sqrt{3}$ in perturbation theory) from fixed poles at the glueball mass scale of 2 GeV will clearly require much more lattice work. The pair of coloured oscillations generated by the energy momentum tensor are clearly the true stable oscillations, but the lattice study of their damping constant clearly can succeed only after one has at least some handle on the mass.

## References

[1] Joseph I. Kapusta, "Finite-temperature field theory", Cambridge University Press, 1989.
[2] N. Attig, F. Karsch, B. Petersson, H. Satz and M.Wolff, Phys. Lett. **209B** (1988) 65.
[3] S. Nadkarni, Phys. Rev. **D34** (1986) 3904.
[4] F. Karsch, "Simulating the quark-gluon plasma on the lattice", CERN Preprint CERN-TH-5498/89, to appear in a review monograph on Quark-gluon plasma, edited by R. C. Hwa, World Scientific.
[5] J. Polonyi and H. W. Wyld, Illinois Report ILL-(TH)-85-23 (1985), unpublished.
[6] J. E. Mandula and M. Ogilvie, Phys. Lett. **201B** (1988) 117.
[7] L. D. McLerran and B. Svetitsky, Phys. Rev. **D24** (1981) 450.
[8] S. Nadkarni, Phys. Rev. **D33** (1986) 3738.
[9] K. Kajantie and J. Kapusta, Ann. Phys. **160** (1985) 477.
[10] U. Heinz, K. Kajantie and T. Toimela, Phys. Lett. 183B (1987) 96.
[11] U. Heinz, K. Kajantie and T. Toimela, Ann. Phys. **176** (1987) 218.
[12] H.-Th. Elze, U. Heinz, K. Kajantie and T. Toimela, Z. Phys. **C37** (1988) 305.
[13] H.-Th. Elze, U. Heinz, K. Kajantie and T. Toimela, Z. Phys. **C37** (1988) 601.
[14] R. Kobes, G. Kunstatter and K. W. Mak, Phys. Lett. **223B** (1989) 433.
[15] M. E. Carrington, T. H. Hansson, H. Yamagishi and I. Zahed, Ann. Phys. **190** (1989) 373.
[16] M. E. Carrington, H. Yamagishi and I. Zahed, Stony Brook Preprint, 1988.
[17] T. Toimela, Phys. Lett. **124B** (1983) 407.
[18] T. H. Hansson and I. Zahed, Nucl. Phys. **B292** (1987) 725.
[19] F. Karsch and H. W. Wyld, Phys. Rev. **D35** (1987) 2518.

# Gluon Response Functions in Noncovariant Gauges [1]

R. Kobes and G. Kunstatter

Physics Department,University of Winnipeg,
Winnipeg, Manitoba, CANADA R3B 2E9

**Abstract:** Nonperturbative dispersion relations for the physical gluon modes are derived from the gluon propagator in a wide class of gauges. A proof is then presented that these dispersion relations yield gauge fixing independent solutions if accurately calculated. This formalism is applied to a one loop calculation of the gluon plasma parameters in a one parameter family of noncovariant gauges which interpolates between the Static Temporal Axial gauge and the Coulomb gauge.

## 1 Introduction

At high temperatures and high pressure hadronic matter is thought to undergo a phase transition to a plasma of essentially free quarks and gluons. This plasma may have experimentally observable consequences for relativistic heavy ion collisions, the astrophysics of neutron stars and early universe cosmology. There is, however, an important theoretical question that has yet to be answered: Can the QCD plasma exist as a stable state of matter? More specifically, a great deal of controversy has recently surrounded the value of the gluon plasma damping constant. There are three distinct, but related issues in this controversy. First of all, questions concerning the correct definition of the gluon plasma response functions have been triggered by the gauge fixing dependence of the damping constant at the one loop level [1-6]. Secondly, as pointed out by several authors [2,7,8], certain higher loop graphs contribute to the damping constant at lowest order, and hence an accurate calculation will probably require summing an infinite subset of all loop graphs. Finally there is the most important question of whether the gluon plasma damping constant, when calculated accurately, is positive or negative; there is a general consensus that a negative result would imply the breakdown of standard perturbation theory.

Clearly the resolution of all three of these issues is of paramount importance. In the present paper, we will address the first issue, namely the correct definition of the damping constant. By resolving this issue we will also be able to make some indirect statements

[1] Talk presented by G. Kunstatter

concerning the validity of the loop expansion. In Section 2 we show how to derive from the gluon propagator nonperturbative dispersion relations for the two physical plasmon modes. In Section 3 we prove that these dispersion relations are gauge fixing independent if accurately calculated[9]. The obvious implication is that gauge fixing independence at the one loop level is a direct consequence of an inconsistent approximation scheme. In Section 4 we apply the formalism introduced in the previous sections to the calculation at the one loop level of the plasma parameters in a one-parameter family of noncovariant gauges which interpolate between the Static Temporal gauge and the Coulomb gauge. Finally, in Section 5, we summarize our results and present some conclusions.

## 2 Linear Response Analysis for Gauge Theories

In the linear response approximation, the stability of a system against external perturbations can be analyzed using one of two approaches. In the operator approach one approximates the Hamiltonian of the perturbed system by

$$\hat{H} = \hat{H}_0 + \hat{H}_{ext} , \tag{2.1}$$

where $\hat{H}_0$ is the unperturbed Hamiltonian and

$$\hat{H}_{ext} = \hat{O}^i J_i \tag{2.2}$$

describes the perturbation in terms of an external source linearly coupled to some well defined operator $\hat{O}^i$. A straightforward analysis then shows that in the linear approximation:

$$\delta\langle\hat{O}^i(x)\rangle = \int d^4y\theta(x_0 - y_0)\langle\left[\hat{O}^i(x), \hat{O}^j(y)\right]\rangle\delta J_j(y) , \tag{2.3}$$

so that the stability of the system under this type of perturbation is determined by the poles in the retarded propagator of $\hat{O}^i$.

The functional approach, on the other hand, assumes the existence of quantum corrected mean field equations of the form:

$$\frac{\delta\Gamma[v]}{\delta v^i(x)} = -J_i(x) , \tag{2.4}$$

where $\Gamma[v]$ is an effective action and $J_i$ is again an external source. In this case, the response of the mean field $v^i$ to a small perturbation in $J_i$ is given by:

$$\delta v^i(x) = -\int d^4y \left[\frac{\delta^2\Gamma}{\delta v^i(x)\delta v^j(y)}\right]^{-1} \delta J_j(y) . \tag{2.5}$$

The connection between the two approaches is obtained by noting that if the effective action $\Gamma[v]$ is obtained via Legendre transformation from a generating functional $W[J]$ in which the source $J$ is linearly coupled to an operator $\hat{O}$, then $v^i = \langle\hat{O}^i\rangle$ and $\left[\delta^2\Gamma/\delta v^i(x)\delta v^j(y)\right]^{-1}$ is equal to minus the time ordered propagator of $\hat{O}$. Now, for finite temperature applications one is interested in thermal averages instead of vacuum expectation values, and as well linear response analysis makes use of the retarded commutator rather than the time ordered propagator. However, if one uses the imaginary time formalism of finite temperature, then analytic continuation of the propagator yields the retarded propagator. Thus, both approaches are expected to yield the same dispersion relations.

In the case of QCD the situation is complicated by the gauge invariance of the classical Lagrangian :

$$L = -\frac{1}{4} F^a_{\mu\nu} F_{a\mu\nu} , \tag{2.6}$$

where

$$F^a_{\mu\nu} = \partial_\mu A^a_\nu - \partial_\nu A^a_\mu - g f^{abc} A^b_\mu A^c_\nu .$$

As before, in this case one would like to start from the basic equation:

$$\delta \bar{A}^a_\mu(x) = \int d^4y D^{ab}_{\mu\nu}(x,y) \delta J^\nu_b(y) , \tag{2.7}$$

where

$$\begin{aligned} \delta \bar{A}^a_\mu(x) &= \delta \langle A^a_\mu(x) \rangle_J \\ &= \langle A^a_\mu \rangle \big|_{J=\delta J} - \langle A^a_\mu \rangle \big|_{J=0} \end{aligned} \tag{2.8}$$

is the variation of the thermal average of the vector potential under the influence of a small perturbation in the external source $\delta J^\nu_a$ and

$$D^{ab}_{\mu\nu}(x,y) \equiv \langle A^a_\mu(x) A^b_\nu(y) \rangle \big|_{J=0} \tag{2.9}$$

is the finite temperature (retarded) gluon correlation function. The problem, of course, is that in order for the gluon propagator to be well defined, the QCD lagrangian must be supplemented by a gauge fixing term and its associated Faddeev-Popov ghost term. The resulting gluon propagator describes unphysical as well as physical modes, and is in general gauge fixing dependent. As we shall show, however, gauge independent dispersion relations for the two physical plasmon modes can be derived directly from the gauge dependent gluon propagator. The remainder of this Section is devoted to deriving these dispersion relations non-perturbatively in a wide class of gauges. The proof of the gauge fixing independence will be presented in Section 3.

In order to simplify the discussion, we will restrict our attention to the class of gauges in which the momentum space gluon propagator can be decomposed as follows (suppressing colour indices):

$$D_{\mu\nu}(k) = \alpha A_{\mu\nu} + \epsilon \frac{\tilde{n}_\mu \tilde{n}_\nu}{\tilde{n}^2} + \beta(\tilde{n}_\mu k_\nu + k_\mu \tilde{n}_\nu) + \delta \frac{k_\mu k_\nu}{k^2} , \tag{2.10}$$

with $\tilde{n}_\mu = P_{\mu\nu} n^\nu$, $P_{\mu\nu} = g_{\mu\nu} - k_\mu k_\nu / k^2$, and $A_{\mu\nu} = P_{\mu\nu} - \tilde{n}_\mu \tilde{n}_\nu / \tilde{n}^2$. Note that $\tilde{n}^\mu k_\mu = 0$ and that $A_{\mu\nu} k^\nu = 0 = A_{\mu\nu} \tilde{n}^\nu$; in the frame $n_\mu = (1,0,0,0)$, $A_{\mu\nu}$ is the usual spatial projection operator:

$$A_{\mu\nu} = \begin{pmatrix} 0 & 0 \\ 0 & \delta_{ij} - \dfrac{k_i k_j}{k^2} \end{pmatrix} .$$

At zero temperature this form of the propagator restricts the (possibly nonlinear) gauge condition to depend at most on derivatives and on one arbitrary, fixed vector $n_\mu$. At finite temperature $O(3,1)$ is broken to $O(3)$ by the presence of the heat bath whose rest frame velocity is $n_\mu$, so that the gauge conditions can depend at most on the velocity of the heat bath and derivatives. One can now use Dyson's equation:

$$D_{\mu\nu}(k) = (\Gamma^{,\mu\nu})^{-1} = \left(D_{(0)}^{-1\ \mu\nu} + \Pi^{\mu\nu}\right)^{-1} , \tag{2.11}$$

with

$$D_{(0)}^{-1\ \mu\nu} = k^2 P^{\mu\nu} + C_{(0)}\frac{\tilde{n}^\mu\tilde{n}^\nu}{\tilde{n}^2} + B_{(0)}(\tilde{n}^\mu k^\nu + \tilde{n}^\nu k^\mu) + D_{(0)}\frac{k^\mu k^\nu}{k^2} \tag{2.12}$$

and radiative corrections

$$\Pi^{\mu\nu} = \Pi P^{\mu\nu} + \chi\frac{\tilde{n}^\mu\tilde{n}^\nu}{\tilde{n}^2} + \Phi(\tilde{n}^\mu k^\nu + \tilde{n}^\nu k^\mu) + \Lambda\frac{k^\mu k^\nu}{k^2} , \tag{2.13}$$

to express the structure functions of $D^{\mu\nu}$ as follows:

$$\alpha^{-1} = k^2 + \Pi \equiv \mathcal{T}(k) \tag{2.14a}$$

$$\epsilon^{-1} = \mathcal{L}(k) \tag{2.14b}$$

$$\beta^{-1} = -\frac{(D_{(0)} + \Lambda)}{B_{(0)} + \Phi}\mathcal{L}(k) \tag{2.14c}$$

$$\delta^{-1} = \frac{(D_{(0)} + \Lambda)}{k^2 + \Pi + C_{(0)} + \chi}\mathcal{L}(k) , \tag{2.14d}$$

with

$$\mathcal{L}(k) \equiv k^2 + \Pi + \chi + \tilde{n}^2 k^2 \left[\frac{B_{(0)}^2}{D_{(0)}} - \frac{(B_{(0)} + \Phi)^2}{D_{(0)} + \Lambda}\right] . \tag{2.15}$$

In Eq. (2.15) we have used the fact that the parameters $B_{(0)}$, $C_{(0)}$ and $D_{(0)}$ are not independent, but satisfy: $B_{(0)}^2 = C_{(0)}D_{(0)}/\tilde{n}^2k^2$. It is clear that apart from kinematical factors the poles in the gluon propagator are determined by the functions $\mathcal{T}$ and $\mathcal{L}$, which in turn determine the dispersion relations of the spatially transverse and spatially longitudinal modes, respectively, of the quark-gluon plasma. Note that although the transverse parts of the propagator are sufficient to determine both dispersion relations, the spatially longitudinal mode $\mathcal{L}$ does in general receive contributions from the nontransverse parts, $\Phi$ and $\Lambda$, of the gluon self-energy.

In the case of linear gauges of the form

$$S_{g.f.} = -\frac{1}{2\alpha}\int dx (F^\mu A_\mu)^2 \tag{2.16}$$

one can use the Ward identity:

$$F^\mu F^\nu D_{\mu\nu} = \alpha \tag{2.17}$$

to relate the nontransverse parts of the self-energy

$$\Lambda(k^2 + \Pi + \chi) = \tilde{n}^2 k^2 \Phi^2 . \tag{2.18}$$

Using this form of the Ward identity, the longitudinal response function takes the form:

$$\mathcal{L} = \frac{\left[k^2 + \Pi + \chi - \tilde{n}^2 k^2 \frac{B_{(0)}}{D_{(0)}} \Phi\right]^2}{\left[k^2 + \Pi + \chi + \tilde{n}^2 k^2 \frac{\Phi^2}{D_{(0)}}\right]} . \tag{2.19}$$

In order to make contact with previous work, it is useful to express the response functions in terms of the components of the proper self-energy $\Pi_{\mu\nu}$. The dispersion relation for the transverse mode $T(k)$ takes the simple form:

$$k_0^2 = f^T(k_0, k) \equiv \frac{1}{2}(\Pi_{ii} - \frac{k_i k_j \Pi^{ij}}{k^2}) . \tag{2.20}$$

The longitudinal mode on the other hand can be written as

$$\begin{aligned} k_0^2 =& f^L(k_0, k) \\ =& \frac{k_i k_j \Pi^{ij}}{k^2} + \tilde{n}^2 \frac{D_{(0)}}{B_{(0)}} \left[1 - \frac{k_0^2 B_{(0)}}{D_{(0)}}\right]^2 \Phi \qquad (B_{(0)} \neq 0) && (2.21a) \\ =& \frac{k_i k_j \Pi^{ij}}{k^2} - 2\tilde{n}^2 k_0 \Phi \qquad (B_{(0)} = 0) . && (2.21b) \end{aligned}$$

Note that when the self-energy is nontransverse, the usual expressions [5,6] for the longitudinal mode are modified.

Before leaving this Section we outline the method for extracting plasma parameters from equations of the general form

$$k_0^2 = f(k_0, k) , \tag{2.22}$$

which will be directly applicable to Eqs.(2.20) and (2.21). One assumes that there is a solution of the form $k_0 = \omega_p - i\gamma$, with $\gamma << \omega_p$. The plasma frequency is then obtained by solving

$$\omega_p^2 = Ref(\omega_p, k) . \tag{2.23}$$

The damping constant will be given by:

$$\gamma = \left. \frac{Imf(\omega, k)}{\omega^2 \frac{\partial}{\partial\omega}\left[\frac{Ref(\omega, k)}{\omega^2}\right]} \right|_{\omega=\omega_p} . \tag{2.24}$$

The convention used is that if such solutions exist, then positive $\gamma$ will result in exponentially damped oscillations, while negative $\gamma$ will result in anti-damping.

By now extensive calculations have been performed at the one loop level using the above equations in a variety of gauges. All such calculations agree on the result for the plasmon mass in the high temperature, long wavelength limit, namely:

$$\omega_p^2 = \frac{1}{9} g^2 N T^2 \tag{2.25}$$

for both the longitudinal and transverse modes. The one-loop damping constant, on the other hand, is gauge fixing dependent. For example, in the usual covariant gauge [3]

$$\gamma = -\left[5 + \frac{3}{2}(\alpha - 1) + \frac{(\alpha - 1)^2}{4}\right]\frac{g^2NT}{24\pi}\,, \tag{2.26a}$$

while in background field gauges [4]

$$\gamma = -\left[11 + \frac{1}{4}(\alpha - 1)^2\right]\frac{g^2NT}{24\pi}\,. \tag{2.26b}$$

Interestingly enough the above values are negative for all real values of the gauge parameter. In the temporal axial gauge and coulomb gauge on the other hand, using correlation functions of the colour electric field, a positive damping constant has been achieved [1-3], namely

$$\gamma = \frac{g^2NT}{24\pi}\,. \tag{2.26c}$$

As we shall see in Section 4 this latter answer can also be obtained in a one-parameter family of noncovariant gauges from the ordinary gluon propagator.

The apparent gauge dependence of the hopefully physical quantity $\gamma$ has resulted in a great deal of controversy in recent years. Pisarski[7] has argued that the gauge dependence at the one- loop level is a consequence of an inconsistent approximation scheme. Despite this, many investigators have been motivated by the gauge dependence in the one loop damping constant to propose alternative definitions of the gluon plasma parameters. These definitions for the most part are based on modified gluon "propagators" that are manifestly gauge fixing independent order by order in perturbation theory. (i.e. gauge dependence cancels algebraically in each order of the loop expansion.) Among the many starting points are: Cornwall's gauge invariant propagator [8,10], the Vilkovisky-DeWitt effective action [4,11], S-matrix elements involving heavy fermions [12], QCD transport equations [13] and correlation functions of non-local operators [14]. It should also be noted that the calculations using the correlation functions of colour electric fields [1-3] were also motivated in part by the problem of gauge dependence. Unfortunately, the above gauge independent, "improved" methods for calculating $\gamma$ still disagree at the one-loop level as to the correct value of the damping constant.

In the next Section we will partially resolve this dilemma by providing a rigorous proof that the dispersion relations derived in this Section from the ordinary gluon propagator do indeed yield gauge independent solutions if calculated accurately[9]. Thus the perturbative modifications to the gluon propagator alluded to above are not necessary from the point of view of gauge dependence, although it may be convenient to consider them for other reasons. As well, since this proof implies the breakdown of the loop expansion for these calculations, the disagreement found among the various gauge independent formalisms at the one loop level is perhaps not surprising.

## 3 Proof of Gauge Fixing Independence

We now outline the proof of gauge fixing independence of physical poles in the gluon propagator. (See ref. [9] and A. Rebhan in these proceedings for more discussion.) It is convenient to use DeWitt's condensed notation [15], in which a generic field is represented by $\phi^i$. The single index $i$ represents all possible indices, internal and spacetime, as well as the spacetime coordinates. For example, in Yang-Mills theory we have $\phi^i \sim A^\mu_a(x)$. Repeated indices imply summation, including spacetime integration with appropriate measure. Consider now a gauge theory with classical action $S[\phi]$ which is invariant under a set of gauge tranformations generated by $K^i_\alpha$,

$$\delta S = \frac{\delta S}{\delta\phi^i} K^i_\alpha \epsilon^\alpha = 0 \tag{3.1}$$

for arbitrary gauge parameters $\epsilon^\alpha$. Again for Yang-Mills theory, the generator $K^i_\alpha$ is the covariant derivative operator

$$K^i_\alpha \sim D^\mu_{ab'}(A)\delta^4(x-y') = \delta_{ab'}\partial^\mu\delta^4(x-y') - gf_{ab'c}A^\mu_c(x)\delta^4(x-y') \,. \tag{3.2}$$

The primed index and coordinate are the group indices, represented by $\alpha$ in the condensed notation.

One can define a gauge fixed generating functional $W[J]$ for the theory using the standard Faddeev-Popov ansatz:

$$\exp\frac{i}{\hbar}W[J] \equiv \int \mathcal{D}\mu[\phi]\mathrm{Det}Q^\alpha_\beta \exp\frac{i}{\hbar}\left(S[\phi] + \frac{1}{2}(F^\alpha[\phi])^2 + J_i\phi^i\right) , \tag{3.3}$$

where

$$Q^\alpha_\beta \equiv \frac{\delta F^\alpha}{\delta\phi^i} K^i_\beta \tag{3.4}$$

is the Faddeev-Popov operator associated with the gauge fixing condition $F^\alpha$. At this stage the only assumptions that we make concerning $F^\alpha$ are that the Faddeev-Popov determinant be non-singular, and that it does not depend on either the mean field or an external background field. The discussion of the so- called background field gauges is presented elsewhere [9].

$W[J]$ generates connected Green's functions for the theory. In particular, the time ordered propagator, in condensed notation, is given by

$$\begin{aligned} D^{ij} &\equiv \langle\langle(\phi^i - \overline{\phi}^i)(\phi^j - \overline{\phi}^j)\rangle\rangle \\ &= \frac{\delta^2 W[J]}{\delta J_i \delta J_j} \,. \end{aligned} \tag{3.5}$$

Here, as in the subsequent proof, it is not necessary to distinguish between vacuum expectation values and thermal averages. In the imaginary time formalism, the difference lies in the temporal boundary conditions placed on the fields in the functional integral. The real time thermal Green's functions are then obtained by analytic continuation from imaginary to real time.

The cornerstone of the proof is the fact that an arbitrary change in the gauge fixing condition, say $F^\alpha \to F^\alpha + \Delta F^\alpha$ can be compensated in the functional integral Eq.(3.3) by a corresponding shift in the quantum field $\phi^i \to \phi^i + \delta^F \phi^i$, where

$$\delta^F \phi^i \equiv K^i_\alpha \mathcal{G}^\alpha{}_\beta \Delta F^\beta \tag{3.6}$$

and $\mathcal{G}^\alpha{}_\beta$ is the Green's function of the Faddeev-Popov operator:

$$\mathcal{G}^\alpha{}_\beta Q^\beta_\epsilon = -\delta^\alpha{}_\epsilon \,. \tag{3.7}$$

It is now straightforward to show that the variation (at fixed $J$) of the generating functional under this change in gauge fixing condition is:

$$\Delta W[J] = J_i \langle K^i_\alpha[\phi] \mathcal{G}^\alpha{}_\beta \Delta F^\beta \rangle \,. \tag{3.8}$$

It should be noted that the analogue of the above for the effective action is

$$\Delta\Gamma[\overline{\phi}] = -\frac{\delta\Gamma}{\delta\overline{\phi}^i} X^i \,, \tag{3.9}$$

where

$$X^i \equiv \Delta\overline{\phi}^i = \langle \delta^F \phi^i \rangle \quad .$$

Eq. (3.9) is in fact generalization of the Nielsen identities determining the gauge parameter dependence of the effective potential:

$$\frac{\partial V}{\partial \alpha} + \frac{d\overline{\phi}}{d\alpha}\frac{\partial V}{\partial \overline{\phi}} = 0 \,. \tag{3.10}$$

The Nielsen identities were used in the '70's to analyze the apparent gauge parameter dependence of particle masses in dynamical symmetry breaking[16].

By differentiating Eq. (3.8) twice with respect to $J$, at fixed $F^\alpha$, one obtains the variation of the propagator under changes in the gauge fixing condition. In the absence of external sources (i.e. at $J = 0$), it is:

$$\begin{aligned} \Delta D^{ij} &= -\frac{\delta X^i}{\delta J_j} - \frac{\delta X^j}{\delta J_i} \\ &= -X^i{}_{,k} D^{kj} - X^j{}_{,k} D^{ki} \,, \end{aligned} \tag{3.11}$$

where

$$X^i{}_{,k} \equiv \frac{\delta X^i}{\delta \overline{\phi}^k} \,,$$

and we have used the fact that

$$\frac{\delta\overline{\phi}^i}{\delta J_j} = \frac{\delta^2 W}{\delta J_i \delta J_j} = D^{ij} \,. \tag{3.12}$$

For the gluon propagator, Eq. (3.11) implies

$$\delta D^{\mu\nu}(k) = D^{\mu\lambda}(k)\delta X^\nu_{,\lambda}(k) + \delta X^\mu_{,\lambda}(k) D^{\lambda\nu}(k) \,. \tag{3.13}$$

Now, returning to the class of propagators described by Eq.(2.10), we find that projecting Eq.(3.13) onto the orthogonal matrices $A_{\mu\nu}$ and $\tilde{n}_\mu \tilde{n}_\nu$, respectively, yields

$$\delta \mathcal{T}(k) = -\mathcal{T}(k) A^\lambda_\nu(k) \delta X^\nu_{,\lambda}(k) \tag{3.14}$$

and

$$\delta \mathcal{L}(k) = -\mathcal{L}(k) 2 \left[ \frac{\tilde{n}_\nu \tilde{n}^\lambda \delta X^\nu_{,\lambda}}{\tilde{n}^2} - \frac{B_{(0)} + \Phi}{(D_{(0)} + \Lambda)} \tilde{n}_\nu k^\lambda \delta X^\nu_{,\lambda} \right] . \tag{3.15}$$

This implies the desired result that a solution to $\mathcal{T}(k) = 0$ is also a solution of $(\mathcal{T} + \delta\mathcal{T})(k) = 0$, and similarly for solutions of $\mathcal{L}(k) = 0$. Hence the dispersion relations derived from the zeroes of $\mathcal{T}$ and $\mathcal{L}$ are gauge fixing independent.

A potential simple zero of $\mathcal{L}$ at $k^2 = 0$ must be excluded from our gauge independence proof because it may be cancelled on the right hand side of Eq. (3.15). It can, in fact, be shown that such a pole does not affect the dynamics of the field strengths. On the other hand, poles in $\delta\mathcal{L}$ coming from the factor $1/(D_{(0)} + \Lambda)$ present no difficulty, since they cannot coincide with the zeroes of $\mathcal{L}$, as is clear from the expression in Eq. (2.15). Moreover, one can use Eq. (3.13) to prove that these poles are in general gauge fixing dependent, and hence unphysical. It is also safe to assume that potential poles in $\delta X$ do not coincide with those given by $1/\mathcal{L}$ and $1/\mathcal{T}$, since $\delta X$ does not contain an external gluon propagator, and the pole structure of the ghost propagator, which essentially determines the pole structure of $\delta X$, is generically different.

Thus, when calculated consistently the dispersion relations derived from the poles of the ordinary gluon propagator are gauge fixing independent. In the next Section we shall examine these relations in a certain class of noncovariant gauges. Although the preceding results indicate that a breakdown of the loop expansion occurs, the calculations of the next Section are done at the one loop level. This is because the use of noncovariant gauges in these calcualations raises certain technical issues even at this level that need to be resolved before a fully selfconsistent calculation can be attempted.

## 4 Noncovariant Gauges

The class of gauges we shall consider in this Section is a one parameter family of noncovariant gauges which interpolate between the static temporal gauge and the Coulomb gauge. This calculation was originally motivated by the work of Heinz, Kajantie and Toimela[1-2], who calculated the plasma parameters from the colour electric field correlation function $\langle E^i(x) E^j(y) \rangle$. They first performed the calculation in the temporal axial gauge (TAG) $A_0 = 0$, since in that gauge the relationship between the colour electric field correlation function and the gluon propagator is particularly simple:

$$\langle E^i(x) E^j(x') \rangle = \partial_0 \partial_0' \langle A^i(x) A^j(x') \rangle . \tag{4.1}$$

Another nice feature of this gauge is the fact that the propagator contains only transverse gluons; there are no ghosts or longitudinal modes. The hope was that in this gauge only physical states would contribute to the thermal average, and this was realized at the one-loop level by the positive damping constant $\gamma = g^2 NT/24\pi$. Moreover, a calculation in the Coulomb gauge, incorporating the vertex corrections necessary to reproduce the colour electric field correlation function from the propagator, yielded precisely the same answer.

The TAG calculation also raised some interesting questions, however. First of all, a characteristic feature of so-called ghost free gauges is the presence of unphysical poles in the propagator. In this case, the pole at $k_0 = 0$ was resolved using the principle value prescription, which was known to break down at two loops at zero temperature[17], and was claimed to yield incorrect answers even at one loop[18]. In addition, there was the issue of the compatibility of the condition $A_0 = 0$ with the imaginary time boundary conditions. Finally, given the linear response analysis and proof of gauge fixing independence in the previous sections, it would be useful to clarify the role of the vertex corrections in the Coulomb gauge calculation. These were necessary, interestingly enough, to yield the correct value for the plasmon mass, but did not contribute directly to the damping constant.

In a recent work [19], these issues were addressed by a calculation of the plasma parameters from the gluon propagator, using a one parameter family of noncovariant gauges which interpolate between the static temporal gauge (STAG) $\partial_0 A_0 = 0$ , and the Coulomb gauge. For ease of reference, we shall call this family the Interpolating Axial gauge (IAG). As we shall see, interesting results emerged even at the one loop level. (For another approach, using the Leibbrandt-Mandelstam prescription in TAG, see refs.[20] and [21].) These gauges, introduced originally by Frenkel and Taylor [22], and independently analyzed more recently by Burnel [23], are implemented by a gauge fixing term

$$S_{g.f.} = -\frac{1}{2\alpha}\int d^4x\,(\partial_0 A_0 - a\partial^\mu A_\mu)^2 \tag{4.2}$$

with associated Faddeev Popov ghost term

$$S_{f.p.} = -\int d^4x\bar{\eta}^b\,(\partial_0 D_0 - a\partial^\mu D_\mu)^{ab}\,\eta^a\ . \tag{4.3}$$

In the above, $\partial_\mu$ denotes ordinary differentiation with respect to $x^\mu$, while $D_\mu$ denotes gauge covariant differentiation. These gauges take on a particularly convenient form if one makes the identification

$$\alpha = a(a-1)\ . \tag{4.4}$$

In this case the bare momentum space propagator is:

$$D^{(0)}_{\mu\nu}(k) = \frac{1}{k^2+i\epsilon}\begin{pmatrix} \dfrac{ak_0^2}{(1-a)k_0^2+ak^2} & 0 \\ 0 & \delta_{ij} - \dfrac{k_ik_j}{(1-a)k_0^2+ak^2} \end{pmatrix}\ . \tag{4.5}$$

Clearly as $a \to 0$, the propagator is equal to that of the TAG and STAG, with unphysical poles at $k_0 = 0$, while as $a \to 1$ the Coulomb gauge propagator is recovered. Moreover, as $a \to \infty$, the IAG becomes the Feynman gauge. These limits are however singular and must be taken only after a physical quantity has been calculated, at which point the gauge parameter $a$ should no longer be present. It should also be noted that for $0 < a < 1$ there are no unphysical poles on the real axis. On the other hand, there are contributions from ghosts and longitudinal gluon modes for all $a > 0$. The Hamiltonian and BRST analysis for this family of gauges can be found in Ref. [24].

We now outline the results of the one loop calculation of the plasma parameters in IAG. Details can be found in Ref. [19]. The transverse mode is obtained by solving the dispersion relation of Eq.(2.20) in the limit $\boldsymbol{k} = 0$. The longitudinal mode of Eq.(2.21) is perhaps the more interesting. In this limit it becomes

$$\begin{aligned} k_0^2 &= f^L(k_0, 0) \\ &= \left.\frac{k_i k_j \Pi^{ij}}{\boldsymbol{k}^2}\right|_{\boldsymbol{k}=0} - \left.\frac{a^2}{(1-a)}\frac{\boldsymbol{k}^2}{k_0}\Phi\right|_{\boldsymbol{k}=0}, \end{aligned} \tag{4.6}$$

where we have used the fact that $\Pi_{ij} \propto \delta_{ij}$ in the long wavelength limit, by rotational invariance. Note that due to the nontransversality of the self-energy, the longitudinal and transverse plasmon modes could in principle differ even in the long wavelength limit. It turns out, however, that

$$\begin{aligned} \left.\frac{2\Phi}{k_0}\right|_{\boldsymbol{k}=0} = -\frac{g^2NT}{3\pi^2 k_0}\int_0^\infty dx \Bigg\{ & \frac{4}{1+b^2x^2} + \\ & \frac{2x^2(1+b^2)^2 + 2(b^2-1)}{x^4(1+b^2)^2 + 2x^2(b^2-1)+1} \Bigg\}, \end{aligned} \tag{4.7}$$

where $b^2 \equiv a/(1-a) > 0$. From Eq.(4.6) this term does not contribute to the plasma parameters to lowest order, so the two dispersion relations are identical in this limit. Thus we have that

$$\begin{aligned} f^{T,L} = {} & g^2N\frac{T^2}{9} + g^2N\frac{k_0T}{3\pi^2}\int_0^\infty dx \frac{1}{1-4x^2+i\varepsilon} \\ & - g^2N\frac{k_0T}{3\pi^2}\int_0^\infty dx \frac{2}{1+4b^2x^2} \\ & + g^2N\frac{k_0T}{3\pi^2}\int_0^\infty dx \frac{x^2(3+4b^2+b^4)+b^2-1}{x^4(1+b^2)^2+2x^2(b^2-1)+1} \\ & - g^2N\frac{k_0T}{3\pi^2}\frac{1}{b^2}\int_0^\infty dx \frac{x^2(b^4-1)+1+3b^2}{x^4(1+b^2)^2+2x^2(b^2-1)+1}. \end{aligned} \tag{4.8}$$

It is crucial that $0 < a < 1$ in order for the pole structure in the last three integrals to be determined. For this range of the gauge parameter, these terms have no imaginary part, and do not contribute to the plasmon mass to leading order. The one loop plasma parameters to leading order are then found to be

$$\omega_p^2 = \frac{g^2NT^2}{9} \tag{4.9}$$

and

$$\gamma_{L,T} = \frac{g^2NT}{24\pi}. \tag{4.10}$$

This is precisely the same result as obtained from the colour electric field correlation function by Heinz et al using the principle value prescription [1-3], and by James[20] and Nachbagauer and Rebhan[21] using the Leibbrandt-Mandelstam prescription, work that is described elsewhere in these proceedings. Remarkably, the damping constant is found to be independent of the gauge parameter $a$ for $0 < a < 1$. This gauge parameter independence of our result leads to the conclusion that the parameters are the same in the STAG and in the Coulomb gauge, without the need for vertex corrections, despite the fact that they are derived from the ordinary gluon propagator which is not the same as the colour electric field correlation function. It should be noted that the ghost contributions do not drop out of our calculation in the singular limit $a \to 0$.

## 5 Summary and Conclusions

We have shown how to extract gauge fixing independent dispersion relations for the physical modes of the gluon plasma from the ordinary gluon propagator. From the point of view of gauge dependence there is no need to consider complicated modifications to the gluon propagator, since the apparent gauge dependence of the damping constant found at the one loop level is necessarily a consequence of an inconsistent approximation scheme. What seems to be required is a self-consistent resummation of an infinite number of diagrams, as suggested by Pisarski. In fact, in a recent paper, Braaten and Pisarski [25] have been able to prove that such a resummation does indeed yield a gauge independent and positive damping constant to order $g^2T$, although they have not as yet calculated its precise numerical value.

We have also shown how to apply these gauge independent dispersion relations at the one loop level to a one parameter family of gauges that interpolates between the static axial gauge and the Coulomb gauge. A positive, gauge parameter independent value was found for the damping constant, in agreement with calculations in the temporal axial gauge and Coulomb gauge using colour electric field correlation functions. This agreement is somewhat remarkable since in our gauges, the ghosts do not decouple even in the STAG limit, nor did we need to encorporate vertex corrections that eem crucial to previous Coulomb gauge calculations. Moreover, for $0 < a < 1$, our propagator has no unphysical poles on the real axis, and thus all prescription problems are avoided.

It should of course be stressed that there is no reason to believe at this stage that this one loop calculation is any more accurate than previous ones, despite the surprising gauge parameter independence of the result. It is however interesting that an unambiguous calculation can be done in noncovariant gauges without the need for vertex corrections. What is now required is a selfconsistent calculation as proposed by Pisarski. It is hoped that the preliminary results presented here indicate that this particular family of noncovariant gauges is worthy of further attention in this regard.

**Acknowledgements** The authors are extremely grateful to K. Mak and A. Rebhan for their collaboration on the work described in Sections 4 and 3, respectively. They would also like to thank R. Pisarski for many useful discussions. Finally, they are grateful to the organizers of this meeting for providing such a useful and enjoyable forum for discussion.

## References

[1] K. Kajantie and J. Kapusta, Ann. Phys. **160**, 477 (1985).
[2] U. Heinz, K. Kajantie, and T. Toimela, Phys. Lett. **B 183**, 96 (1987).
[3] U. Heinz, K. Kajantie, and T. Toimela, Ann. Phys. **176**, 218 (1987).
[4] T.H. Hansson and I. Zahed, Phys. Rev. Lett. **58**, 2397 (1987).
[5] T.H. Hansson and I. Zahed, Nucl. Phys. **B 292**, 725 (1987).
[6] H.-Th. Elze, U. Heinz, K. Kajantie, and T. Toimela, Z. Phys. **C 37**, 305 (1988).
[7] R. Pisarski, Phys. Rev. Lett. **63**, 1129 (1989).
[8] S. Nadkarni, Phys. Rev. Lett. **61**, 396 (1988).
[9] R. Kobes, G. Kunstatter, and A. Rebhan, Univ. Winnipeg preprint 10/89. See also A. Rebhan, these Proceedings.
[10] J.M. Cornwall, Phys. Rev. **D 26**, 1453 (1982).
[11] R. Kobes and G. Kunstatter, Phys. Rev. Lett. **61**, 392 (1988).
[12] U. Kraemmer, M. Kreuzer, and A. Rebhan, Ann. Phys., in press.
[13] H.-Th. Elze, CERN preprints CERN-TH 5223/88 and 5423/89.
[14] S. Catani and E. d'Emilio, Princeton Univ. preprint PUTP 1134/89.
[15] B. DeWitt, Phys. Rev. **162**, 1195 (1967).
[16] K.K. Nielsen, Nucl. Phys. **B101**, 173 (1975). See I.J.R. Aitchison and C.M. Fraser, Ann. Phys. **156**, 1 (1984) for a review.
[17] V.F. Muller and W. Ruhl, Ann. Phys. **133**, 240 (1981); S. Caracciolo, G. Curci and P. Menotti, Phys. Lett. **B113**, 311 (1982).
[18] M.E. Carrington, H. Yamagishi, I. Zahed, Stony Brook preprint, 1988.
[19] R. Kobes, G. Kunstatter, and K. Mak, Phys. Lett. **B223**, 433 (1989); Z. Phys. **C45**, 129 (1989).
[20] K. James, DAMTP preprint 89-20, and in these Proceedings.
[21] H. Nachbagauer and A. Rebhan, unpublished.
[22] J. Frenkel and J.C. Taylor, Nucl. Phys. **B 109**, 439 (1976).
[23] A. Burnel, Phys. Rev. **D 36**, 1852 (1987)
[24] A. Burnel, R. Kobes, G. Kunstatter, and K. Mak, manuscript in preparation.
[25] E. Braaten and R. Pisarski, Northwestern Univ. preprints NUHEP-TH-89-7, NUHEP-TH-89-11.

# Towards a Consistent Calculation of the QCD Plasmon Damping Constant[1]

U. Kraemmer [1], M. Kreuzer [2], A. Rebhan [1], and H. Schulz [2]

[1]Institut für Theoretische Physik, Technische Universität Wien, Wiedner Hauptstr. 8–10, A–1040 Wien, Austria
[2]Institut für Theoretische Physik, Universität Hannover, Appelstr. 2, D–3000 Hannover, FRG

**Abstract:** One-loop results for high-temperature QCD plasma parameters are recapitulated with particular attention being paid to the hitherto gauge fixing dependent results for the plasmon damping constant. In the case of the temporal axial gauge, where superficially the result is satisfactory, it is shown that the one-loop damping constant is in principle dependent on the pole prescription with an accidental coincidence of the LM and the PV prescription in the long wavelength limit. It is then argued that the conventional perturbative expansion fails because higher loop graphs can contribute to the same order $g^2T$. Therefore it is necessary to employ an improved perturbative expansion of which a comparatively simple prototype is elaborated. This provides for the first time a naturally gauge independent framework to derive the plasmon parameters from an *on-shell* gluon propagator.

## 1 Introduction

At sufficiently high temperature quantum chromodynamics (QCD) is expected to be in a deconfined phase, the quark-gluon plasma, where due to asymptotic freedom perturbation theory should be applicable to some extent [1,2]. The existence of this form of matter would have possible implications for heavy ion collisions as well as for the early universe or also for the core of a neutron star.

Of central interest for the theoretical understanding of the quark-gluon plasma is the dispersion relation for colour oscillations which either could propagate as collective gluonic modes or could be screened dynamically by the plasma. In a linear response theory [3] the basic quantity to be investigated is the gluon propagator

$$\begin{aligned}\langle \mathrm{T} A^a_\mu(x) A^b_\nu(y)\rangle_\beta &= \mathrm{Tr}\left[e^{-\beta H}\mathrm{T}A^a_\mu(x)A^b_\nu(y)\right] / \,\mathrm{Tr}\left[e^{-\beta H}\right] \\ &= \delta^{ab}\Delta_{\mu\nu}(x-y),\end{aligned} \tag{1}$$

evaluated at high temperatures $T = \frac{1}{\beta} \gg \Lambda_{QCD}$.

[1] Talk presented by A. Rebhan

At finite temperatures, there is a preferred coordinate system, the rest frame of the plasma with velocity $n_\mu$ ($= \delta_{\mu 0}$ in the following), and unless a non-covariant choice of gauge further reduces Lorentz symmetry, the general form of the momentum space propagator is given by

$$\Delta_{\mu\nu}(Q) = \frac{1}{a(Q)} A_{\mu\nu} + \frac{1}{b(Q)} B_{\mu\nu} + \frac{1}{c(Q)} C_{\mu\nu} + \frac{1}{d(Q)} D_{\mu\nu}. \tag{2}$$

Here we use the notation $Q_\mu = (q_0, q_i)$ and we employ the following tensorial basis

$$A_{\mu\nu} \equiv g_{\mu\nu} - \frac{Q_\mu Q_\nu}{Q^2} - \frac{\tilde{n}_\mu \tilde{n}_\nu}{\tilde{n}^2}, \tag{3}$$

$$B_{\mu\nu} \equiv \frac{\tilde{n}_\mu \tilde{n}_\nu}{\tilde{n}^2}, \tag{4}$$

$$C_{\mu\nu} \equiv \tilde{n}_\mu Q_\nu + Q_\mu \tilde{n}_\nu, \tag{5}$$

$$D_{\mu\nu} \equiv \frac{Q_\mu Q_\nu}{Q^2}, \tag{6}$$

where

$$\tilde{n}_\mu \equiv \left( g_{\mu\nu} - \frac{Q_\mu Q_\nu}{Q^2} \right) n^\nu. \tag{7}$$

Only $A_{\mu\nu}$ and $B_{\mu\nu}$ are transverse with respect to the 4-vector $Q_\mu$, and they correspond to *spatially* transverse and longitudinal modes, respectively.

In the leading order ($g^2T^2$) of a high-$T$ expansion, only $a(Q^2)$ and $b(Q^2)$ receive corrections to the classical piece $Q^2$ which read [1]

$$a(Q^2) \equiv Q^2 + \Pi_T = Q^2 - \frac{3}{2} m_{pl.}^2 \left\{ \frac{q_0^2}{q^2} + \left(1 - \frac{q_0^2}{q^2}\right) \frac{q_0}{2q} \ln\left(\frac{q_0 + q}{q_0 - q}\right) \right\}, \tag{8}$$

$$b(Q^2) \equiv Q^2 + \Pi_L = Q^2 - 3m_{pl.}^2 \left(1 - \frac{q_0^2}{q^2}\right) \left(1 - \frac{q_0}{2q} \ln\left(\frac{q_0 + q}{q_0 - q}\right)\right), \tag{9}$$

where $q = |\mathbf{q}|$ and

$$m_{pl.}^2 = e^2T^2/9, \qquad e^2 \equiv (N + N_F/2)g^2. \tag{10}$$

The dispersion relation for spatially transverse and spatially longitudinal gluonic modes is given by the poles of $a$ and $b$, respectively, and is displayed in Fig. 1.

It is made evident by the quadratic scales used in Fig. 1 that both dispersion relations are not Lorentz invariant, for which they would have to be straight lines parallel to the diagonal. Further, propagating modes exist only for frequencies $q_0 > m_{pl.}$, the "plasmon mass", below which there is dynamical screening. In the static limit $q_0 \to 0$, the screening of spatially longitudinal modes is described by the color-electric mass $m_e^2 = e^2T^2/3$, whereas a corresponding magnetic mass is thought to be generated non-perturbatively at the order $g^2T$, thereby solving the problem of infra-red divergencies arising at higher loops [4].

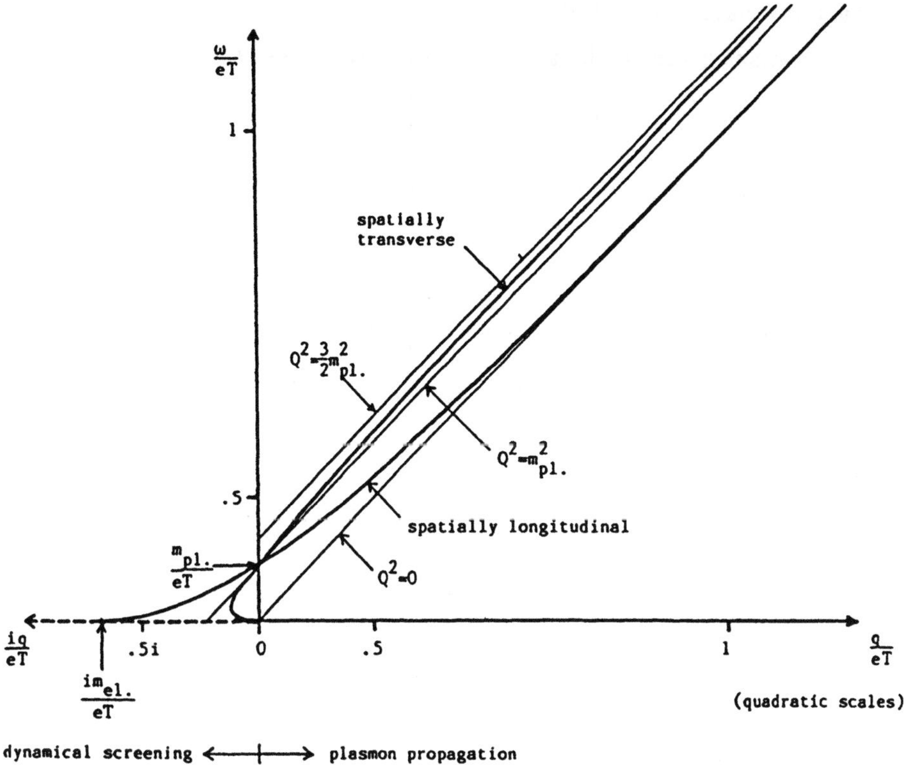


**Fig. 1.** High temperature dispersion relation of the spatially longitudinal and transverse gluonic modes. The enveloping straight lines give the mass hyperboloids (note the quadratic scales) corresponding to the asymptotic values of mass 0 and $\sqrt{\frac{3}{2}}m_{pl.}$. The line in between corresponds to the approximation introduced in (29).

## 2 Overview of results for the one-loop plasmon damping constant

To leading order in $g^2T^2$ there is no imaginary part in the results (8) and (9) at the location of the dispersion relation, hence any damping of plasma oscillations is weak in the high temperature expansion, where one assumes $q_0, q \sim gT \ll T$.

The one-loop plasmon damping constant which is defined at $q_0 = m_{pl.}$, $q = 0$, turns out to be

$$\gamma = c\frac{g^2TN}{24\pi}, \tag{11}$$

where in contrast to $m_{pl.}$ only the gluons are contributing, the quarks being Pauli suppressed by another factor $g^2$. However, while the results (8) and (9) are gauge independent as one expects them to be, the constant $c$ in (11) is found to be gauge dependent, indicating an incomplete calculation.

### 2.1 Ordinary covariant gauges [1,5]

In ordinary covariant gauges with gauge parameter $\alpha$, the result for $c$ is

$$c = -\left[5 + \frac{3}{2}(1-\alpha) + \frac{1}{4}(1-\alpha)^2\right] < 0 \tag{12}$$

which is negative for all values of the gauge parameter, seemingly signalling anti-damping of plasma oscillations and consequently instability of the QCD plasma itself.

### 2.2 Temporal axial gauge [5-10]

On the other hand, calculations based on non-covariant gauges made it clear that not even the negative sign in (12) is a gauge independent answer. In the temporal axial gauge, using the PV prescription, $c$ turned out to be positive [6],

$$c = +1, \tag{13}$$

and the temporal axial gauge is in fact a gauge which seems to be most suited to the problem at hand. In this gauge there is a linear relation between gauge potentials and (colour-electric) field strenghts,

$$E_i = \partial_0 A_i, \tag{14}$$

and second, in the rest frame of the heat bath the time direction is already singled out so that the gauge does not spoil any symmetries.

However, the temporal gauge is problematic in that it is excluding certain *physical* field configurations [7,11]. In fact, the imaginary time formulation of the temporal gauge is not well understood [7,8,10].

Nevertheless, the positive result (13) has been confirmed by circumventing both problems by means of a regulated static temporal gauge [5] which interpolates the static temporal gauge and the Coulomb gauge. Most recently, James [8] has shown that the temporal axial gauge can be implemented consistently within a real time formulation, and that employing the generalized LM [12] prescription there again gives the same result (13).

There is in fact a simple reason for the coincidence of the results obtained in the PV and in the LM prescription [10] which is peculiar to the limit $q \to 0$ usually studied: The integrals one ends up evaluating in the LM prescription typically read

$$\int d^3k\, f(q_0, \boldsymbol{q}, \boldsymbol{k}, \tfrac{1}{p_0+i\eta}), \quad \eta \equiv \varepsilon \operatorname{sgn} p_3, \quad p_0 = q_0 \pm k, \quad \boldsymbol{p} = \boldsymbol{q} + \boldsymbol{k} \tag{15}$$

and in the limit $\boldsymbol{q} \to 0$ one therefore obtains expressions like

$$\int d^3\boldsymbol{k}\, g(q_0, |\boldsymbol{k}|, \eta = \varepsilon \operatorname{sgn} k_3) =$$
$$= 2\pi \int_0^\infty k^2 dk \int_{-1}^1 dz\, g(q_0, |\boldsymbol{k}|, \eta = \varepsilon \operatorname{sgn} z) = 2\pi \int_0^\infty k^2 dk \left[g(\eta = \varepsilon) + g(\eta = -\varepsilon)\right], \tag{16}$$

thus reducing to the PV results.

However, for $q \neq 0$, which no less is relevant for plasmon damping, one can expect dependences on the pole prescriptions. Indeed, already with $q = 0$ a result different from (13) has been reported in [9], allegedly by employing the PV prescription, but in fact by following a procedure similar to Lee's "no prescription" [13].

Beyond this dependence on the pole prescription, there is still the problem of gauge dependence. Although the propagator in the axial gauge is directly related to the correlation function $\langle E^i(x)E^j(y)\rangle$, the latter is not really gauge fixing independent. The Ward identity which describes its variation under a change of the gauge condition $f_\mu A^\mu \to (f_\mu + \delta f_\mu)A^\mu$ reads

$$\delta\langle E^a_j(x)E^e_k(y)\rangle = -gf^{abc}\int d^4z\langle E^b_j(x)\bar{c}^c(x)c^d(z)\delta f_\mu A^{d\mu}(z)E^e_k(y)\rangle + (a,j,x \leftrightarrow e,k,y), \tag{17}$$

where $\bar{c}$, $c$ are Faddeev-Popov ghost fields, and the r.h.s. of (17) vanishes only in the Abelian case. Indeed, although the same result (13) has been obtained from $\langle E^i(x)E^j(y)\rangle$ evaluated in the Coulomb gauge [6], a different, gauge dependent result arises in covariant gauges [14].

### 2.3 Background field gauges and other attempts

Historically the next attempt to arrive at a consistent definition of the plasmon damping constant was the proposal [16] to use background covariant gauges which yield a gauge invariant effective action.[2] However, this background gauge field invariance does not exclude gauge parameter dependences, and indeed, the result turned out to be gauge fixing dependent and negative definite as in ordinary covariant gauges:

$$c = -11 - \frac{1}{4}(1-\alpha)^2 < 0. \tag{18}$$

This persisting gauge dependence of the one-loop plasmon damping constant finally gave rise to a series of proposals aiming at manifest gauge fixing independence. In [6] it was proposed to invoke the off-shell gauge independent Vilkovisky-DeWitt effective action [18] which in effect singles out $\alpha = 0$ in (18), hence $c = -45/4$. Afterwards it was pointed out in [19] that compatibility with linear response theory requires a modification of the Vilkovisky-DeWitt effective action, leading to $c = -27/4$, which still is negative. In [20] the use of Cornwall's "gauge invariant propagator" [21] was advocated, which gives $c = -11$, and this result was interpreted as corroboration of the conjecture of [22] that the negative sign is bound up with asymptotic freedom. While this last calculation has been criticized in [23] as being ambiguous, the result $c = -11$ has also been derived from an on-shell scattering amplitude using superheavy test fermions [24]. But not all manifestly gauge independent schemes agree on the sign if not on the magnitude of the plasmon damping: recently an improved, gauge independent analogue of the correlation function of colour-electric field strenghts has been put forward [25], which reproduces the positive result (13) of the temporal axial gauge.

---

[2] It has been overlooked in [16] that already the conventional temporal axial gauge can be interpreted [17] as being a background covariant gauge in a trivial sense.

## 3 Higher loop corrections

All the aforementioned proposals try to obviate the gauge fixing dependence of the gauge boson propogator in non-Abelian theories and to achieve manifest gauge independence order by order *in the loop expansion.*

However, most recently a general proof has been given by Kobes, Kunstatter and one of the present authors [26] that the structure functions of the gluon propagator $a(Q)$, $b(Q)$ (cf. eq. (2)), from which the dispersion relations of the spatially transverse and longitudinal modes are derived, exhibit the following gauge fixing dependence: A variation of the gauge by $f_\mu A^\mu \to (f_\mu + \delta f_\mu)A^\mu$ induces the variations

$$\delta a(Q) = a(Q)\delta Y(Q) \tag{19}$$

$$\delta b(Q) = b(Q)\delta Z(Q) \tag{20}$$

with some complicated functions $\delta Y$ and $\delta Z$. This entails that after changing the gauge the dispersion relation $a(Q)+\delta a(Q) = 0$ has the same solutions as before in $a(Q) = 0$, and similarly with $b(Q)$. By the way, no such relations hold for the other structure functions $c(Q)$ and $d(Q)$ of (2).

Now we have just seen in the previous chapter that only the leading contributions $\sim g^2T^2$ to the one-loop dispersion relations are gauge independent. The identities (19) and (20) thus imply that the sub-leading contributions (like $\gamma$) cannot be given by one loop results alone, and that gauge independence will indeed be restored by higher loop corrections.

What then happens to the identities (19) and (20)? Being identities they must hold also loop order by loop order. Expanding e.g.

$$a = a^{(0)} + a^{(1)} + a^{(2)} + \ldots, \qquad \delta Y = \delta Y^{(0)} + \delta Y^{(1)} + \delta Y^{(2)} + \ldots, \tag{21}$$

one finds that at lowest order $\delta a^{(0)} = a^{(0)}\delta Y^{(0)}$ and, since at the classical level $a(Q)$ is trivially gauge independent, that $\delta Y^{(0)} \equiv 0$. Therefore at the one loop level

$$\delta a^{(1)} = a^{(0)}\delta Y^{(1)}, \tag{22}$$

which has to be considered on the one-loop "mass shell" $a^{(0)} + a^{(1)} = 0$. Now if in the latter equation $a^{(1)}$ cannot be neglected in comparison with $a^{(0)}$, the r.h.s. of (22) will not vanish and a gauge dependence appears in general. Precisely that happens in the high temperature expansion: there a mass of order $gT$ is generated at one loop which cannot be considered small but instead sets the scale for the dispersion relation. Assuming that the coupling constant $g$ is a good expansion parameter, the perturbative order is now determined by the excess of $g$'s over $T$'s rather than by the powers of $g$ alone, corresponding to a mixing of different loop orders. In the example of the plasmon damping constant, a two-loop result with imaginary part proportional to $g^4T^3/q_0$, evaluated at $q_0 = m_{pl.} \sim gT$, would be of the same order of magnitude as the one-loop result $\sim g^2Tq_0$.

Technically, what happens in the case of the imaginary part of the dispersion relations is that higher loop contributions can be viewed as acting as mass insertions which shift the location of the poles of internal propagators on which the imaginary parts depend. The already gauge independent results $\sim g^2T^2$, on the other hand, are not changed because they are determined by "hard" loop momenta $\sim T \gg gT$, as has been noted by

Pisarski [27,28] who has proposed to use an improved expansion scheme which resums all the leading terms $\sim g^2T^2$.

In order to understand how such an improved expansion scheme can be established, consider a scalar field theory, assuming that the only effect of order $gT$ is the generation of a mass in the self energy

$$\Pi(K) = cg^2T^2 + dg^2TK + \ldots \equiv m^2 + agmK + \ldots \tag{23}$$

Mass insertions can then be resummed in the following way. Rewrite

$$\mathcal{L} = \tfrac{1}{2}(\partial\varphi)^2 + \text{interactions} = \tfrac{1}{2}(\partial\varphi)^2 - \tfrac{1}{2}m^2\varphi^2 + \tfrac{1}{2}m^2\varphi^2\ell\Big|_{\ell=1} + \text{interactions} \tag{24}$$

and identify $\ell$ with a loop counting parameter, i.e. whose powers give the loop order, and which will be set to one only after having performed the loop expansion to a given order. The r.h.s. of (24) then means to do perturbation theory with a massive propagator of mass $m$ and to compensate all mass insertions at one loop higher. This does not alter the theory itself, for with infinite loop order nothing has changed.

Since $m$ is generated by hard loop momenta $\sim T \gg gT$, $m$ itself does not change and the improved one-loop self energy will be given by

$$\breve{\Pi}(K) = m^2 + \breve{a}gmK + \ldots. \tag{25}$$

This guarantees that after this resummation higher loops are indeed suppressed by powers of $g$, for the ratio of the one-loop part of the self energy to the lowest order self energy now is given by

$$\frac{\breve{\Pi}(K) - m^2}{K^2 - m^2} = \frac{\breve{a}gmK}{K^2 - m^2} \sim g, \tag{26}$$

where the subtraction of $m^2$ from $\breve{\Pi}$ is carried out by the 'one-loop-higher' counterterm in (24).

## 4 Partially resummed high-$T$ QCD

A resummation of the order $g^2T^2$-contributions in QCD is of course incomparably more complicated as is evident from the form of the leading order contributions to the dispersion relations (8) and (9).

A possible analogue of the Lagrangian (24) which would replace the bare self energy $Q^2(A+B)_{\mu\nu}$ by $a(Q)A_{\mu\nu} + b(Q)B_{\mu\nu}$ could be

$$\mathcal{L} = -\tfrac{1}{4}F_{\mu\nu}^2 + \left\{-\tfrac{1}{4}F_{\mu\nu}\frac{a(D)-D^2}{D^2}F^{\mu\nu} - \tfrac{1}{4}F_{ij}\frac{b(D)-a(D)}{D^2}F^{ij}\right\}(1-\ell), \tag{27}$$

where $D_\mu$ is the covariant derivative. The resummation of the loop expansion could be again achieved by identifying $\ell$ with the loop counting parameter and setting $\ell = 1$ in the end. That (27) is non-local presents no particular difficulty, however, in order to fix the ordering of the covariant derivatives in $a(D)$ and $b(D)$ and to decide whether one should not choose more complicated covariantizations of the self energy one would have to take into account also the leading order corrections to $n$-point vertex functions.

Since (27) is gauge invariant for all $\ell$, gauge fixing can proceed as usual by adding e.g.

$$\mathcal{L}_{g.f.} = \frac{1}{2\alpha}(\partial_\mu A^{a\mu})^2 + \mathcal{L}_{gh.}. \tag{28}$$

However, the structure functions $a$ and $b$ are so complicated that until now a fully resummed one-loop calculation has remained a mere gedankenrechnung. We shall thus content ourselves for the time being with a partially resummed loop expansion based on the approximation

$$a(D) = b(D) = D^2 + m_{pl.}^2, \tag{29}$$

which shifts the improved bare mass shell so that it is correct at the point $q_0 = m_{pl.}$, $q = 0$ at which we will calculate the thus modified plasmon damping constant (cf. Fig. 1). Note that for any functions $a$ and $b$ (27) still represents the true theory, but, as we shall see, with (29) we are able to achieve gauge independence order by order in the redefined perturbation series.

Hence we start from a Lagrangian

$$\mathcal{L} = -\tfrac{1}{4}F_{\mu\nu}^2 - \tfrac{1}{4}F_{\mu\nu}\frac{m_{pl.}^2}{D^2}F^{\mu\nu}(1-\ell) + \mathcal{L}_{g.f.}. \tag{30}$$

The improved Feynman rules are obtained at $\ell = 0$ so that the propagator now reads

$$\Delta_{\mu\nu}^{ab}(Q) = \delta^{ab}\left\{\frac{-1}{Q^2 - m_{pl.}^2}\left(g_{\mu\nu} - \frac{Q_\mu Q_\nu}{Q^2}\right) - \alpha\frac{Q_\mu Q_\nu}{Q^4}\right\}. \tag{31}$$

Despite the oversimplification (29) the new vertices are rather complicated, and they contain non-local pieces, i.e. propagator-like contributions.

For the recalculation of the one-loop gluon self energy we need the 3-gluon vertex

$$\begin{aligned}
V[-Q,K,Q-K]_{\mu\sigma\lambda}^{acb} = g f^{abc}\Bigg\{ & g_{\mu\sigma}(K+Q)_\lambda + g_{\sigma\lambda}(Q-2K)_\mu + g_{\lambda\mu}(K-2Q)_\sigma \\
& + m_{pl.}^2\Bigg[-g_{\mu\sigma}\left(\frac{K_\lambda}{K^2} + \frac{Q_\lambda}{Q^2}\right) + g_{\sigma\lambda}\left(\frac{K_\mu}{K^2} + \frac{(K-Q)_\mu}{(K-Q)^2}\right) \\
& + g_{\lambda\mu}\left(\frac{(Q-K)_\sigma}{(Q-K)^2} + \frac{Q_\sigma}{Q^2}\right) + (Q+K)_\lambda\frac{g_{\mu\sigma}Q\cdot K - K_\mu Q_\sigma}{Q^2K^2} \\
& + (2K-Q)_\mu\frac{g_{\sigma\lambda}K\cdot(Q-K) - (Q-K)_\sigma K_\lambda}{K^2(Q-K)^2} \\
& + (K-2Q)_\sigma\frac{g_{\lambda\mu}Q\cdot(Q-K) - Q_\lambda(Q-K)_\mu}{Q^2(Q-K)^2}\Bigg]\Bigg\},
\end{aligned} \tag{32}$$

and from the still more complicated 4-gluon vertex ‘only’ the following part

$$
\begin{aligned}
-\frac{i}{2}\delta^{cd} \times V[-Q,Q,-K,K]^{abcd}_{\alpha\beta\gamma\delta} = g^2 C_A \delta^{ab} \frac{1+(\alpha\leftrightarrow\beta)}{2} \Bigg\{ & -g_{\alpha\beta}g_{\gamma\delta} + g_{\alpha\gamma}g_{\beta\delta} \\
& - \frac{m^2_{pl.}}{(Q-K)^2}\Bigg[ g_{\alpha\gamma}g_{\beta\delta} - g_{\alpha\beta}g_{\gamma\delta} \\
& + \frac{1}{Q^2}\left(g_{\alpha\beta} - \frac{Q_\alpha Q_\beta}{Q^2}\right)\{g_{\gamma\delta}(Q-K)^2 - (2Q-K)_\gamma(2Q-K)_\delta\} \\
& + \frac{1}{K^2}\left(g_{\gamma\delta} - \frac{K_\gamma K_\delta}{K^2}\right)\{g_{\alpha\beta}(Q-K)^2 - (Q-2K)_\alpha(Q-2K)_\beta\} \\
+(1+(\gamma\leftrightarrow\delta))\Big\{ & (2Q-K)_\delta \frac{g_{\alpha\beta}Q_\gamma - g_{\alpha\gamma}Q_\beta}{Q^2} + (Q-2K)_\beta \frac{g_{\alpha\gamma}K_\delta - g_{\gamma\delta}K_\alpha}{K^2} \\
& + \frac{g_{\alpha\gamma}Q\cdot K - K_\alpha Q_\gamma}{Q^2K^2}\left[g_{\beta\delta}(Q-K)^2 - (Q-2K)_\beta(2Q-K)_\delta\right]\Big\}\Bigg]\Bigg\},
\end{aligned} \tag{33}
$$

whereas the Feynman rules involving Faddeev-Popov ghosts are the usual ones.

The most pressing question is of course whether the new one-loop result will again be invalidated by gauge dependences. Extracting only the gauge parameter ($\alpha$) dependent contributions to the structure function $b$ of the gluon self energy yields

$$
\begin{aligned}
iC_A g^2 \int_T d^4K \Bigg\{ & \alpha^2 (Q^2 - m^2_{pl.})^2 \frac{[k_0 Q^2 - q_0 K\cdot Q]^2}{q^2 Q^2 K^4 (Q-K)^4} \\
& + \alpha \frac{Q^2 - m^2_{pl.}}{q^2 Q^2 (Q-K)^4}\Bigg[ q_0^2 \frac{q^2k^2 - (q\cdot k)^2}{k^2}\left(1 - \frac{Q^2 - m^2_{pl.}}{K^2 - m^2_{pl.}}\right) \\
& + \left(q_0^2\frac{(q\cdot k)^2}{k^2} - q^4\right)\frac{K^2 - Q^2}{K^2 - m^2_{pl.}} + \frac{[k_0Q^2 - q_0 K\cdot Q]^2}{K^2}\frac{2K^2 - Q^2 - m^2_{pl.}}{K^2 - m^2_{pl.}}\Bigg]\Bigg\} \\
& = 0 \quad \text{at} \quad q_0 = m_{pl.}, q = 0,
\end{aligned} \tag{34}
$$

so gauge independence arises at the location of the plasmon pole already *before* performing the integration resp. the thermal sum implicit in $\int_T d^4K$. This is in perfect agreement with the identities (19) and (20) guaranteeing gauge independence of the complete dispersion relation. Gauge dependences remain, however, for $q \neq 0$, where due to the simplifications of (29) we are still off the plasmon dispersion relation.

We now proceed with the calculation of the plasmon damping constant at $q = 0$, although we cannot really expect that our result, although gauge independent, will be the final answer, since by using the approximation (29) we must still admit that higher loop corrections of the redefined perturbation series can contribute to the same order $g^2T$. What we can hope for, however, is that our partial resummation will give an idea of how the bare one-loop result is changed.

At $q = 0$ the dispersion relations for the spatially transverse and longitudinal modes coincide, $a(m_{pl.}, 0) = b(m_{pl.}, 0)$, and this can be shown to hold also for the one-loop results of our partially resummed perturbation theory, hence it suffices to recalculate only one structure function.

For the modified gluon self energy we obtain the following contributions for $\Pi_L$ at $q_0 = m_{pl.}$, $q = 0$. From the gluon loop:

$$ig^2 C_A \int_T d^4K \frac{1}{[K^2 - m_{pl.}^2][(K-Q)^2 - m_{pl.}^2]} \frac{k^2}{3} \Bigg[ 5 + \frac{(K-Q)^2 + 8m_{pl.}^2}{K^2} + \frac{2m_{pl.}^2 (K-Q)^2}{K^4} + 4\frac{m_{pl.}^4}{K^2(K-Q)^2} + \frac{m_{pl.}^8}{K^4(K-Q)^4} \Bigg] \Bigg|_{\mathbf{q}=0}, \tag{35}$$

from the gluon tadpole:

$$ig^2 C_A \int_T d^4K \frac{1}{K^2 - m_{pl.}^2} \Bigg\{ 5 + \frac{3m_{pl.}^2 - \frac{1}{3}k^2}{K^2} - \frac{2m_{pl.}^2}{(K-Q)^2} + \frac{3m_{pl.}^2 k^2}{K^2(K-Q)^2} \Bigg\} \Bigg|_{\mathbf{q}=0}, \tag{36}$$

and from the ghost loop

$$-ig^2 C_A \int_T d^4K \frac{k^2}{3K^2(K-Q)^2}, \tag{37}$$

where the latter is the same as in the conventional perturbation theory.

A straightforward evaluation of the imaginary parts gives the 'improved' result

$$c = -\frac{13}{4} \tag{38}$$

for the plasmon damping constant — which is still negative. However, we do not have control over how far our partial resummation still is from a complete one taking into account all contributions $\sim gT$. The modest outcrop of our calculation is that (38) is at least less negative than the bare covariant gauge results. Most recently arguments have been put forward [28] that in a fully resummed calculation $c$ would have to be positive.

What really is significant in our partially resummed perturbation theory is the explicit demonstration that the resummation of terms $\sim gT$ is indeed necessary and that gauge independence follows naturally from the fact that then the calculation is 'on-shell' with respect to the leading order dispersion relations.

**Acknowledgements:** A.R. is indebted to R. Kobes and G. Kunstatter for their generous hospitality during a stay at the University of Winnipeg and for many valuable discussions. M. K. has been supported by Deutsche Forschungsgemeinschaft (DFG).

## References

[1] J.C. Collins, M.J. Perry, Phys. Rev. Lett. **34** (1975) 1353; O.K. Kalashnikov, V.V. Klimov, Yad. Fiz. **31** (1980) 1357 [Sov. J. Nucl. Phys. **31** (1980) 699]; D.J. Gross, R.D. Pisarski, L.G. Yaffe, Rev. Mod. Phys. **53** (1981) 43; H.A. Weldon, Phys. Rev. **D26** (1982) 1394; O.K. Kalashnikov, Fortschr. Phys. **32** (1984) 525.

[2] S. Nadkarni, "Asymptotic freedom and the running charge in high-$T$ QCD", U.o.Kentucky report no. UK/89-07.

[3] A.L. Fetter, J.D. Walecka, *Quantum Theory of Many-Particle Systems*, New York: McGraw-Hill, 1971.
[4] A.D. Linde, Phys. Lett. **B96** (1980) 289.
[5] R. Kobes, G. Kunstatter, this volume; R. Kobes, G. Kunstatter, K.W. Mak, Z. Phys. **C45** (1989) 129.
[6] K. Kajantie, J. Kapusta, Ann. Phys. (NY) **160** (1985) 477; U. Heinz, K. Kajantie, T. Toimela, Phys. Lett. **B183** (1987) 96; Ann. Phys. (NY) **176** (1987) 218.
[7] K. Kajantie, this volume.
8. K. James, this volume.
[9] M.E. Carrington, H. Yamagishi, I. Zahed, "The timelike axial gauge at zero and finite temperature", Stony Brook preprint (1988).
[10] H. Nachbagauer, A. Rebhan (unpublished).
[11] S. Nadkarni, Phys. Rev. **D33** (1986) 3738.
[12] P. Gaigg, M. Kreuzer, Phys. Lett. **B205** (1988) 530.
[13] H.C. Lee, this volume.
[14] K. Kajantie, private communication.
[15] For an overview see "Thermal Field Theories and Their Applications", eds. K.L. Kowalski, N.P. Landsman, Ch.G. van Weert, Physica **A158** (1989) 1-558.
[16] T.H. Hansson, I. Zahed, Phys. Rev. Lett. **58** (1987) 2397; Nucl. Phys. **B292** (1987) 725.
[17] A. Rebhan, G. Wirthumer, Z. Phys. **C28** (1985) 269.
[18] G.A. Vilkovisky, in *Quantum Theory of Gravity*, S. Christensen (ed.), p. 169, Bristol: Hilger, 1984; Nucl. Phys. **B234** (1984) 125; B.S. DeWitt, in *Quantum Field Theory and Quantum Statistics*, Vol. 1, I.A. Batalin, C.J. Isham, G.A. Vilkovisky (eds), p. 191, Bristol: Hilger, 1987; E.S. Fradkin, A.A. Tseytlin, Nucl. Phys. **B234** (1984) 509; A. Rebhan, Nucl. Phys. **B288** (1987) 832; Nucl. Phys. **B298** (1988) 726; R. Kobes, G. Kunstatter, D.J. Toms, in *TeV Physics*, G. Domokos, S. Kovesi-Domokos (eds), p. 73, Singapore: World Scientific, 1988.
[19] R. Kobes, G. Kunstatter, Phys. Rev. Lett. **61** (1988) 392.
[20] S. Nadkarni, Phys. Rev. Lett. **61** (1988) 396.
[21] J.M. Cornwall, Phys. Rev. **D26** (1982) 1453; J.M. Cornwall, W.-S. Hou, J.E. King, Phys. Lett. **B153** (1985) 173.
[22] J.A. Lopez, J.C. Parikh, P.J. Siemens, Texas A&M preprint (1985).
[23] M.E. Carrington, T.H. Hansson, H. Yamagishi, I. Zahed, Ann. Phys. (NY) **190** (1989) 373.
[24] U. Kraemmer, M. Kreuzer, A. Rebhan, "QCD plasma parameters from the S-matrix", U.o.Hannover preprint ITP-UH 8/89, to appear in Ann. Phys. (NY).
[25] S. Catani, E. d'Emilio, "Gauge invariant chromoelectromagnetic field and hot gluon plasma oscillations", Princeton Univ. preprint PUTP 1134 (1989).
[26] R. Kobes, G. Kunstatter, A. Rebhan, "Gauge independence of the gluon propagator poles and QCD plasma parameters", U.o.Winnipeg preprint (1989).
[27] R. Pisarski, Phys. Rev. Lett. **63** (1989) 1129.
[28] E. Braaten, R.D. Pisarski, "Resummation and gauge invariance of the gluon damping rate in hot QCD", NUHEP-TH-89-12; "Soft amplitudes in hot gauge theories: a general analysis", FERMILAB-PUB-89/152-T .

# Nonlocality, Coexistence of UV and IR Asymptotic Freedom and the Landau Damping of Hot QCD Plasma[1]

Emilio d'Emilio and Stefano Catani

[1]Dipartimento di Fisica dell'Universitá di Pisa and
Istituto Nazionale di Fisica Nucleare, Sezione di Pisa

[2]Istituto Nazionale di Fisica Nucleare, Sezione di Firenze

The obvious fact that there is something wrong with the description of the hot QCD plasma is well known. The so many disagreeing calculations of its Landau damping width [1-12] (in particular its gauge-fixing dependence and, in most cases, its negative sign)are *per se* a sufficient justification to the above statement. But there is more to it.

On the one side, in fact, the general folklore is that PT should be applicable to the casehand at hand. Indeed, thanks to the temperature $T \gg \Lambda_{QCD}$ and to UV-asymptotic freedom, one expects that the effective coupling is $\alpha_s(T) \ll 1$ (although we are not aware of any systematic investigation implementing the RG at high $T$). On the other side one computes, in the above framework, the plasmon frequency $\omega_p$ obtaining the result

$$\omega_p^2 = \frac{1}{9} g^2 C_A T^2, \tag{1}$$

about which there is general agreement. Obviously it comes out that $\omega_p \simeq gT \ll T$.

Now, to be consistent with the philosophy of the RG, one should have found an energy $\omega_p$ higher that the fiducial scale $T$ at which PT starts being reliable. In other words it seems that PT could be trusted only if a situation like QED were realized, *i.e.*, only if $g = 0$ behaved for this particular problem as an *IR fixed point*.

Stated in this way, the problem looks awkward - to put it mildly.

Nonetheless it is exactly the aim of this talk (based on the results of [13]) to outline the reasons why this fact does indeed take place, without spoiling the statement that $g = 0$ remains an UV-fixed point, as far as the colour singlet sector of QCD is concerned.

The gauge invariance and positivity of the $\gamma$ will be side results of the discussion we give below. Stronger motivations for the construction we are going to sketch are provided

[1] Talk presented by E. d'Emilio

by the necessity (or, at least, by our wish) of giving an algorithm for constructing, at least in PT, locally gauge invariant fields that carry nontrivial global colour indices.

In more explicit terms, we will be looking for locally gauge invariant substitutes for the usual vector potential $\mathbf{A}_\mu$ and the Noether colour current $\mathbf{J}_\mu$, say $\mathbf{a}_\mu$ and $\mathbf{j}_\mu$, that, in the $T=0$ quantum field theory, share all the symmetries fixing the rules of the game: translations and, notably, Lorentz symmetry. Switching on the temperature is something postponed to the final part of this note.

Three steps are necessary to make our wish come true.

(i) One introduces a *nonlocal* string-like operator $V(x;n,c)$ whose Fourier transform with respect to $x$:

$$\hat{V}(k;n,c) = \int d^4x\, e^{ikx}\, V(x;n,c)$$

is given by the formal power series

$$\hat{V}(k;n,c) = (2\pi)^4\delta^{(4)}(k) + \sum_{\ell=1}^{+\infty}(-g)^\ell \int \frac{d^4k_1}{(2\pi)^4}\cdots\cdots\frac{d^4k_\ell}{(2\pi)^4}(2\pi)^4\delta^{(4)}(k-\sum_{i=1}^{\ell}k_i)\cdot$$

$$\cdot\frac{n\cdot\hat{\mathbf{A}}(k_1)}{\left[n\cdot(k_1+k_2+\ldots+k_\ell)\right]_c}\cdot\ldots\cdot\frac{n\cdot\hat{\mathbf{A}}(k_\ell)}{\left[n\cdot k_\ell\right]_c}\ . \tag{2}$$

Here $n^\mu$ is an arbitrary four vector and the symbol $1/[n\cdot k]_c$ denotes a linear convex combination of the $\pm i\varepsilon$ regularizations of the $(n\cdot k)^{-1}$ singularity, normalized in such a way that $c=0$ denotes the Cauchy Principal Value prescription.

(ii) One then defines the non local composite fields

$$\mathcal{A}_\mu(x) = V^\dagger(x)\left[\mathbf{A}_\mu(x) + \frac{i}{g}\partial_\mu\right]V(x)\,, \tag{3a}$$

$$\mathcal{F}_{\mu\nu}(x) = V^\dagger(x)\,\mathbf{F}_{\mu\nu}(x)\,V(x)\,, \tag{3b}$$

$$\mathcal{J}_\nu(x) = g\left[\partial^\mu(\mathcal{A}_\mu\wedge\mathcal{A}_\nu) + \mathcal{A}^\mu\wedge\mathcal{F}_{\mu\nu}\right]\,. \tag{3c}$$

The $(n,c)$-dependence on both sides of (3) is understood. $\mathcal{A}_\mu, \mathcal{F}_{\mu\nu}$ and $\mathcal{J}_\nu$ respectively generalize the gauge field $\mathbf{A}_\mu$, the field-strength tensor $\mathbf{F}_{\mu\nu} = \partial_\mu\mathbf{A}_\nu - \partial_\nu\mathbf{A}_\mu + g\,\mathbf{A}_\mu\wedge\mathbf{A}_\nu$ and the Noether current $\mathbf{J}_\nu$ in a sense to be clarified shortly.

(iii) The gauge invariant fields $\mathbf{a}_\mu$ and $\mathbf{j}_\nu$ are obtained by first Fourier transforming $x\to k$ in (3), then setting $n=k$ and, finally, anti-Fourier transforming. For the case of $\mathbf{a}_\mu$ we get

$$\mathbf{a}_\mu(x;c) = \int\frac{d^4k}{(2\pi)^4}\,e^{-ikx}\int d^4y\,e^{iky}\mathcal{A}_\mu(y;k,c)\,, \tag{4}$$

and likewise for $\mathbf{j}_\nu$.

Several comments are in order.

1. The $\delta^{(4)}$ in (2) exposes the fact that translation invariance has been maintained in the construction of the fields (3).

2. The introduction of nonlocal operators such as $V$ is unavoidable [14, 15] in trying to construct locally gauge invariant fields carrying nontrivial global colour numbers.

3. The relationship between the operators $V$ and standard string operators introduced in gauge theories [16] emerges, *e.g.*, by first choosing the $\pm i\varepsilon$ regularization prescription and then by representing the denominators in (2) in the well known way $(a \pm i\varepsilon)^{-1} = \mp i \int_{-\infty}^{+\infty} \theta(\pm\alpha)\, exp[i\alpha(a \pm i\varepsilon)]$. It turns out that the operators $V(x; n, \pm 1)$ are nothing but rectilinear path-ordered strings extending from $x$ to infinity in Minkowsky space, along the direction selected by the four-vector $n$.

No such straightforward comparison is possible for different values of $c$, $c = 0$ in particular. Nevertheless, independently of the regularization prescription, the operators $V$ defined in (2) are unitary and, under a local gauge transformation $\delta \mathbf{A}_\mu(x) = (\partial_\mu + g\,\mathbf{A}_\mu \wedge)\,\mathbf{C}(x)$ vanishing at infinity, $\mathbf{C}(t, \mathbf{x}) \to 0$ for $(\mathbf{x}^2 + t^2)^{1/2} \to \infty$, they transform according to $\delta V(x) = ig\,\mathbf{C}(x)\,V(x)$. This transformation property guarantees that the nonlocal fields defined in (3) are locally gauge invariant, *i.e.* they automatically satisfy $\delta \mathcal{A}_\mu = \delta \mathcal{F}_{\mu\nu} = \delta \mathcal{J}_\nu = 0$.

The composite fields (3), although gauge invariant, are not Lorentz covariant, due to the $n$-dependence of $V(x; n, c)$. The choice for $n$ performed in the step (iii) above is suggested by the necessity of restoring Lorentz symmetry. To this stage it is neither compulsory nor self evident. We thus invite the reader to make up his mind about the opportunity of such a choice on the basis of the results to which it leads. For the moment we can only partly give a justification for it by mentioning the analogy with QED, where the same choice leads to the introduction of charged fields with a near mass shell behaviour mild enough as to allow for the extension of LSZ formalism to the physical charged sectors [17].

In any case, once the choice $n = k$ involved in (iii) is done, it is evident from (4) that one is no longer dealing with a single rectilinear string along $n$. Rather, there is a string for any momentum component of $\mathcal{A}_\mu$. Then, these strings are integrated upon to give rise to $\mathbf{a}_\mu$.

Going back to the QCD plasma dynamics, we suggest to use the current $\mathbf{j}_\mu$ in computing nonabelian correlation functions. This suggestion is due to the fact that $\mathbf{j}_\mu$, besides of being a gauge invariant and Lorentz covariant operator, has the correct properties of a physical colour current: (a) it is a conserved current and, (b) its associated charges $\mathbf{q} = \int d^3x\, \mathbf{j}_0(x)$ can be identified with the colour charges $\mathbf{Q}$ associated with the Noether current $\mathbf{J}_\mu$.

The property (a) follows by noting that the weak YM equations of motion

$$\partial^\mu(\partial_\mu \mathbf{A}_\nu - \partial_\nu \mathbf{A}_\mu) + \mathbf{J}_\nu \simeq 0\,, \tag{5}$$

imply the weak Maxwell equations

$$\partial^\mu \mathbf{f}_{\mu\nu} + \mathbf{j}_\nu \simeq 0\,, \tag{6}$$

relating what we call the chromoelectromagnetic field $\mathbf{f}_{\mu\nu} = \partial_\mu \mathbf{a}_\nu - \partial_\nu \mathbf{a}_\mu$ and the colour current $\mathbf{j}_\nu$. Incidentally, any temptation of abandoning (5) in favor of (6) would be deceiving: (6) is not an equation for $\mathbf{a}_\mu$ as (5) is for $\mathbf{A}_\mu$. We have in fact spoiled the equivalence on effecting the choice $n = k$ involved in step (iii) of the construction. But, and this is the real important thing, the Maxwell equations (6) provide a physical interpretation of the nonabelian plasma dynamics that is quite parallel to that of the QED plasma and is, therefore, quite solid - a feature that is not shared, in our opinion, by the approaches of refs. [1-12]. Needless to add, this is the main reason for our suggestion.

Concerning (b) we have to be more specific. Obviously the identification is crucial if (6) is claimed to provide a description of chromodynamics in terms of gauge invariant operators. Likewise crucial are the elements that make this identification possible. The details (for which we evidently have to refer the reader to [13]) show that two elements play a key role. The first is the choice $n = k$ in point (iii) above. The second is that denominators in (2) must be regulated by the $c = 0$, *i.e.*, by the PV, prescription.

We are now ready to examine the role of nonlocality in the solution of our puzzle. Consider, as an example, the lowest order term in the expansion of $\mathbf{j}_\mu$:

$$j_\mu^m(x) = g \int dy_1 \int dy_2 O_{\mu\alpha\beta}^{mab}(x - y_1, x - y_2) A_a^\alpha(y_1) A_b^\beta(y_2) + O(g^2)\,, \tag{7}$$

where the convolutions expose the preserved translation invariance and the kernel $O$ could be given in terms of both the $n^\mu/(n \cdot k)_c$ vertices appearing in (2) and the Yang-Mills vertices, but its explicit form is not needed here. The above formula makes it evident that several time orderings are possible. The first is:

(i) The T-ordering of the *external* $x^0$ times (let it be either the usual Schwinger-Dyson ordering required, *e.g.*, in the LSZ reduction formalism, or the retardation required by causality in applying linear response theory to QCD plasma).

These functions must in general be constructed from the Wightman functions of the ordinary vector potential $\mathbf{A}_\mu$. As for the two-point functions which are relevant in the case of the plasma, they have to be constructed starting from the spectral weight $\rho(k^2)$, appearing in

$$\delta^{mn} \rho_{\mu\nu}^+(k) = \delta^{mn}(k_\mu k_\nu - g_{\mu\nu}k^2)\theta(k^0)\rho(k^2) = \int d^4x e^{ikx} \langle 0|j_\mu^m(x) j_\nu^n(0)|0\rangle\,, \tag{8}$$

through the Lehman representation. It is evidently no surprise that $\rho$ comes out positive (in one loop $\rho^{(1)}(k^2) = 1/6 \cdot \alpha_s \theta(k^2)$, expectedly gauge-fixing independent and insensitive to the $c$-prescription of the $p \cdot k$ denominators of (2) demanded by gauge invariance), since, much as in QED, $\mathbf{j}_\mu$ only couples the vacuum to positive norm states.

The second ordering suggested by (7) is:

(ii) The ordering of the *internal* $y_i^0$ times, that we will refer to as the Causal, $C$, one.

The correlation functions of (i) and (ii) no more need to coincide, as they do in the case of local fields (notably the fields that describe the colour singlet sector). Indeed the calculation of the discontinuity of

$$-i\delta^{mn}\pi_{\mu\nu}^C(k) = -i\delta^{mn}(g_{\mu\nu}k^2 - k_\mu k_\nu)\,\pi^C(k^2) = \int d^4x\, e^{ikx}\, \langle 0|C j_\mu^m(x) j_\nu^n(0)|0\rangle\ , \tag{9}$$

can be done in such a way as to explicitly expose the positive contribution of the Wightman function (8), on the one hand, and the negative contribution from the nonlocal denominators $1/(p \cdot k)_c$ of (2) and the sensitivity to the $c$-prescription, on the other hand. The results are, first, that $\pi^C$ has the usual analiticity properties (*i.e.* a cut along the positive axis of the complex $k^2$-plane) only if $c = 0$ (in agreement with the classical argument, mentioned above, coming from the identification of $\mathbf{j}_\mu$ as the colour current); and, second, the overall discontinuity along the cut is the celebrated $-11/6 \cdot \alpha_s C_A \theta(k^2)$.

In other words our $g^2$-order calculation establishes the fact that the gauge invariant field $\mathbf{a}_\mu$ renormalizes in two different ways, according to whether one has to compute either $T$- or $C$-ordered products, (i) and (ii) above.

While the latter ordering evidently gives rise to the well known UV asymptotic freedom result

$$\alpha_s^{UV}(k^2) = \alpha_s(-\mu^2) \cdot \left[1 + \frac{11}{12}\frac{\alpha_s(-\mu^2)}{\pi} C_A \, ln\left(\frac{k^2}{-\mu^2}\right)\right]^{-1}, \quad k^2 \to -\infty, \tag{10}$$

the former, on the contrary, suggests

$$\alpha_s^{IR}(k^2) = \alpha_s(-\mu^2) \cdot \left[1 - \frac{1}{12}\frac{\alpha_s(-\mu^2)}{\pi} C_A \, ln\left(\frac{k^2}{-\mu^2}\right)\right]^{-1}, \quad k^2 \to -0, \tag{11}$$

similar to the QED case. The only real difference is in the fact that, due to the masslessness of gluons and to the impossibility of normalizing $\alpha_s$ at $k^2 = 0$, $\alpha_s^{IR}(0) = 0$ (by the way: has the vanishing of the on-shell residue of the Schwinger-Dyson time-ordered gluon propagator anything to do with confinement?).

Let us now switch on the temperature.

Which of the two orderings is required by linear response theory?

Evidently (i), *i.e.* the thermal trace $\pi^R_{\mu\nu}(t, \mathbf{x})$ of the commutator of $j^m_\mu(t, \mathbf{x})$ retarded in the external $t$ time. Unless one does not wish to violate causality.

How do we compute it? From the thermal Wightman function (*i.e.*, from (8), in which the VEV has been replaced with the Boltzmann weighted trace) and through the Lehman representation (let us restrict ourselves to only the longitudinal modes):

$$Im\, \pi^R_{00}(\omega, \mathbf{k}, \beta) = \frac{1}{2}\left(\rho^+_{00}(\omega, \mathbf{k}, \beta) - \rho^+_{00}(-\omega, \mathbf{k}, \beta)\right), \tag{12}$$

$$Re\, \pi^R_{00}(\omega, \mathbf{k}, \beta) = -\int_{-\infty}^{+\infty} \frac{d\omega'}{\pi}\frac{PV}{\omega - \omega'} Im\, \pi^R_{00}(\omega', \mathbf{k}, \beta). \tag{13}$$

Before turning to the results of the perturbative calculation, we only want to mention that in the formula for the width

$$\gamma(\mathbf{k}) = \frac{Im\, \pi^R_{00}}{\frac{\partial}{\partial\omega} Re\, \pi^R_{00}}, \tag{14}$$

the numerator (which is the one whose sign fluctuates in refs. [1-12] and is therefore the origin of the controversy) is easily seen to have the same sign as $\omega$, independently of any perturbative approximation. The above statement provides, therefore, an argument for the positivity of Landau damping that is valid under quite general assumptions, namely translation invariance and the positivity of the metric of the physical subspace. To recall how this comes about one writes (12) in the form

$$Im\,\pi^R_{00} = \tag{15}$$

$$= \sum_{M,N} C_A^{-1}(2\pi)^4\delta^{(3)}(\mathbf{k}-\mathbf{P}_M+\mathbf{P}_N)\delta(\omega-E_M+E_N)[e^{-\beta E_N}-e^{-\beta E_M}]\sum_m |\langle N|j_0^m(0)|M\rangle|^2 .$$

In the above equation we have exploited translation invariance (comment 1 above) and the sum over $M$, $N$, denotes a sum restricted to a complete set of physical, positive norm, eigenstates of four-momentum belonging to the eigenvectors $E_M$, $\mathbf{P}_M$. The restriction is allowed because our colour current is a gauge invariant operator carrying zero ghost number, so that the analysis of [18] applies and guarantees the decoupling of all the unphysical states from the sum. It will be now noted that for positive $\omega$ only those states for which $E_M > E_N$ contribute to the sum, so that the factor in the square bracket, which by now is the only dangerous one, is positive. In the sum, therefore, each term is positive definite, in the sense of distributions. Likewise for $\omega < 0$.

As for the perturbative calculation, in the high $T$ asymptotics (*i.e.* retaining only the most singular terms as $\beta \to 0$) we find (1) for plasmon energy (and we are now glad!) and

$$\gamma(\mathbf{k}) = \frac{g^2 C_A T}{24\pi}\left[1 + \frac{353}{320}\left(\frac{\mathbf{k}^2}{\omega_p^2}\right)^2 + ...\right], \quad |\mathbf{k}| \ll \omega_p , \tag{16}$$

which gives, for $\gamma(0)$, the same result as physical gauge calculations. A coincidence? Well, roughly speaking, before effecting the choice $n = k$ involved in (iii), one is sitting in the axial gauge (indeed $n \cdot \mathcal{A} = 0$ easily follows from (2). For a two point function (and only for two points) there just are one momentum $k$ and one string direction $n$: this means that the two point function of interest, *i.e.* the one obtained by doing step (iii), can be obtained from the two point function of the usual YM fields in the gauge $n \cdot \mathbf{A} = 0$ by setting $n = k$. In addition the kinematics of the plasmon problem $k \to (\omega_p, 0)$ hints at some relation between our calculation and those in the temporal gauge. But there are some subtleties connected with the fact that in time-like gauges the Lehman representation, assumed in refs. [1-3], just does not exist [19]. Some more details about this are to be found, again, in [13].

It remains to be see whether the conjectures of [20] and [21] (who claim the inadequacy of PT for the computations of quantities $O(g^2T)$, like $\gamma$) are true.

In this respect, the two loop calculation of both $\omega_p$ and $\gamma$ certainly looks, in our dispersive approach, complicated - but not impossible. The two loop calculation may also serve to check in a concrete example whether the coexistence of the different renormalizations of $\mathbf{a}_\mu$ corresponding to the $T$- and $C$-products does consistently carry over to higher orders - an issue that besides being challenging is also interesting, given the fact that physics seems not to ignore the former (remember the retardation in applying the linear response theroy to the plasma).

We shall report on these subjects in the near future.

## References


[1] K. Kajantie, J. Kapusta, Ann. Phys. (N.Y.) **160** (1985) 477.

[2] U. Heinz, K. Kajantie, T. Toimela, Phys. Lett. **183B** (1987) 96.

[3] U. Heinz, K. Kajantie, T. Toimela, Ann. Phys. (N.Y.) **176** (1987) 218.

[4] R. Kobes, G. Kunstatter, K. W. Mak, Phys. Lett. **223B** (1989) 433.

[5] O.K. Kalashnikov, V. V. Klimov, Sov. J. Nucl. Phys. **31** (1980) 699.

[6] T.H. Hansson, I. Zahed, Nucl. Phys. **B292** (1987) 725.

[7] H.-Th. Elze, K. Kajantie, T. Toimela, Z. Phys. **C37** (1988) 305.

[8] R. Kobes, G. Kunstatter, Phys. Rev. Lett. **61** (1988) 392.

[9] S. Nadkarni, Phys. Rev. Lett. **61** (1988) 396.

[10] M.E. Carrington, T.H. Hansson, H. Yamagishi, I. Zahed, Ann. Phys. (N.Y.) **190** (1989) 373.

[11] J. Kapusta, T. Toimela, Phys. Rev. **D39** (1989) 3197.

[12] J. Milana, Phys. Rev. **D39** (1989) 2418.

[13] S. Catani, E. d'Emilio, *Gauge Invariant Chromoelectromagnetic Field and Hot Gluon Plasma Oscillations*, Princeton University preprint PUTP 1134; *Gauge Invariant Description of the Plasmon in Hot QCD*, Pisa University preprint IFUP-TH 49/89.

[14] O. Steinmann, Ann. Phys. (N.Y.) **157** (1984) 232, Helv. Phys. Acta **58** (1985) 995, *Physical States in Gauge Theories*, lectures delivered at the workshop *Gauge Theories and Fundamental Interactions*, Stefan Banach International Mathematical Center, Warsaw, 1988.

[15] R. Ferrari, L.E. Picasso, F. Strocchi, Comm. Math. Phys. **35** (1974) 25, Nuovo Cimento **A39** (1977) 1.

[16] P.A.M. Dirac, Canad. J. Phys. **33** (1955) 650;
I. Bialynicki-Birula, Bull. Acad. Polon. Sci. **XI** (1963) 135;
S. Mandelstam, Ann. Phys. (N.Y.) **19** (1962) 1, Phys. Rev. **175** (1968) 1580;
K. Wilson, Phys. Rev. **D10** (1974) 2445.

[17] E. d'Emilio, M. Mintchev, Nuovo Cimento Lettere **34** (1982) 545, Phys. Rev. **D27** 1840, Fortschr. Phys. **32** (1984) 473, 503.
N.N. Bogoliubov, A. Logunov, I.T. Todorov, *General Principles of Quantum Field Theory* (Nauka, Moscow, 1987; in Russian).

[18] H. Hata, T.Kugo, Phys. Rev **D21** (1980) 333.

[19] J. Frenkel, J.C. Taylor, Nucl. Phys. **D109** (1976) 439; G.B. West, Phys. Rev. **D27** (1983) 1878.

[20] R. D. Pisarski, *Scattering Amplitudes in Hot Gauge Theories*, Fermilab preprint 88/123-T.

[21] A. Rebhan, this conference.

# The Temporal Axial Gauge at Finite Temperature

K.A.James

Department of Applied Mathematics and Theoretical Physics,
University of Cambridge, Silver Street, Cambridge, CB3 9EW, U.K.

**Abstract:** A consistent set of Feynman rules for the temporal axial gauge is derived within the Real Time Formalism (RTF); these rules lead to finite temperature dependent terms. In the Imaginary Time Formalism (ITF) we encounter temperature dependent divergences; these problems are discussed.

## 1 Introduction

The temporal axial gauge when used within the ITF gives rise to temperature dependent divergences. In an effort to resolve these problems we use the more physical RTF to derive a consistent set of Feynman rules for the Temporal gauge, which can be used for perturbation theory at finite temperature. This may also give an understanding of what, if anything, can be done in the Imaginary Time Formalism and why.

I shall use the *Vienna* prescription [1],[2] for the $1/k_0^2$ poles at zero temperature. This is defined by

$$\frac{1}{k_0^2} = \frac{1}{(k_0 + i\eta/k_3)^2} , \tag{1.1}$$

where $\eta \to 0$ at the end of a calculation. This is a generalisation of the Leibbrandt-Mandelstam prescription [3]. Firstly I shall describe, briefly, the Hamiltonian formulation of Landshoff [4] for deriving this prescription. Next I shall use this formulation to derive the necessary Feynman rules at $T = 1/\beta > 0$ for the temporal gauge. Using these rules I calculated the plasmon damping constant $\gamma$ of an electrically perturbed plasma to one loop order and the two loop interaction pressure $P_{(2)}^{glue}$ of the plasma. The latter is a gauge invariant object, so is a good check. I shall state my results obtained. Finally I shall give a simple derivation of the ITF and briefly discuss the problems that arise with the temporal gauge and why.

## 2 Hamiltonian Formulation

The essential ideas of the Hamiltonian formulation of Landshoff [3] for deriving the *Vienna* prescription are as follows. Set

$$A_0 = \frac{\eta}{\partial_3} \frac{1}{\partial^2} \partial \cdot \mathbf{A} \ , \tag{2.1}$$

where we take the limit $\eta \rightarrow 0$ at the end, giving $A_0 = 0$. The longitudinal and transverse operators are defined by

$$L_{ij}(\partial) = \frac{\partial_i \partial_j}{\partial^2} \ , \qquad T_{ij}(\partial) = \delta_{ij} - L_{ij}(\partial) \ . \tag{2.2}$$

We can decompose the gluon field $A_i(\mathbf{x}, t)$ and the free Hamiltonian $H_0$ as

$$A_i = A_i^T + A_i^L \ , \qquad H_0 = H_0^T(A^T) + H_0^L(A^L) \ . \tag{2.3}$$

where

$$A_i^T = T_{ij}(\partial) A_j \ , \qquad A_i^L = L_{ij}(\partial) A_j \ . \tag{2.4}$$

$H_0^T$, $H_0^L$ are functionals of the transverse and longitudinal fields respectively. The transverse field is quantised using the normal transverse field operators, $a(\mathbf{k}, s)$, $a^\dagger(\mathbf{k}, s)$ ; $s = 1, 2$ giving the two physical transverse polarizations. Their time evolution is governed by $H_0^T$.

The longitudinal modes do not carry energy. Thus, Landshoff uses "*ghost*" operators to describe $A_i^L(\mathbf{x}, t)$:

$$A_i^L(\mathbf{x}, t) = -i \int \frac{d^4k}{(2\pi)^3} \frac{k_i}{\mathbf{k}^2} \theta(k_3) \left( r(\mathbf{k}, t) e^{-i(\frac{\eta}{k_3} t - \mathbf{k}\cdot\mathbf{x})} - r^\dagger(\mathbf{k}, t) e^{i(\frac{\eta}{k_3} t - \mathbf{k}\cdot\mathbf{x})} \right) \ , \tag{2.5}$$

where

$$r(\mathbf{k}, t) = q(\mathbf{k}) + t p(\mathbf{k}) \ .$$

"*Ghost*" states in the Heisenberg picture are created by acting with $q^\dagger(\mathbf{k})$ and $p^\dagger(\mathbf{k})$ on the vacuum. The only nonzero commutation relation between the q's and p's is

$$[q(\mathbf{k}), p^\dagger(\mathbf{k}')] = i(2\pi^3)\mathbf{k}^2 \delta^{(3)}(\mathbf{k} - \mathbf{k}') \ . \tag{2.6}$$

We also have the longitudinal field operators commuting with the transverse ones:

$$[q, a] = [q^\dagger, a] = [q, a^\dagger] = [q^\dagger, a^\dagger] = 0 \ , \tag{2.7}$$

similarly for the p's. Gauss's law in this approach is defined to be

$$p(\mathbf{k}) | > = 0 \ . \tag{2.8}$$

With this definition of the Gauss law the vacuum is defined by

$$q(\mathbf{k}) | 0 >= 0 \ . \tag{2.9}$$

The physical states are those states which satisfy the Gauss law and have positive norms. These are the transverse states, $\{|\psi_T>\}$; $|\psi_T>$ is obtained by acting with the transverse creation operator $a^\dagger(\mathbf{k}, s)$ on the vacuum.

## 3 Feynman Rules and Results

In the RTF, the propagator required to do perturbation theory at finite temperature $T = 1/\beta$, has a $2 \times 2$ matrix structure [5]. In an operator formalism this is given by

$$i\mathbf{D}_{ij}(\mathbf{x},t) = \begin{pmatrix} < T(A_i(\mathbf{x},t)A_j(0)) >_0 & < A_j(0,-\frac{1}{2}i\beta)A_i(\mathbf{x},t) >_0 \\ < A_i(\mathbf{x},t-\frac{1}{2}i\beta)A_j(0) >_0 & < T^*(A_i(\mathbf{x},t)A_j(0)) >_0 , \end{pmatrix} \tag{3.1}$$

where $T^*$ is the anti-time ordering operator and $< ... >_0$ is the free thermal average defined by

$$< \mathcal{O}(\mathbf{x},t) >_0 = \frac{Tr(e^{-\beta H_0}\mathcal{O}(\mathbf{x},t))}{Tr(e^{-\beta H_0})} . \tag{3.2}$$

In the temporal gauge the trace is over the physical states of the system:

$$Tr(\mathcal{O}(\mathbf{x},t)) = \sum_T < \psi_T|\mathcal{O}(\mathbf{x},t)|\psi_T > . \tag{3.3}$$

Decompose $\mathbf{D}_{ij}$ into its transverse and longitudinal parts as

$$\mathbf{D}_{ij}(\mathbf{x},t) = \mathbf{D}^T_{ij}(\mathbf{x},t) + \mathbf{D}^L_{ij}(\mathbf{x},t) , \tag{3.4}$$

where

$$i\mathbf{D}^W_{ij}(\mathbf{x},t) = \begin{pmatrix} < T(A^W_i(\mathbf{x},t)A^W_j(0)) >_0 & < A^W_j(0,-\frac{1}{2}i\beta)A^W_i(\mathbf{x},t) >_0 \\ < A^W_i(\mathbf{x},t-\frac{1}{2}i\beta)A^W_j(0) >_0 & < T^*(A^W_i(\mathbf{x},t)A^W_j(0)) >_0 , \end{pmatrix} \tag{3.5}$$

$$W = (L,T) .$$

In Fourier space we can write $\mathbf{D}^T_{ij}$ as

$$\mathbf{D}^T_{ij}(\mathbf{k},k_0) = \mathbf{D}_T(\mathbf{k},k_0)T_{ij}(\mathbf{k}) . \tag{3.6}$$

$\mathbf{D}_T(\mathbf{k},k_0)$ can be found straightforwardly to give the expected result:

$$i\mathbf{D}_T(\mathbf{k},k_0) = \begin{pmatrix} \Delta(k) & 0 \\ 0 & \Delta^*(k) \end{pmatrix} + 2\pi\delta(k^2)N_k \begin{pmatrix} 1 & e^{\beta|k_0|/2} \\ e^{\beta|k_0|/2} & 1 , \end{pmatrix} \tag{3.7}$$

as derived by Niemi and Semenoff [5], where

$$\Delta(k) = \frac{i}{k^2+i\epsilon} \qquad N_k = \frac{1}{e^{\beta|k_0|}-1} , \tag{3.8}$$

and the $\delta(k^2)$ is $\epsilon$-regularized [4,5], i.e.

$$2\pi i\delta(k^2) = \lim_{\epsilon\to 0}\left(\frac{1}{k^2-i\epsilon} - \frac{1}{k^2+i\epsilon}\right) , \tag{3.9}$$

wherever a pinch singularity occurs.

From (1.7) it is clear that

$$q(\mathbf{k})|\psi_T> = p(\mathbf{k})|\psi_T> = 0\ . \tag{3.10}$$

This implies that, given any two functionals $F(A^T)$ and $G(A^L)$ of transverse and longitudinal fields respectively, the matrix element of their product between any two physical states factorises as

$$<\psi'_T|\,F(A^T)G(A^L)\,|\psi_T> = <\psi'_T|\,F(A^T)\,|\psi_T><0|\,G(A^L)\,|0>\ . \tag{3.11}$$

Consider a functional $\mathcal{O}(A^L)$ of the longitudinal fields only; its free thermal average is

$$<\mathcal{O}(A^L)>_0 = \frac{\sum_T <\psi_T|e^{-\beta H_0^T}e^{-\beta H_0^L}\mathcal{O}(A^L)|\psi_T>}{\sum_T <\psi_T|e^{-\beta H_0^T}e^{-\beta H_0^L}|\psi_T>}\ , \tag{3.12}$$

since $H_0^T$ commutes with $H_0^L$. From (3.11) this reduces to

$$\begin{aligned} <\mathcal{O}(A^L)>_0 &= \frac{\sum_T <\psi_T|e^{-\beta H_0^T}|\psi_T><0|e^{-\beta H_0^L}\mathcal{O}(A^L)|0>}{\sum_T <\psi_T|e^{-\beta H_0^T}|\psi_T><0|e^{-\beta H_0^L}|0>} \\ &= <0|\mathcal{O}(A^L)|0>\ , \end{aligned} \tag{3.13}$$

where the vacuum is normalised to one. Thus in the canonical approach of [4] the free thermal average of operator functionals of the longitudinal fields only is just the vacuum expectation value of the operator.

From (3.13) the components of the longitudinal part of the propagator $\mathbf{D}_{ij}^L$ in (2.4) are given by

$$[\mathbf{D}_{ij}^L]^{11}(\mathbf{x},t) = -i<0|T(A_i^L(\mathbf{x},t)A_j^L(0))|0>\ , \tag{3.14a}$$

$$[\mathbf{D}_{ij}^L]^{22}(\mathbf{x},t) = -i<0|T^*(A_i^L(\mathbf{x},t)A_j^L(0))|0>\ , \tag{3.14b}$$

$$[\mathbf{D}_{ij}^L]^{12}(\mathbf{x},t) = -i<0|A_j^L(0,-\frac{1}{2}i\beta)A_i^L(\mathbf{x},t)|0>\ , \tag{3.14c}$$

$$[\mathbf{D}_{ij}^L]^{21}(\mathbf{x},t) = -i<0|A_i^L(\mathbf{x},t-\frac{1}{2}i\beta)A_j^L(0)|0>\ . \tag{3.14d}$$

In Fourier space $\mathbf{D}_{ij}^L$ can be written as

$$\mathbf{D}_{ij}^L(\mathbf{k},k_0) = \mathbf{D}_L(\mathbf{k},k_0)L_{ij}(\mathbf{k})\ , \tag{3.15}$$

where the components of $\mathbf{D}_L$ are as follows:

$$\mathbf{D}_L^{11}(\mathbf{k},k_0) = \frac{1}{(n\cdot k)^2} = \frac{1}{(k_0-(\eta-i\epsilon')/k_3)^2}\ , \tag{3.16}$$

which, as described in [4] is equivalent to the *Vienna* prescription (1.1) after a continuation to an imaginary value for $\eta$.

It is straightforward to show that the (22) component is given by

$$\mathbf{D}_L^{22}(\mathbf{k},k_0) = -\frac{1}{(n\cdot k)^{*2}}\ , \tag{3.17}$$

where n is given in (3.16); this is expected from the definition of the $T^*$-product. The (12) component can be found to be as follows

$$iD_L^{12}(\mathbf{k}, k_0) = 2\pi\theta(-k_3)\delta'(k_0 - \eta/k_3)e^{\beta k_0/2} . \tag{3.18}$$

Similarly the (21) component is

$$iD_L^{21}(\mathbf{k}, k_0) = -2\pi\theta(k_3)\delta'(k_0 - \eta/k_3)e^{-\beta k_0/2} . \tag{3.19}$$

Here, as before, wherever pinches occur we use

$$2\pi i\delta'(a) = \lim_{\epsilon'\to 0}\left(\frac{1}{(a+i\epsilon')^2} - \frac{1}{(a-i\epsilon')^2}\right) . \tag{3.20}$$

In the limit $T \to 0$, the (12) and (21) components must both give zero to reproduce the $T = 0$ theory. From an inspection of these components we find that we need $\eta$ real and positive. Thus we must take the limit $\eta \to 0^+$ at the end of a calculation.

Before stating the results obtained, I would like to make a comment on the free thermal propagator $D_{ij,\beta}(\mathbf{x}, t)$ found in this approach. It is defined at temperature $\beta^{-1}$ by

$$\begin{aligned} D_{ij,\beta}(\mathbf{x}, t) &= < T(A_i(\mathbf{x}, t)A_j(0)) >_0 \\ &= D^>_{ij,\beta}(\mathbf{x}, t)\theta(t) + D^<_{ij,\beta}(\mathbf{x}, t)\theta(-t) . \end{aligned} \tag{3.21}$$

The longitudinal part of $D_{ij,\beta}$ is given by

$$D^L_{ij,\beta}(\mathbf{x}, t) = < T(A^L_i(\mathbf{x}, t)A^L_j(0)) >_0 . \tag{3.22}$$

From (3.13) this reduces to

$$D^L_{ij,\beta}(\mathbf{x}, t) = < 0|T(A^L_i(\mathbf{x}, t)A^L_j(0))|0 > , \tag{3.23}$$

which is temperature independent. Thus the longitudinal part of the propagator clearly violates the periodic or KMS condition [4] defined by

$$D^>_{ij,\beta}(t - i\beta) = D^<_{ij,\beta}(t) , \tag{3.24}$$

which in Fourier space reads

$$D^<_{ij,\beta}(\mathbf{k}, k_0) = e^{-\beta k_0} D^>_{ij,\beta}(\mathbf{k}, k_0) . \tag{3.25}$$

It can be easily checked that the transverse part of the propagator $D^T_{ij,\beta}$ satisfies this periodicity condition (2.24). As we shall see this violation is the source of the problems for the ITF.

Using the Feynman rules above I found the following results.

(*i*) The one loop plasmon damping constant, $\gamma$ of an electrically perturbed gluon plasma is

$$\gamma = \frac{Ng^2T}{24\pi} \; .$$

This contribution came solely from the following diagram:

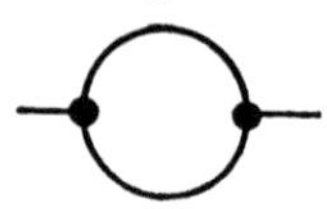


i.e from the transverse modes $T$ only, in agreement with what was found by Kajantie et al. [7].

(*ii*) The two loop interaction pressure, $P_2^{glue}$ of the gluon plasma is

$$P_2^{glue} = \frac{-N(N^2-1)g^2T^4}{144} \; ,$$

which is identical to the results obtained using a covariant gauge within the RTF [5] and using the Feynman gauge within the ITF [8].

In the above two calculations I had no problems with the order of any limits to be taken nor with the product of the various distribution functions. There were singularities encountered in the integrals; all temperature dependent terms were finite unlike in the ITF. Thus at this stage, it would seem that this set of Feynman rules allows one to do consistent calculations at finite temperature using the temporal gauge.

## 4. The Imaginary Time Formalism

In the ITF we use the contour $C$ drawn below:

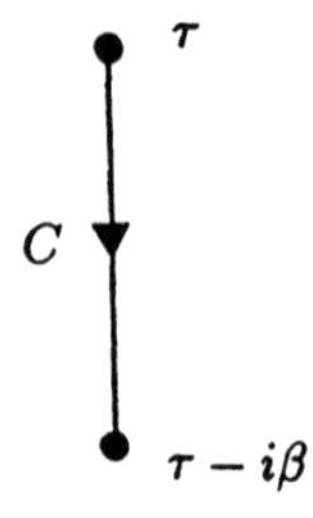

where $\tau$ is real and finite; the standard choice for $\tau$ is zero. The object required for doing perturbation theory in the ITF is the euclidean propagator defined as

$$\begin{aligned}\Delta_{ij}(\mathbf{x},t) &=< T_c(A_i(\mathbf{x},t)A_j(0)) >_0 \\ &= \Delta^{>}_{ij}(\mathbf{x},t)\theta(\text{-Im}\,t) + \Delta^{<}_{ij}(\mathbf{x},t)\theta(\text{Im}\,t)\end{aligned} , \tag{4.1}$$

where $T_c$ is the time ordering along the contour $C$; $t$ is pure imaginary and takes the values $-\beta < \text{Im}\,t < \beta$.

The KMS condition (3.24) when applied to $\Delta_{ij}$ leads to the periodic condition

$$\Delta_{ij}(t - i\beta) = \Delta_{ij}(t) . \tag{4.2}$$

Defined over the finite interval $C$ this condition leads to the Fourier sum

$$\Delta_{ij}(\mathbf{p},t) = \frac{i}{\beta}\sum_{n=-\infty}^{\infty} \Delta_{ij,n}(\mathbf{p})e^{\omega_n t} , \tag{4.3}$$

where

$$\omega_n = 2n\pi/\beta \tag{4.4}$$

are the Matsubara frequencies.

The inverse is given by

$$\begin{aligned}\Delta_{ij,n}(\mathbf{p}) &= \int_0^{-i\beta} dt\Delta^{>}_{ij}(\mathbf{p},t)e^{-\omega_n t} \\ &= \int_{-\infty}^{\infty} d\omega \frac{\rho_{ij}(\mathbf{p},\omega)}{i\omega_n - \omega} .\end{aligned} \tag{4.5}$$

$\Delta_{ij,n}(\mathbf{p})$ is like the $T = 0$ propagator with $i\omega_n$ instead of $p_0$. Thus the Feynman rules of the ITF are just like at $T = 0$ except we have

$$k_0 \to i\omega_n ,$$

$$\int \frac{d^4k}{(2\pi)^4} \to \frac{i}{\beta}\int \frac{d^3\mathbf{k}}{(2\pi)^3}\sum_n ,$$

$$(2\pi)^4\delta^{(4)}(\sum k) \to (2\pi)^3\frac{\beta}{i}\delta^{(3)}(\sum \mathbf{k})\delta(\sum \omega) .$$

As is clea. from above the KMS or periodic boundary condition (2.24) is crucial in setting up the ITF. Using the RTF I found that only the longitudinal modes violate this periodic condition; this, however, does not pose any problems in the RTF as far as temperature dependent divergences are concerned. In the ITF the temperature dependent divergences come directly and only from the longitudinal modes.

To understand these problems let us go back to the beginning and look at the Hamiltonian approach of Landshoff [4]. In the temporal gauge, the Faddeev-Popov ghosts decouple from the theory and so we must remove the unphysical longitudinal modes by imposing a weak condition on the states . When we trace over states, we trace over the physical states; in our case we can identify these easily as the transverse states, since the unphysical longitudinal modes cannot attain thermal equilibrum with the heat bath. This leads to a temperature independent propagator for the longitudinal modes (as one would expect) and hence violation of the KMS condition.

Clearly the violation of the KMS condition by the longitudinal modes is the reason for the problems encountered when trying to set up a perturbative framework in the Imaginary Time Formalism using the Temporal gauge. How one can get around this obstacle, if at all possible, is yet unsolved to the knowledge of the author.

## References

[1] P. Gaigg and M. Kreuzer, Phys. Lett **205B** (1989) 530.

[2] H.Hüffel, P.V. Landshoff and J.C.Taylor, Phys.Lett. **217B** (1989) 147.

[3] G.Leibbrandt, Rev. Mod. Phys. **59** (1987) 1067.

[4] P.V. Landshoff, Phys. Lett. **227B** (1989) 427.

[5] N.P.Landsman and Ch.G.van Weert, Phys. Rep. **145** (1987) 141.

[6] J.Niemi and G.Semenoff, Ann. Phys. **152** (1984) 105.

[7] U.Heinz, K.Kajantie, T.Toimela, Ann.Phys. **176** (1987) 218.

[8] J.I.Kapusta, Nucl. Phys. **B148** (1979) 461.

This book was processed by the author using the TEX macro package from Springer-Verlag.

GPSR Compliance

*The European Union's (EU) General Product Safety Regulation (GPSR) is a set of rules that requires consumer products to be safe and our obligations to ensure this.*

*If you have any concerns about our products, you can contact us on ProductSafety@springernature.com*

In case Publisher is established outside the EU, the EU authorized representative is:

Springer Nature Customer Service Center GmbH
Europaplatz 3
69115 Heidelberg, Germany

**Batch number: 10620881**

Printed by Printforce, the Netherlands